Breeding for robustness in cattle

EAAP – European Federation of Animal Science

The European Association for Animal Production wishes to express its appreciation to the *Ministero per le Politiche Agricole e Forestali* and the *Associazione Italiana Allevatori* for their valuable support of its activities

Breeding for robustness in cattle

EAAP publication No. 126

edited by

Marija Klopčič

Reinhard Reents

Jan Philipsson

Abele Kuipers

Wageningen Academic
P u b l i s h e r s

ISBN 978-90-8686-084-5

ISSN 0071-2477

First published, 2009

Wageningen Academic Publishers
The Netherlands, 2009

Introduction

Abele Kuipers[1] and Reinhard Reents[2]
[1]President of Cattle Commission of EAAP, Wageningen University and Research Centre, the Netherlands
[2]Chairman of INTERBULL, United Information Systems for Animal Production, VIT Verden, Germany

Robustness is a rather novel concept in cattle husbandry. In poultry production, robustness of the animal has received special attention for several decades now; a review of the developments in poultry breeding has been published by Flock *et al.* (2005). In pig production, the interest in robustness of the animal stems from a much more recent time - a session was devoted to this topic at EAAP 2005, and Knap (2008) has given an overview of the developments that are currently taking place in pig breeding. Interest in the robustness of the animal has arisen in recent years in cattle husbandry as well, most especially in dairy. However, for many years a range of functionally important traits in dairy cows have been considered to a varying extent in breeding programmes of many countries or breeds. The present discussion about robustness adds more dimensions to this.

The past decade has revealed unfavourable trends in fertility, udder health and locomotion in some major dairy cattle breeds. This results in economic losses. The enlargement of herds leads to less available labour time per individual cow. This requires cows that are easy to handle. At the same time, society is demanding a higher welfare standard of livestock. These developments have increased the desire for more "robust" animals. This was the reason for the organisation of a joint session of EAAP Commissions and INTERBULL in August 2007 in Dublin to review the state of the art of robustness research, its definitions and applications, in cattle.

This book covers a series of articles presented at this seminar as well as additional articles from authors who were later found to have delivered significant work in dealing with the concept of robustness. The book contains 24 contributions of 86 authors from 14 countries all over the world. It is the first attempt to summarise the available know-how concerning this topic in cattle. That makes this book unique. The active input of the authors was very encouraging. The editing team reviewed all articles carefully and revisions were made when needed. We would like to thank all authors very much for their participation in the process that led to this book.

Chapter 1 deals with the general concept of robustness and its merits. A variety of definitions of robustness by the various authors can be found in this book, including:
- 'a robust dairy cow is a cow that is able to maintain homeostasis in the commonly accepted and sustainable herds of the near future';
- 'the ability of the cow to function well in the environment she lives in, being resilient to the changes in the micro-environment that she encounters during her life';
- 'a robust cow can be defined as one that adapts well to a wide range of environmental conditions or in genetic terms expresses a reduced genotype by environment interaction'; and
- 'a cow may be said to be robust if she is capable of coping well with, sometimes unpredictable, variations in her operating environment with minimal damage, alteration or loss of functionality'.

More descriptions of the concept of robustness can easily be found in the book. It shows that the concept of robustness is not exactly defined, and that some differences may exist if looking at the individual cow over her lifetime in a given environment versus the variation among cows within a population across environments.

Chapter 1 also deals with basic concepts like evolution, genetics, environment, animal health and welfare, and integrity as aspects related to robustness. The concept of robustness is mostly associated with secondary traits, like health, fertility and longevity traits. But several authors also recommend including other non-production traits in the breeding goal, such as behaviour and traits meeting specific ethical considerations. Several approaches to the impact of disturbances are discussed, like the control model which is characterised by maintaining stability by means of keeping away disturbances and the adaptation model, which is characterised by minimising disturbances using the intrinsic capacity of the animal to adapt where possible. The adaptation model looks at methods such as the use of reaction norms and selection for minimal residual variation in offspring.

In *Chapter 2* the link between robustness and longevity is described and in *Chapter 3* between fertility and health traits. Besides fertility and udder health, special focus is given in several articles in this and other chapters to claw disorders and body condition score as indicators of robustness.

The energy balance of the cow is another biological phenomenon that can be linked to the robustness of the cow. The possible contribution of the instrument of energy balance to breed a more sustainable cow is described in *Chapter 4.* This chapter is more exploratory in its content.

The link between robustness and hot climatic conditions is tackled in *Chapter 5.* It is stated that 'dairy farmers in hot climates, whose dairy cattle experience heat stress, should value traits differently than dairy farmers in temperate climates'. It was concluded that 'tropically adapted breeds are generally more robust for changes in thermal environments than temperately adapted breeds'.

The attitude and input of stakeholders towards robustness as part of the breeding program are discussed in *Chapter 6.* The attitudes of farmers towards robustness were measured in a country in transition in Eastern Europe. With regard to the input of farmers, the observation is made that 'when studying literature the impression arises that the young organic sector gives, relatively to the traditional sector, quite some attention to the opinions of farmers in formulating the breeding goals, whereas it may not be as well documented in the traditional sector'. It is advised to 'choose a diverse composition of working groups to be involved in the preparation of breeding goals to obtain a well balanced sustainable breeding program'. In this chapter the inclusion of robustness as a fourth type trait of the cow in a national breeding program is described.

We conclude that robustness is not yet an exactly defined concept, although all contributions in this book describe it as a combination of functional traits and other non-production traits, like behaviour and energy balance of the cow and traits meeting ethical standards emerging from society. The combination of traits, i.e. characteristics of the cow chosen to express robustness, differs from author to author, but it seems to be a general idea that robustness expresses the ability of the cow or breed to be able to cope with variable environments. The 'other' non-production traits are generally not (yet) part of the breeding programs, because of priority ranking and non-availability of measurements and data at wider scale.

To make the concept of robustness operational in practice requires agreement on the definition of this concept and the characteristics to be included in it. In fact, robustness must be considered as a composite trait. Moreover, to make it worthwhile to include in a breeding program, it should add an advantage over the existing traits and indices in use.

A weighted combination of some selected fertility, health, longevity and type traits can be considered as a composite robustness trait. If any reliable measure related to the energy balance becomes accepted, then this criteria can also be added. In this respect, body condition score is presently the most likely candidate. Other non-production traits need further development to become operational in practice.

In relation to the various definitions given in this book on robustness, and the understanding of these for practical purposes, it would be important to devote more research efforts on studies of genotype by environment interactions for a number of longevity and functional traits. For that purpose INTERBULL evaluations, considering countries or regions representing quite different environments and production systems would be valuable to use.

The above discussion is perhaps a starting point for a deeper understanding of what robustness really means. More analysis and discussion is needed. Consideration may be given to establishing an international working group to analyse the pros and cons of the inclusion of some additional characteristics of robustness into the breeding goals of cattle. The definition and composition of such a trait would be part of the task of such a working group. Moreover, the utilisation of reaction norms and breeding for minimal genotype by environment interaction ('generalists' vs 'specialists') may be part of the study. INTERBULL and EAAP could take the lead together to realize such an initiative.

When talking about robustness, the focus is presently on farm animal or breed level as illustrated in this book. This is also the reality in the field. But in future 'robustness' of the whole production system requires attention as well. The role of improved disease resistance in ruminants and the cross-breeding systems, as described in this book, may actually already be considered as examples of contributions to support the robustness of the whole production system.

References

Flock, D.K., K.F. Laughlin and J. Bentley, 2005. Minimizing losses in poultry breeding and production: how breeding companies contribute to poultry welfare. World's Poultry Science Journal 61, 227-237.
Knap, P.W., 2008. Robustness. In: Rauw, W.M. (ed.). Resource allocation theory applied to farm animal production. CABI, Wallingford, UK, p.288-301.

Table of contents

Part 6: Application and attitude

Part 1
General concepts

The role of environmental sensitivity and plasticity in breeding for robustness: lessons from evolutionary genetics

E. Strandberg
Swedish University of Agricultural Sciences, Department of Animal Breeding and Genetics, P.O. Box 7023, 75007 Uppsala, Sweden

Abstract

A dictionary definition of robust is: having or exhibiting vigorous health, strength or stamina; resilient; sturdy; capable of performing without failure under a wide range of conditions. This definition highlights two perspectives of robustness: (1) the ability of an individual to function well in the environment she lives in, being resilient to the changes in the 'microenvironment' that she encounters during her life; and (2) the ability of individuals (genotypes) to function well over a wide range of macroenvironments, e.g. production systems or herds. Both of these perspectives are important and environmental sensitivity plays an important role for both. Given that the breeding goal is correctly and broadly defined, it seems reasonable that animals with high and even performance over environments in the total breeding goal, i.e. flat reaction norms, are desirable. In a global perspective, however, it is unlikely that truly robust animals should be the goal, i.e. generalists that are the best everywhere. Over such a wide range of environments, specialisation would be expected to be a better option. However, it could also be of importance for an animal to have a high and even performance within its environment over the lifetime. It has been seen from studies of genotype by environment interaction that the genetic variation varies across environment. This could be modeled as genetic variation in residual variance, where a high variation would be undesirable for the producer. Having a flat reaction norm for the breeding goal, does not necessarily mean that all component traits have a high and even performance. In general terms, one can envision several ways that an individual can achieve a high and flat reaction norm.

Keywords: evolution, plasticity, environmental sensitivity

Introduction

In a session named 'Breeding for robustness in cattle' the first question that comes to mind is: *what is robustness?* I think you will find that among the authors in this volume, there are as many definitions as authors. In Longman dictionary of the English language (1984) *robust* is defined as: 'having or exhibiting vigorous health, strength or stamina; resilient; sturdy; capable of performing without failure under a wide range of conditions'. If we translate this to cow language we might say that robustness is:
- the ability of a cow to function well in the environment she lives in, being resilient to the changes in the 'microenvironment' that she encounters (during her life);
- the ability of a cow to function well over a wide range of macroenvironments, e.g. climates, production systems or herds.

If this is our definition of a robust cow, then the next important question is: *do we want cows to be robust?* This might seem like a no-brainer but let's keep that question in mind for the time being. The rest of this paper will focus on what we can learn from evolutionary genetics, an area which for a long time has struggled with the issues of plasticity in relation to adaptation. An adaptation can be described as a feature that has evolved by natural selection and that increases the fitness of the individual. Adaptations can be of all kinds, biochemical, physiological, morphological, behavioural, and so on. I think it is important to remember that an adaptation is something that evolves in a population as a result of environmental cues, whereas when an individual 'adapts', this is through plasticity, i.e. the ability to change the phenotype in response to a change in the environment.

The structure of this paper, which is based on some lecture notes from a course on 'Genes and Environment' in connection with EAAP 2005, is as follows. First a brief history of plasticity, followed by a section on how genetics can explain plasticity. The next topic is whether plasticity is adaptive and when plasticity is expected to evolve. Finally, I try to draw some conclusions from this from an animal breeding point of view.

A brief history of phenotypic plasticity[1]

Around the turn of the 19th century, and really before the rediscovery of the results by Mendel, Darwinism was experiencing some serious challenges from other schools, notably Lamarck, mainly because Darwin was unable to suggest a model for the inheritance of phenotypic differences. In a paper entitled 'A new factor in evolution', Baldwin (1896) claimed that application of Darwin's principle of natural selection was enough to explain the origin of new phenotypes – one did not need to invoke any Lamarckism (i.e. inheritance of acquired characters). His new factor was that he argued that individuals do not only differ in their phenotypes but they also differ in the way these phenotypes are changed by altered environmental conditions (i.e. in their reaction norms, although he did not use the word). He argued that individuals that are more adaptable (more plastic, in our current vocabulary) to new environmental conditions will get more offspring and will therefore contribute more to the future generation(s) – a perfect Darwinian argument. (This phenomenon could otherwise have been used by Lamarckists – new environments 'give rise to new genes' that are inherited to the offspring.) Baldwin elaborated on this idea in a later book (Baldwin, 1902).

Quite soon after the rediscovery of the work by Mendel of 'genetic factors', what we now call genes, one started to realise that there was not a clear one-to-one link between genes and the phenotype. The first type of deviation was the discovery of non-additive effects, dominance and epistasis. The other discovery was that genetically homozygous lines would still show phenotypic variation. Even in experimental settings there is still some kind of microenvironmental 'noise', which was shown to give phenotypic variation. This led researchers to realise the necessity to distinguish the phenotype from the genotype. It also became clear that if microenvironmental variation could lead to phenotypic variation, then one should expect the large variation existing in nature to lead to even larger phenotypic variation.

This was the background which led to the concept of the reaction norm. This term *per se* was introduced by Woltereck (1909). He used the term to describe a phenomenon he saw in clones of the small crustacean *Daphnia*. He noticed differences in head size and shape and he plotted this against a measure of the environment (food availability) for each clone. These 'phenotypic curves' are the first published reaction norms and they are illustrated in Figure 1. They show that the clones react differently to the environmental change and it would seem that is also a good example of the distinction between genotype and phenotype which was being developed at the same time by Johannsen (more about him later).

However, Woltereck focused on something else that made him doubt the usefulness of distinguishing between phenotype and genotype, as suggested by Johannsen. He saw that distinct clones showed (almost) the same phenotype under certain environmental conditions. Woltereck therefore concluded that there is no distinction between genotype and phenotype, because differences between supposedly distinct genotypes could disappear only as a result of changes in the environment.

---

[1] Most of this treatment is from Schlichting and Pigliucci (1998) and Pigliucci (2001).

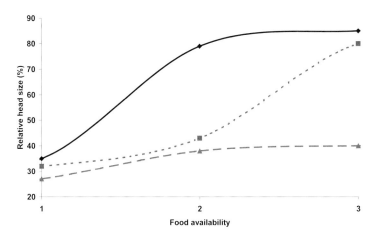

Figure 1. Relative head size in three clones of Daphnia as a function of three environments (low, medium and high food (algae) availability). The three distinct points are connected with a line. Results from Woltereck (1909) as presented in Schlichting and Pigliucci (1998).

Johannsen (1911), in responding to Woltereck, first pointed out the similarity between the *genotype* and the *reaction norm*: 'The 'Reaktionsnorm' emphasises the diversity and still the unity in the behaviour of the individual organism… Hence, the notion of 'Reaktionsnorm' is fully compatible with the genotype-conception'. Having said that, he criticised Woltereck's misinterpretation that just because the genotypes can produce the same phenotypic values in some environments, that does not mean that the genotypes are not reacting to the environment, and in fact sometimes they do not show the same phenotypes. With our modern eyes, it is quite difficult not to see that the three clones have different reaction norms, isn't it?

However, the ideas of Johanssen were not an over-night success, rather people tended to adopt the more gene-centered approach and sent the effects of environment into the obscure backroom of residual noise, something to be avoided or at least minimised by experimental design.

The next advance in the history of plasticity was made in the middle of the twentieth century by the book of Schmalhausen, who, in spite of his name, was a Russian geneticist. The book was originally in Russian and had probably never been read by a larger audience if it had not been translated to English in 1949 by a population geneticist friend of the author: Theodosius Dobzhansky (Schmalhausen, 1949).

Several ideas in this book have become key concepts in modern theory of phenotypic plasticity. He clearly distinguished between *external and internal factors* and considered both as potentially equally important in evolution. Later biologists have usually had a tendency to advocate either one or the other as the factor of main (or only) importance. Early paleontologists believed in some inner force directing evolution, rather than Darwin's natural selection. The Neo-Darwinists showed that it was not necessary to have such an internal force, natural selection could actually explain the trends seen in fossil records. The impact of Neo-Darwinist research was, however, so strong that the role of internal factors was ignored completely.

Schmalhausen also introduced the concept of reaction norm and discussed how reaction norms may develop when individuals are exposed to a variety of environments. He hypothesised what would happen when a change in the environment occurs (say by a change in climate or that individuals move to a new environment). That part of the reaction norm which was used before is now not observed, and that part can be subject to random events (mutation and drift) and may therefore change. A new part of the reaction norms is exposed to the environment and comes under selection and may therefore change. Both these factors can lead to a new (average) reaction norm which may be different from that of the original population. Schmalhausen called this stabilising selection, as it gives as a result a new 'stabilised' reaction norm (note that this use is different from the common meaning of stabilising selection – selection towards the mean). Schmalhausen also brought forward the idea of *integration through common regulation* as an answer to the question how different parts that need to work together evolve. This was long before any knowledge on regulatory genes was available. Good as it was, Schmalhausen's book, it still did not have an enormous impact on the research community. With the rise of Lysenkoism in the Soviet Union, it became impossible for Schmalhausen to continue his work. Also, the perspective of the book differed too much with the prevailing Neo-Darwinian thoughts at that time to be fully accepted.

Another important researcher, Conrad Waddington, suggested a mechanism he called 'genetic assimilation' which has a strong similarity to the idea of 'stabilising selection' suggested by Schmalhausen (Waddington, 1942, 1952). Genetic assimilation is most easily explained by the experiment by Waddington (1952) on Drosophila reactions to heat shock. In the base population a certain fraction showed a novel phenotype as a reaction to heat shock at larval stage. Selection for this new phenotype resulted in a selection response, i.e. a larger proportion of flies showed the new phenotype. That is not earth shattering. However, the surprising result was that after some generations flies showed the novel phenotype *even without the heat shock*! At first sight, this would be in accordance with Lamarckian theory, i.e. acquired characters are inherited! However, a less controversial interpretation is that the old reaction norm had changed so much that where before an extreme environment was needed to show the new phenotype, now a normal environment was enough to trigger the new phenotype, i.e. the flies had become more sensitive.

The next major step forward was a review by Bradshaw (1965). Up till now, most evolutionary biologists considered plasticity to be a nuisance, a measurement error that should be reduced as much as possible by experimental design or adjustment in statistical models. For instance, Falconer (an animal breeder) wrote in The American Naturalist (an evolutionary journal) a paper entitled 'The Problem of Environment and Selection' (Falconer, 1952). And of course it was a problem for Falconer as an animal breeder. This paper led to the approach of treating genotype by environment interaction (*GxE*) as across-environment genetic correlations. Bradshaw had a much more evolutionary perspective on this phenomenon – even though he was a plant breeder – and saw plasticity as an alternative for the plants (organisms) to adapt to changing environmental condition.

Bradshaw covered many aspects in his paper – here just a few main points are mentioned briefly. He interpreted the existence of genetic variation within and across species as an indication that plasticity for a trait is (at least partly) under different genetic control than the trait 'itself' – i.e. the average level of the trait. He also stressed that a genotype can be plastic for one trait and not for another – one cannot talk of a plastic genotype, one must define the trait (and the environment) too. Therefore, plasticities of different traits also must be controlled by (partly) different genes. On the other hand he noted that plasticities of developmentally related traits also are related. Furthermore, he emphasised that plasticity for *fitness* must be selected against (in nature) – the ideal situation must be to maintain as high fitness as possible under all environmental conditions (that the individual is likely to meet). However, plasticity for a given trait may result in fitness *homeostasis* (i.e. stable fitness). Bradshaw also considered what will happen with plasticity under disruptive, directional and stabilising selection and various environmental fluctuations.

Now we are coming closer to our own time and it might be wise to stop here, so as not to inadvertently say that still active researchers are 'history'. However, I think it is quite clear that most of the theory on plasticity is coming from people with an evolutionary and natural selection perspective – not from animal or plant breeding. This is, as mentioned, probably quite as expected: the aim of agriculture and perhaps especially animal husbandry has been to try to make the environment less variable by controlling it, thereby making the production of food more reliable and efficient. Plasticity, or at least *GxE*, is then something that should not exist.

How can genetics explain phenotypic plasticity?

Before venturing into the various genetic models or mechanisms that have been suggested to explain phenotypic plasticity, it is important to remember (as should be clear from the brief history of plasticity) that most of the development regarding plasticity has come from the world of evolutionary biology/genetics. One of the main questions in that field is to understand how organisms (both individuals and populations) adapt to complex environments and how these (populations of) organisms evolve over (long) time.

Most of the work has been done on *developmental plasticity*, i.e. usually studies of how organisms change their morphology in response to changes in the environment. However, all other traits, including *physiological and behavioural* reactions, also fall into the general theory of plasticity. There are, of course, some differences between the plasticity of these three types of traits. Both physiological and behavioural responses are short-term (sometime very much so) whereas developmental responses take longer time. The latter are also often irreversible. The physiological and behavioural plasticity usually exists throughout life, whereas developmental plasticity often has a certain window of opportunity, if the triggering environment comes before or after that window, the change in morphology does not occur.

Genetic models

In order to try to explain both the existence of and the variation in plasticity three main genetic models have been suggested (Scheiner and Lyman, 1991; Pigliucci, 2001).
1. *Overdominance*. This model assumes that plasticity is a function of heterozygosity – the more heterozygous an individual, the less plastic it is. This model stems from the idea that heterozygotes have the highest fitness, which was suggested by Lerner (1954) in his book on homeostasis. In order for this theory to make sense (in my mind) it should therefore be the *plasticity of fitness* which is dependent on heterozygosity, not necessarily the plasticity of other traits.
2. *Pleiotropy*. This model assumes that the same genes affect the traits expressed in the different environments but not to the same degree. Perhaps expression of the genes is different or the actual effect of the gene products is (physically) altered. In an animal breeding situation this is usually accounted for by using a multiple trait model.
3. *Epistasis*. This model has been used to try to explain why the mean of the trait can be genetically (at least somewhat) independent from the plasticity of a trait. Different set of genes affect the mean and the slope of the reaction norm and they also interact. I would suggest that also regulatory genes that are sensitive to environmental cues, which then affect the expression of (possibly different) sets of genes, fall into this category.

One should note that in this type of description, the terms overdominance, pleiotropy, and epistasis refer to effects that are shown *over environments*, not within environments (as is otherwise the common use).

Canalisation, homeostasis and plasticity

There are some terms that have been used in the evolutionary literature that need some explanation because they are sometimes referred to as the opposite of plasticity.

Canalisation means that a genotype can *buffer against small changes in the environment and in its own genotype* (mutations). The buffering means that even if small changes occur, the resulting phenotype is more or less the same. These terms are most commonly used for developmental traits, e.g. even if the environment is somewhat variable, the developmental plan is still followed and the resulting body, say, is still the same. It is, for instance, canalisation that makes sure that cows have four legs whether they are born in Sweden or in Africa. In developmental genetics, if the environment has an effect on the phenotype, it is usually an either-or effect, i.e. one type or another, and not a gradual effect.

Homeostasis is the outcome of canalisation. Homeostasis measures the degree of (in)variance in the phenotype when the individual is exposed to changes either in the environment or in the genome (by mutation). A more canalised genotype has higher homeostasis (is less affected). In his definition of canalisation, Waddington (1942) defined it for *minor* variations in conditions, what we might call microenvironmental variation. As an example of canalisation, he used the environmentally triggered metamorphosis of axolotls (salamanders). They produce one of two distinct phenotypes – which one is defined by a large change in environment (a reaction norm, or plasticity). Once the developmental pathway is chosen, however, small variations in environment do not affect the outcome, i.e. within each phenotypic outcome we have canalisation.

Homeostasis for fitness can be expected to be favourable, i.e. variations in the environment should preferably not give a large effect on fitness. If we expand that to large variation in environments, it would also mean that plasticity for fitness should not be favourable.

Allelic sensitivity and regulatory plasticity genes

Some researchers have suggested that one could distinguish between two types of genetic control of plasticity: *allelic sensitivity* and *regulatory plasticity genes*, sometimes referred to as *regulatory switches* (Schlichting and Pigliucci, 1993; Schlichting and Pigliucci, 1995). These two factors have been suggested as explanation for two plasticity phenomena that have been observed, so-called *phenotypic modulation* and *developmental conversion*, respectively. Phenotypic modulation is a gradual change in the phenotype, whereas developmental conversion changes the phenotype from one type to another (Figure 2).

Allelic sensitivity can come about from enzyme kinetics. For instance, an enzyme coded for by a structural gene probably shows some relation between its activity and the temperature where it is functioning. If the trait we are interested in is affected by the activity of this enzyme, then a change in temperature may also give a change in the trait. This will result in plasticity. Different alleles at this locus may code for enzymes with different activity curves which could give rise to genetic variation in plasticity. This may lead to a genetic correlation between environments that is <1, even though the same genes are involved, it is their (relative) effects that have changed. The activity of enzymes may be affected also by other factors, such as pH or nutrient availability, which also can be seen as allelic sensitivity (Schlichting and Pigliucci, 1998; Pigliucci, 2001).

One very old example of allelic sensitivity was given by Schmalhausen (1949; in Schlichting and Pigliucci, 1998) – the effect of two alleles on wing reduction in *Drosophila* in relation to the temperature (Figure 3). The reaction to temperature is quite different for the two homozygotes, whereas the heterozygotes is somewhat intermediate. There was also some difference between the sexes (only male shown here).

Allelic sensitivity was shown in *E. coli* mutants in relation to nutrient environment. The fitness of the mutant S1 relative to the wild type was severely reduced in one environment but not in two other (Hartl and Dykhuizen, 1981).

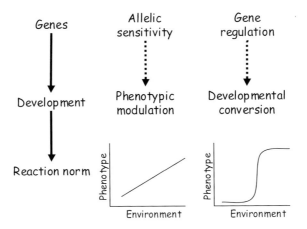

Figure 2. Schematic description of the hypothesised relation between allelic sensitivity and regulatory plasticity genes on one hand and phenotypic modulation (a gradual change in the phenotype) and developmental conversion (a marked phenotypic change). Based on Figure 3.16 in Schlichting and Pigliucci (1998).

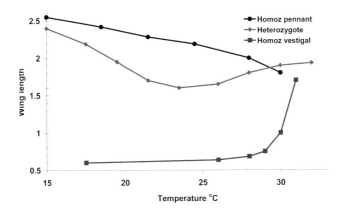

Figure 3. Reaction norms for wing length in relation to temperature for the homozygotes for the alleles pennant and vestigal, and the heterozygote, in male Drosophila melanogaster. Results from Schmalhausen (1949) presented in Figure 2.6 in Schlichting and Pigliucci (1998).

Regulatory plasticity which creates more drastic changes in phenotype, sometimes morphological changes during development, are often described in terms of an environmentally dependent 'switch'. Plasticity genes are then defined as regulatory loci that directly respond to a specific environmental stimulus by triggering a specific series of (usually morphogenic) changes. There are numerous examples of regulatory switches. Sex determination in reptiles is determined by temperature, shade-avoidance in plants is determined by phytochrome genes, photo-period induced changes between winged and wingless forms occur in insects, the *lac* operon which is turned on when lactose is present in the media for *E. coli*, etc.

As I already alluded to, the same environmental changes may give rise to both allelic sensitivity and regulatory plasticity. Growth rate in many organisms is related to temperature, usually with an intermediate optimum. This may be the result of allelic sensitivity and it gives rise to phenotypic modulation, in the terminology just introduced. However, some (large) changes in temperature can trigger a specific reaction called heat-shock response, which helps the organism to cope with high temperatures (that it 'expects' in the future). The response includes the production of special proteins that prevent other proteins to denature when exposed to high temperature. This change would qualify a developmental conversion, a large change in phenotype.

Molecular mechanisms underlying phenotypic plasticity

How do the genes know about the environment?

In order for the environment to interact with the genes there must be some kind of signaling system from the environment into the cells that are the targets. We usually think of the environment as the 'external' environment, however, the 'internal' environment can also be envisioned as having an interaction with the genotype. The internal environment could be rather far away from the target cells or it could be the more near-by environmental conditions.

Regardless of the type of environmental signal, there must be some signal reception, a signal transduction to the target cells, a translation of the signal, and then the effect on the phenotype must be mediated in some way. Once the signal has reached the cell, it could affect the transcription of one or more genes. However, it could also affect the translation of already created transcripts, or the processing of already translated products (Figure 4; Schlichting and Smith, 2002).

A *nervous system* is a very refined way of sending signals very quickly over long distances within the body. This type of signaling system is obviously of great importance for very fast reactions to environmental signals (such as flight as a reaction to the appearance of a lion). The response to this type of signal (various kinds of physiological reactions and behaviour) is generally also very fast. Another example of a signaling system is *hormones*. Although not as fast as the nervous system, hormones may also act over long distances.

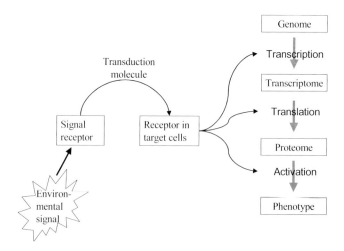

Figure 4. The steps necessary in a pathway from environmental signal to a changed phenotype. Transcriptome is the genetic material that has been transcribed but not yet translated, the proteome is the products of translation that are available for cellular metabolism into functionally active products. The environmental signal could be external to the organism or internal. Based on Schlichting and Pigliucci (1998) and Schlichting and Smith (2002).

Breeding for robustness in cattle

The diet available has been shown to give rise to phenotypic (developmental) differences in several organisms. The southern spadefoot toad tadpole can evolve to be either omnivores or carnivores. If there is a lot of macroscopic prey such as shrimp in their environment, they develop to be carnivores as these are more efficient in catching the prey and develop more quickly. In an experiment where tadpoles were exposed to a diet that would normally give rise to omnivores, but the diet was spiced with the hormone thyroxin, they instead developed into the larger carnivorous form. The suggestive conclusion from the experiment was that this hormone might be the internal signal, somehow triggered by the environment, to produce the carnivorous rather than the omnivorous form (Pfennig, 1992).

A certain butterfly (*Bicyclus anynana*) exists in two forms, one dry-season and one wet-season form. The wet-season form has a different wing colouration and an increased size of the eyespot. In an experiment two selection lines were created under a constant temperature environment, always producing the dry-season or the wet-season form, respectively. Differences were found in the levels of some steroid hormones among the lines – the wet-season form showed an earlier increase in the hormones and the end level was twice as high as for the dry-season form. When the dry-season line was injected with increasing levels of the hormones, the wing pattern and the eyespot size changed towards those existing for the wet-season form. So, again it seems that the internal signal is the hormone, but how the external environment triggers the signal is not known (Koch *et al.*, 1996).

Sex determination in some reptiles (turtles, snakes, lizards, crocodilians) is determined by the temperature the egg is exposed to. Commonly there is a threshold temperature, on each side of which the animals become males or females. In the snapping turtle, temperatures between 23 and 27 °C give only males, mixed sex ratios are produced between 27 and 29.5, and only females are produced above 29.5 °C. This sex determination seems to be the effect of the kinetic properties of two different enzymes in relation to temperature (Figure 5). Both 5α-aromatose and 5α-reductase can use the hormone testosterone as substrate. At high temperatures 5α-aromatose is better at competing for the substrate. This leads to the conversion of testosterone into estradiol, which in turn leads to a female individual. At lower temperatures, 5α-reductase can compete successfully and converts testosterone into dihydrotestosterone, which is involved in the development of a male (Crews, 1994; Rhen and Lang, 1995; Crews, 1996). This would then seem to be an example of when allelic sensitivity can result in developmental conversion.

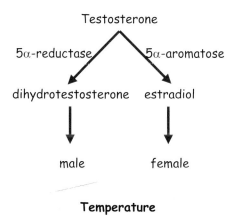

Figure 5. Schematic description of the temperature-dependent sex determination in reptiles. At low temperatures 5α-reductase competes successfully for the substrate testosterone and the outcome is a male. At higher temperatures 5α-aromatose is a better competitor and the outcome is a female.

A hormone-induced effect on gene expression of a specific locus (A1494) has been shown in *Arabidopsis*. Wilting (drought) increases the expression of A1494 and at the same time the hormone ABA was seen to increase. However, experiments with mutants showed that expression increased both with wilting and with exogenous ABA (in ABA-deficient mutants), so it seems expression can influenced by both. However, when exposed to low temperature, the ABA signal is responsible for expression of A1494. Heat shock does not lead to expression of A1494, so it is not a general temperature stress locus (Williams *et al.,* 1994). A similar role for ABA was found in a heterophyllous plant (*Marsilea quadrifolia)*. Heterophylly means that the plant produces different types of leaf depending on the circumstances, e.g. if it grows under or above water. Lin and Yang (1999) showed that both blue light and ABA, independently, could result in the aerial left type – so there must be independent pathways of signaling. Increased nutrient availability seemed to give a rise in ABA and induce the aerial leaf type, too.

An interesting example of a signaling system is the regulation of development and behaviour (!) in plants in response to light availability (or shade). This regulation is mediated by the phytochrome genes, which encode for a family of protein photoreceptors. In *Arabidopsis thaliana* there are five phytochrome genes, *phyA* to *phyE*. This gene family seems to be evolutionarily very old, and has been found in photosynthetic eubacteria, but in more developed organisms it has evolved by gene duplication. The divergence of *phyB* and *phyD* in *A. thaliana* is relatively recent (evolutionary speaking). The phytochrome gene product is combined with a light absorbing 'chromphore' inside the cell, and this product can absorb light and trigger a signal transduction. The molecule has a conserved region (the part that receives the light signal) and a more variable regulatory part, which may give different signals depending on the gene family member. The various phytochrome family members therefore seem to have slightly different roles. The following example is from *A. thaliana*:
- *Germination* of seeds is often dependent on light. When exposed to light *phyA* signals the start of germination.
- *Seedling establishment*. The seedling grows differently under the soil, compared with above ground. Once it breaks through the surface, the light signals picked up mainly by *phyB* gives rise to the inhibition of stem elongation, expansion of leaves instead, and the development of the photosynthetic machinery.
- *Shade avoidance*. When there are many neighboring plants (i.e. competition for resources), the light reflected from them (the wavelength change) can be detected by, especially, *phyB*. This signal results in the shade-avoidance syndrome, which e.g. leads to stem elongation growth (to find more light), different resource allocation and changed flowering time.
- *Photoperiodism*. The relative lengths of day and night detected by *phyA* and *phyB* is vital for the induction of flowering or the initiation of dormancy (for winter).

More molecular evidence

Recent studies using microarray techniques have shown that a seemingly simple environmental signal can give rise to a large response in changed expression of many genes. For instance, a specific light treatment (most likely mediated through the phytochromes already mentioned) gives rise to the expression of a third of all genes in *Arabidopsis*. Some of the increased expression occurred within one hour, but some the expression of some genes were delayed several hours up to days. Pathogen defense responses gave a changed expression in more than 2000 genes and exposure to drought or cold stress affected the expression of about 1300 genes (summarised in Schlichting and Smith, 2002).

In *Arabidopsis* two alcohol fermentation pathways exist. Different variants of two enzymes alcohol dehydrogenase (ADH) and pyruvate decarboxylase (PDC) are used, being the result of different genes. One of the pathways is always active but the other is activated by low oxygen levels, and specifically in the root system. At least seven genes are involved in each pathway (Dolferus *et al.,*

1997; in Pigliucci, 2001). This is an example of a environmental stress which results in a specific reaction, a reaction which most likely is adaptive, i.e. that increases the fitness of the individual. In this example, it is not known how the hypoxia signals to the cells that they should start transcription of the alternative genes.

Another kind of stress is the exposure to high salt concentrations. A comparison of the expression rates of genes between the salt-sensitive cultivated wheat and a wild relative, when exposed to salt stress showed that the expression of 11 genes was elevated in the wild grass. These genes existed and were transcribed also in the wheat, but to a much lower degree. So, in this case it is not a different set of genes that are expressed (as in the alcohol fermentation example), but that the regulation of these genes is somehow affected by the salinity – again unclear how.

Yet another environmental stressor is heat. The heat-shock response, which occurs in both plants and animals, includes the creation of a special protein that protects other proteins from denaturation. In *Arabidopsis*, 4 genes have been found that are involved in tolerance to heat shock. One of these, *Hot1* exists in a multitude of other organisms, including yeast and bacteria. The heat-shock response then seems to be of very old date, and one can therefore presume that it has been of great evolutionary value, to all organisms. One interesting note is that some cold-water fishes expressed a heat-shock induced protein at the blistering heat of 5 °C!

Is plasticity adaptive?

Does plasticity give the individual higher fitness? This is in general quite difficult to prove experimentally. One reason is actually that it is difficult to define fitness in a proper and unambiguous way. Therefore the evidence of adaptive plasticity is mainly along the lines of reasons and arguments.

It would seem rather natural to assume that plasticity for fitness is unfavourable. We would like fitness to be high even if the environment changes. So flat reaction norms for fitness would seem desirable. However, intuitively there must be some relation to the distribution of the environments over time and space – it doesn't make sense to have a very high fitness for an environment the individual never encounters. For instance, it would be very costly for humans to have a parallel system of getting oxygen (gills), even though we sometimes might fall in the sea. So, flat reaction norms for fitness are probably desirable over a reasonable range of environments.

Now, how does one achieve such flat reaction norms for fitness? The most intuitive way is to be very *plastic in other traits* that contribute to the overall fitness. A nice example of this was given by Sultan (1995) who exposed plants to two different lighting conditions and then looked at the resulting seedling biomass. She found that that the reaction norm in question was almost flat, even though one of the light treatments (the low light) should have been very stressful for the parent plant. The reaction norm for the weight of the fruit produced (achene) was also almost flat, although with a slight decrease. However, when looking at the pericarp, the thick shell surrounding the embryo and its nutrition, there was a clear decrease in that. So, the strategy of the parent plant in response to the low light availability was to decrease the less important outer shell, but try to keep the amount of nutrients for the embryo more or less the same. So by being plastic for shell thickness, the plant could stay non-plastic for fitness (keep homeostasis) (Figure 6).

One case of plasticity which is intuitively easy to understand is the adaptive value of predator-induced changes in morphology of the prey. There are several examples of such alterations. Actually, the first example of phenotypic plasticity described (by Woltereck, 1909) was the change in body shape in *Daphnia* in response to the presence of a predator. When the predator is present (or the chemicals released by the predator) the *Daphnia* individuals grows a 'helmet' on the head, which lowers the predation rate.

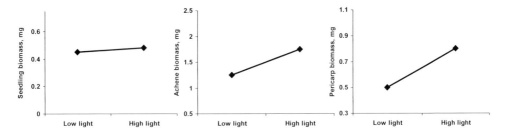

Figure 6. Schematic description of reaction norms of seedling, achene, and pericarp mass, as function of low and high light conditions for the parents. The stable seedling mass is due to a marginal change in fruit mass (achene) and a drastic decrease in pericarp (shell) mass.
Source: Summarised results from Sultan (1995).

A similar change in body conformation was found in a certain carp fish who increased its body depth in the presence of a predator (pike) (Bronmark and Miner, 1992). Since the pike has a limit to how much it can open its mouth, this size change can prevent predation. However, there must be a cost to increase size if it is not necessary, so there is some kind of balance between the benefits and costs of expressing the alternative phenotype. And it should only be expressed in the presence of the predator, so it is important that the environmental cue is reliable.

The acorn barnacle exists in two forms. The so-called bent form is more efficient in defending the animal from its predator, however, this advantage comes with a cost. It has slower growth rate and reduced fecundity, so having this form when there is no predator around is not a good idea (Lively, 1986).

A striking example of plasticity which is difficult to understand unless it is adaptive is the environmentally induced body polymorphism in a certain caterpillar, also related to predator avoidance. Caterpillars that develop during spring on oak catkins (a cluster of flowers) mimic these. Other caterpillars (of the same species) that develop during the summer, when the catkins are gone, mimic the tree twigs instead. Interestingly, the environmental cue is not the visual one, but rather a diet one. The larvae raised on catkins sense the low tannin levels in this food and develop to mimic catkins. Larvae raised on twigs sense the high tannin levels and develop to mimic twigs. Larvae experimentally raised on catkins with added tannins actually developed to mimic twigs.

Also behaviour can change as a result of a predator signal. *Daphnia* (again) have been shown to react to the presence of a fish predator by migrating to greater depths. However, there was also genetic (clonal) variation in this plasticity (Figure 7; De Meester, 1993). This reaction occurred not only when the fish was actually there, but also when a chemical released by the fish was added to the water tank. This indicates that this migrating behaviour may be anticipatory, i.e. the *Daphnia* can feel the presence of the fish before the predator threat has become too large, and act accordingly.

But this game can be played from both sides. Also predators can show plasticity and react to the existence of (various type of) prey. When a certain fish (threespine stickleback, *Gasterosteus* spp.) was exposed to a certain type of prey, they changed morphologically to better handle this type of prey, and changed their behaviour to better find the prey (Day and McPhail, 1996).

Another example of behavioural plasticity, which is likely to be adaptive, is different nest-building behaviour in mice. This was compared in five natural strains, in several replicated lines selected for high or low nesting behaviour, and in four inbred lines. All genotypes showed plasticity (with more nest-building material assembled at the low temperature), but also variation in plasticity. However,

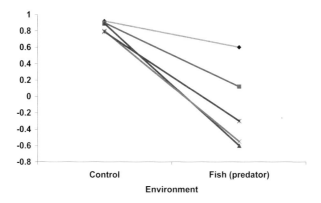

Figure 7. Behavioural plasticity for phototaxis in five clones of Daphnia magna as a response to predator exposure. When exposed to a fish predator (or the chemical released by the fish) they migrate to greater depths, but the response varies between clones. Adapted from De Meester (1993).

there was no tendency for crossing of reaction norms. The natural populations and the selected lines showed more variability under low than under high temperature (Lynch, 1994).

Given all these examples, and the plethora of more examples available, I think it is reasonable to assume that – yes, plasticity can be adaptive. But then, indeed, the opposite question arises: if plasticity is so good, why bother to adapt (as a population) when you can be plastic instead?

When does plasticity evolve?

Considering that environments change and never are stable, it would seem natural to expect that plasticity would always exist. However, environmental variability is a necessary but not sufficient condition for plasticity to evolve and there has been a lot of theoretical and simulation work trying to work out when plasticity actually would be favoured by natural selection. Another way of phrasing this is to ask what conditions would favour the development of *generalist* or *specialist* genotypes. The generalists are thought of as being plastic; they can change their phenotypes (usually morphological traits) to adapt to various environmental conditions, whereas the specialists are non-plastic (or at least less plastic). However, this relates to plasticity for various traits, not for fitness. Specialists would be plastic for fitness if exposed to environments they are not adapted for, whereas generalists would be more non-plastic for fitness. Note that there might be some confusion of the terms 'generalist' and 'specialist' between the animal breeding world and the evolutionary biology world, e.g. cows that keep a high production in many environments have been called 'generalists' in animal breeding literature.

The general outcome of all these studies can be summarised as follows (Bradshaw, 1965; Levins, 1968; Schlichting and Pigliucci, 1998; Schlichting and Smith, 2002). Plasticity is *selected against* if:
- the environmental change is rare;
- environmental cues are not reliable;
- the environment fluctuates faster than the typical response time;
- a single phenotype is optimal in both (all) environments;
- the environment is spatially coarse-grained, and the organism can choose its habitat.

The first two conditions are fairly easy to understand, there is no need for plasticity to evolve if changes are rare or if the environmental changes are not easily predictable. Also, if the environment changes (back and forth) quicker than the needed phenotypic change can take place (e.g. morphology) then by the time the phenotype has changed, the environment has already changed again.

Coarse-grained environments mean that the environment changes across a wide spatial or temporal range (i.e. over a long distance or over generations). The result is that one individual experiences only one environment. On the other hand, *fine-grained* means that changes occur within a short distance or within a generation (each individual can be exposed to more than one environment). If an individual can choose a habitat for which it is better adapted, then plasticity is not needed.

The conditions when plasticity is *selected for* are if:
- environmental change is frequent;
- environmental cues are reliable;
- environmental variation is temporally or spatially fine-grained;
- environmental variation is temporally coarse-grained with predictive cues (leads to polyphenism);
- environmental variation is temporally fine-grained in a predictable sequence (leads to heteroblasty).

If one and the same individual experiences more than one environment, plasticity is expected to evolve. Actually, if there are no constraints (or costs) to plasticity, the population will converge on a genotype that results in the optimal phenotype in all environments. This would mean that there would be no genetic variation for plasticity. The prediction has led to a lot of scientific discussion, in relation to whether plasticity can play a role in the maintenance of genetic variation or not.

If the environmental cues are predictable but the changes do not occur within the generation (e.g. short-lived organisms and seasonal variation) then *polyphenism* can evolve, i.e. the individual displays discrete phenotypes adapted to that specific season. If the sequence of environmental changes is predictable, but temporally fine-grained (happens within generation), then each individual may have a fixed sequence of phenotypes during its development (*heteroblasty*) and each phenotype adapts to its specific environment (e.g. season).

Costs and limits of plasticity

Plasticity would seem to be the optimal solution to cope with changing environments, and one might expect that natural selection would create organisms that show the best phenotype in all environments encountered. Such organisms are sometimes called 'Darwinian monsters'. However, in reality, this does not happen. Why not? One answer to that question is that plasticity does not come free – there is also a cost associated with it. Furthermore, there may be factors limiting the benefit of plasticity, some of which we have already discussed. The following costs and limitations have been suggested (DeWitt *et al.*, 1998):

Costs of plasticity

As costs of plasticity can be mentioned:
- Maintenance of machinery – energetic costs of sensory and regulatory mechanisms.
- Production cost – cost of producing structures plastically (compared to creating them by fixed genetic responses).
- Information acquisition – energy used for sensing the environment (which could have been used for other activities).
- Development instability – plasticity may imply reduced canalisation within each environment.
- Genetic – deleterious effects of plasticity genes, either through linkage, pleiotropy, or epistasis with other genes.

Limits of plasticity

As limits of plasticity can be mentioned:
- Information reliability – environmental cues may be unreliable or changing too fast.
- Lag time – if the lag between the cue and the response is too long the response may be maladaptive, e.g. if the environment has already changed again.
- Developmental range – plastic genotype may not be able to express as large a range of phenotypes as would be possible by 'specialists' (ecotypes).
- Epiphenotype problem – the plastic response may be an 'add-on' to the basic system, rather than an integrated system that is expected in a genetically adapted organism.

Some of these limits have already been raised as factors leading to plasticity *not* evolving, e.g. when cues are unreliable or too fast-changing. Sometimes a different cue than the actual environmental change can be used. For instance, plants preparing for the winter might have a problem if they try to follow the temperature changes (which may go up and down quite drastically during the fall). By instead using the photoperiod change as a cue, which is much more continuous than the change in temperature, the plant can better predict the on-coming winter (Schlichting and Pigliucci, 1998).

Even though the suggested reasons for costs and limitations in most cases make very good sense, it is not easy to measure these costs. Some results are however available. Flies (*Drosophila*) conditioned to high temperatures (thus turning on the heat-shock response) survived severe heat stress much better than unconditioned flies. However, these conditioned flies produced fewer offspring than nonconditioned flies when *not* exposed to heat. This shows that in normal environments the heat-shock response is maladaptive (Krebs and Loeschcke, 1994; Kreps and Simon, 1997).

The costs and limits to plasticity have also led to the hypothesis called 'Jack-of-all-trades and master-of-none'. What this basically says is that there is no generalist as good as the best specialist for a given environment. Again, no Darwinian monster. This is illustrated in Figure 8 where real data on growth habits of two specialist varieties of *Plantago lanceolata* (Van Tienderen, 1990) are complemented by hypothetical reaction norms for two generalists, one perfect (that does not exist) and one imperfect. The (imperfect) generalist would be more plastic for the growth habit than either of the specialists, but would not achieve the same high fitness as either of the specialists in their intended environment.

Consequences for animal breeding

What then can we learn from studies of plasticity and adaptation in evolutionary genetics? First of all, the concept of 'one size fits all' is not a reasonable one. If you have too large an environmental range in nature, specialists evolve. If we look at animal breeding from a truly global perspective, I think that we need to come to the same conclusion. For instance, there are too many diseases in the world for all cows to be especially resistant to all of them, and even if it were possible it would be extremely costly to achieve such a resistance (e.g. to diseases that cows in a certain region never encounter). Specialisation is also expected to give rise to higher genetic diversity as a between-population component.

Having said that, within a smaller range of environments, animals should be robust, i.e. achieve a high value of the function related to 'fitness' regardless of 'micro-environmental' changes. What is then fitness from an animal breeding perspective? The closest counterpart that I can think of is the profit function that is used to develop the total merit index (TMI). This profit function, or production fitness function, should be broad and include both production and functional traits otherwise it will be in too much conflict with the natural fitness function, diseases and fertility will become severely limiting, and thus the breeding program will not be sustainable in the long term. This profit function,

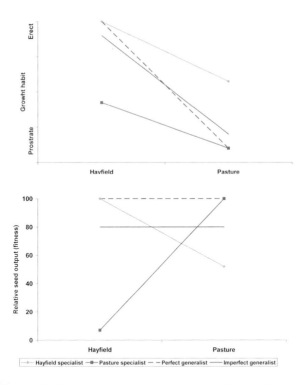

Figure 8. Growth habit (erect or prostrate, upper panel) and relative seed output (a measure of fitness, lower panel) for two specialist varieties of Plantago lanceolata *(after Van Tienderen, 1990), complemented with hypothesised values for a perfect and an imperfect generalist. Note that the imperfect generalist (Jack of all trades) would have lower fitness than either of the specialists in the environment they are adapted to. The perfect generalist (Darwinian monster) would never exist in reality. The generalists are more plastic for the growth habit than the specialists.*

or rather a predictor of the TMI, is used to select animals for reproduction and the genetic change in any trait is proportional to the covariance between that trait and the *predictor* of TMI. Therefore, it is important to ensure that the structure of the breeding program allows for good prediction of all traits, otherwise the genetic response pattern will be very different from that indicated by the profit function and TMI.

Finally then, back to the posed question in the beginning: do we want cows to be robust? In good scientific tradition, the answer is both yes and no!

References

Baldwin, J.M., 1896. A new factor in evolution. Am. Nat., 30: 441-451; 536-553.

Baldwin, J.M., 1902. Development and Evoluton. New York, Macmillan.

Bradshaw, A.D., 1965. Evolutionary significance of phenotypic plasticity in plants. Advances in Genetics, 13: 115-155.

Bronmark, C. and J.G. Miner, 1992. Predator-induced phenotypical change in body morphology in crucian carp. Science, 258: 1348-1350.

Crews, D., 1994. Temperature, steroids, and sex determination. J. Endocrinology, 142: 1-8.

Crews, D., 1996. Temperature-dependent sex determination: the interplay of steroid hormones and temperature. Zool. Sci., 13: 1-13.

Breeding for robustness in cattle

Day, T. and J.D. McPhail, 1996. The effect of behavioural and morphological plasticity on foraging efficiency in the threespind stickleback (*Gasterosteus* sp.). Oecologica, 108: 380-388.

De Meester, L., 1993. Genotype, fish-mediated chemicals and phototactic behavior in *Daphnia magna*. Ecology, 74: 1467-1474.

DeWitt, T.J., A. Sih and D.S. Wilson, 1998. Costs and limits to benefits as constraints on the evolution of phenotypic plasticity. Trends Ecol. Evol., 13: 77-81.

Dolferus, R., M. Ellis, G. de Bruxelles, B. Trevaskis, F. Hoeren, E.S. Dennis and W.J. Peacock, 1997. Strategies of Gene Action in *Arabidopsis* during Hypoxia. Annals of Botany, 79: 21-31.

Falconer, D.S., 1952. The Problem of Environment and Selection. Am. Nat., 86: 293-298.

Hartl, D.L. and D.E. Dykhuizen, 1981. Potential for selection among nearly neutral allozymes fo 6-phosphogluconate dehydrogenase in *Escherichia coli*. Proc Natl Acad. Sci. USA, 78: 6344-6348.

Johannsen, W., 1911. The genotype conception of heredity. Am. Nat., 45: 129-159.

Koch, P.B., P.M. Brakefield and F. Kesbeke, 1996. Ecdysteroids control eyespot size and wing color pattern in the polyphenic butterfly *Bicyclus anynana* (Lepidoptera: Satyridae). J. Insect Phys., 42: 223-230.

Krebs, R.A. and V. Loeschcke, 1994. Costs and benefit of activation of the heat-shock response in *Drosophila melanogaster*. Functional Ecology, 8: 730-737.

Kreps, J.A. and A.E. Simon, 1997. Environmental and genetic effects on circadian clock-regulated gene expression in *Arabidopsis*. Plant Cell, 9: 297-304.

Lerner, I.M., 1954. Genetic Homeostasis. London, Oliver and Boyd.

Levins, D.A., 1968. Evolution in Changing Environments. Princeton, NJ, USA, Princeton Univ. Press.

Lin, B.L. and W.L. Yang, 1999. Blue light and abscisic acid independently induce heterophyllous switch in *Marsilea quadrifolia*. Plant Phys., 119: 429-434.

Lively, C.M., 1986. Competition, comparative life histories, and maintenance of shell dimorphism in a barnacle. Ecology, 67: 858-864.

Lively, C.M., 1986. Predator-induced shell dimorphism in the acorn barnacle *Chthamalus anisopoma*. Evolution, 40: 232-242.

Lynch, C.B., 1994. Evolutionary inferences from genetic analyses of cold adaptation in laboratory and wild populations of the house mouse. Chicago, Univ. of Chicago Press.

Pfennig, D.W., 1992. Proximal and functional causes of polyphenism in an anuran tadpole. Functional Ecology, 6: 167-174.

Pigliucci, M., 2001. Phenotypic plasticity: beyond nature and nurture. Baltimore, The Johns Hopkins University Press.

Rhen, T. and J.W. Lang, 1995. Phenotypic plasticity for growth in the common snapping turtle: effects of incubation temperature, clutch, and their interaction. Am. Nat., 146: 726-747.

Scheiner, S.M. and R.F. Lyman, 1991. The genetics of phenotypic plasticity. II. Response to selection. J. Evol. Biol., 4: 23-50.

Schlichting, C. and M. Pigliucci, 1993. Control of phenotypic plasticity via regulatory genes. Am. Nat., 142: 366-370.

Schlichting, C. and M. Pigliucci, 1995. Gene regulation, quantitative genetics and the evolution of reaction norms. Evolutionary Ecology, 9: 154-168.

Schlichting, C. and M. Pigliucci, 1998. Phenotypic evolution: a reaction norm perspective. Sunderland, MA, Sinauer Associates.

Schlichting, C. and H. Smith, 2002. Phenotypic plasticity: linking molecular mechanisms with evolutionary outcomes. Evolutionary Ecology, 16: 189-211.

Schmalhausen, I.I., 1949. Factors of evolution: the theory of stabilizing selection. Chicago, University of Chicago Press.

Sultan, S.E., 1995. Phenotypic plasticity and plant adaptation. Acta Botanica Neederlandica 44: 363-383.

Van Tienderen, P.H., 1990. Morphological variation in *Plantago lanceolata*: Limits of plasticity. Evol. Trends Plants, 4: 35-43.

Waddington, C.H., 1942. Canalization of development and the inheritance of acquired characters. Nature, 150: 563-565.

Waddington, C.H., 1952. Selection of the genetic basis for an acquired character. Nature, 169: 278.

Williams, J., M. Bulman, A. Huttly, A. Phillips and S. Neill, 1994. Characterization of cDNA from *Arabidopsis thaliana* encoding a potential thiol protease whose expression is induced independently by wilting and abscisic acid. Plant Molecular Biol., 25: 259-270.

Woltereck, R., 1909. Weitere experimentelle Untersuchungen über Artveränderung, speziell über das Wesen quantitativer Artunterschiede bei Daphniden. Versuche Deutsche Zoologische Gesellschaft, 19: 110-172.

Genetic concepts to improve robustness of dairy cows

J. ten Napel, M.P.L. Calus, H.A. Mulder and R.F. Veerkamp
Animal Breeding and Genomics Centre, Wageningen UR, P.O. Box 65, 8200 AB Lelystad, the Netherlands

Abstract

The past decade has revealed unfavourable trends in udder health, fertility and locomotion in dairy cattle. More recently, dairy herds are increasing and the availability of skilled labour per animal is decreasing. This has increased the interest in more robust dairy cows. Robustness of a cow was defined as 'the ability to maintain homeostasis in commonly accepted and sustainable dairy herds of the near future'. Robustness is largely an acquired characteristic. Adaptive systems never cease to develop. Success of adaptation also depends on available resources and opportunities from the environment. Genotype merely gives an individual an advantage or disadvantage in becoming robust. Several concepts of breeding for robustness exist. These include avoiding inbreeding depression, utilising heterosis, using multi-trait selection, reducing environmental sensitivity in the target range dairy production systems and breeding for group performance. Environmental sensitivity may be reduced through natural selection, using genetic correlations between environments, using reaction norms or using genetic variation in residual variance within progeny groups of sires.

Keywords: adaptability, animal breeding, genotype environment interaction

Introduction

Cannon (1932) first used the term homeostasis to indicate that a body continuously acts to maintain a stable internal environment by responding to external environmental stimuli. In the last two decades, there have been concerns that high-yielding dairy cows struggle to maintain homeostasis. Several studies reported unfavourable genetic correlations between milk yield and reproductive problems, locomotive problems and udder health problems (Pryce *et al.,* 1997; Rauw *et al.,* 1998; Royal *et al.,* 2000), and there is general consensus that selection for milk fat and milk protein yield alone may give an unfavourable correlated response in these traits (Dechow *et al.,* 2002). The magnitude of these correlated responses is rather small, compared with the effects of environmental disturbance, albeit when the effects of breeding are accumulated across years these might be substantial. The gradual reduction in genetic levels for fertility and health will put more pressure on management to maintain performance at acceptable levels. Such cows require more management attention.

At the same time, the level of management is increasingly coming under pressure from other directions. For example, due to economic pressure, herd size is increasing and therefore the amount of labour available per animal is decreasing. The shortage of labour is aggravated by the fact that it is increasingly difficult to find suitably skilled labour. Also, pressure on management increases because previously simple and effective management tools, such as the use of antibiotics, are now perceived as potential risks for human health and therefore regulated much stronger.

These two trends, i.e. negative effects from selection for yield and increasing pressure on management, have fuelled the demand for more robust cows. These two trends may amplify each other in practice, as correlations with yield vary across environments (Windig *et al.,* 2005, 2008). A popular description of robust cows could be something like productive dairy cows that:
1. maintain homeostasis in an increasingly dynamic production environment;
2. are suitable for a wider range of dairy production systems;

3. only need the basic individual care; and
4. are easy to manage.

However, here we describe a more formal concept for robust cows first, and then describe the contribution that genetics can make to robustness. Robustness of a system is dependant on the chosen system level (e.g. cow, farm or sector), but here we restrict ourselves to the level of the cow.

The concept of robust cows

Robustness: control or adaptation

From a farm management point of view, there are essentially two ways to control the impact of disturbances on an animal (Ten Napel *et al.,* 2006). They are not mutually exclusive, but go together well. The one approach is called the Control Model and is characterised by maintaining stability through keeping away disturbances. Typically a strategy is used of protecting animals from disturbances as much as possible, constantly monitoring animals, whether a disturbance occurs, and interventions targeted at the disturbance, when it does occur. When taken to the extreme, the homeostasis of the animal is dependent on proper and timely functioning of humans and technical equipment.

The other approach is called the Adaptation Model. This approach is characterised by maintaining stability through minimising the impact of disturbances in the presence of the disturbance. The design of such a production system seeks to utilise the intrinsic capacity of animals to adapt where possible, and use the Control Model approach where necessary (Ten Napel *et al.,* 2006).

Definition of robustness

In the Netherlands, we gradually developed a concept of robustness of farm animals in the course of three to four years, based on discussions with many groups of stakeholders. This process resulted in the following definition of a robust dairy cow: 'A robust dairy cow is a cow that is able to maintain homeostasis in the commonly accepted and sustainable dairy herds of the near future'. It is clear from this definition that robustness is not just a matter of the average level of management being suitable for the cow. Dairy herds are dynamic and fluctuations in temperature, air speed, humidity, disease pressure, fodder quality, stocking density, social interaction with other cows, aggression, interaction with stockmen, among other factors, occur. Over time or across herds, common fluctuations largely fit in a certain band width. A cow that is robust is able to maintain homeostasis in a range of production environments with a band width that is wider than the common band width of fluctuations. It does not mean that a robust cow must be able to cope with anything. It is acceptable to look for ad-hoc solutions to unlikely disturbances and freak incidents.

To interpret the current situation, as described in the Introduction, in terms of band widths, it appears that the band width that cows can handle is gradually reducing, while the band width of fluctuations in dairy production systems is gradually increasing again, after decades of reducing the bandwidth by reducing the standardisation and improving the level of management following the Control model.

Intrinsic factors for robustness of cows

The genotype of a cow is a major determinant of the robustness of a cow. There are many examples that illustrate the role of genetics in susceptibility to disease (Morris, 2000; Owen *et al.,* 2000) and traits of the immune system (Detilleux *et al.,* 1994). In some cases, the genotype merely gives an individual an advantage or disadvantage in becoming robust. In other cases, the genotype conveys complete resistance or susceptibility to particular threats.

However, apart from genotype, there is another important factor for robustness, i.e. the development of adaptive systems, such as the immune system and the nervous system. These adaptive systems develop in response to environmental signals throughout an animal's life, and this development is unique for every individual. As illustrated by (Tada, 1997): 'These systems engender their own elements from a single progenitor. The diverse elements thus generated form relationships by mutual adaptation and co-adaptation, and thus create a dynamic self-regulating system through self-organisation. They are closed self-satisfied systems, yet open to the environment, receiving outside signals to transduce them into internal messages for self-regulation and expansion'.

The adaptive systems of a newborn are primed for the environment in which the calf is born, for example through maternal antibodies and circulating maternal cortisol. In humans, maternal antibodies attenuate most infections in early life and turn them into effective vaccines. If this 'natural vaccination' does not occur, the infections may be severe, unless the child is actively vaccinated in synchronisation with the immune system's maturation (Zinkernagel, 2003).

The intrinsic robustness of a cow depends on the genotype, the early development of adaptive systems in the body and the challenges experienced in life.

Expression of robustness

Robustness of cows not only depends on the intrinsic robustness of the individual animals, but also on the opportunity for a cow to use the ability at farm level. For example, a diseased cow may have a different feed and temperature requirement and may want to temporarily isolate itself from the social group. Expression of robustness is not so relevant in the design of systems based on the Control Model, as it is focused on keeping away disturbances. Therefore, using the Adaptation model may require a re-design of the common dairy production systems, to allow for utilising the intrinsic capacity of animals to adapt.

Another aspect is the temporary vulnerability of a cow. For example, a cow that experiences a severely negative energy and protein balance during the peak of lactation may not have the resources to successfully combat an infection. So whether a cow is successful in maintaining homeostasis not only depends on the genotype and the development of adaptive systems, but also on the resources and opportunities offered by the environment to mount a response.

Breeding for robust animals

Robustness of an animal in a production environment is related to the genetic fitness in a natural environment. Genetic fitness of an individual is the contribution of genes that it makes to the next generation, or the number of its progeny represented in the next generation in absence of artificial selection (Falconer and Mackay, 1996). The characteristics that determine the contribution to the next generation in a natural environment may largely make up robustness in a livestock production environment. Animal breeding aims to use the genetic variation present for fitness and different strategies to do so will be discussed here.

Heterosis and inbreeding depression

Overcoming or avoiding inbreeding depression and maximising heterosis is a relatively easy way to improve genetic fitness and is widely utilised in pig and poultry breeding programmes. These effects come from the observation that characteristics associated with genetic fitness often reveal overdominance, that is the phenotype of heterozygotes is superior to the phenotype of any of the two types of homozygotes. In studies of national subpopulations of 2.2 to 3.7 million heads of the largest dairy breed in the world, the Holstein, the effective population size has been estimated to be between 46 and 68, as reviewed by Taberlet (2008) The consequence is a high rate of inbreeding, associated with possible reduced fitness due to inbreeding depression. Widening the pool from which bulls are selected and limiting the use of individual bulls would reduce the rate of inbreeding significantly. This might lead to a loss of genetic progress, but this can be limited by optimising gain and inbreeding (Meuwissen and Sonesson, 1998). Although common in pig and poultry breeding, crossbreeding is rarely practiced in a structured manner in dairy cattle breeding. The first main reason is that a breeding programme with a pyramid structure is not possible, except when using super-ovulation and embryo transfer. Without these reproductive technologies, the average number of replacement heifers available per cow per year is between 0.4 and 0.5. So only when the replacement rate is around 20% per year, can a purebred herd support a crossbred herd of the same size. The second main reason is the unfavourable recombination that counteracts the favourable heterosis for milk production traits (Dechow *et al.*, 2007, Pedersen and Christensen, 1989), which reduced the need to look for feasible ways to utilise heterosis in dairy cattle breeding. However, for maintaining overall robustness, crossbreeding has become of interest again (Heins *et al.*, 2008). Rotational crossing, as sometimes practiced in beef cattle breeding (Marshall *et al.*, 1990), may be suitable for this purpose (McAllister, 2002).

Multi-trait selection

The most common way to breed for robustness, is to include fitness traits in the breeding goal, breeding value estimation and the selection index. This has been practiced in many dairy countries (Miglior *et al.*, 2005). One example of multi-trait selection being effective is shown in Figure 1, where the unfavourable trend in calving interval disappeared in progeny-tested bulls of both the Dutch Black and White and Red and White Holstein breed after inclusion of the trait in the selection index in 2001 (Van Drie, 2007). Philipsson and Lindhé (2003) provided similar examples for mastitis and fertility in the Nordic countries.

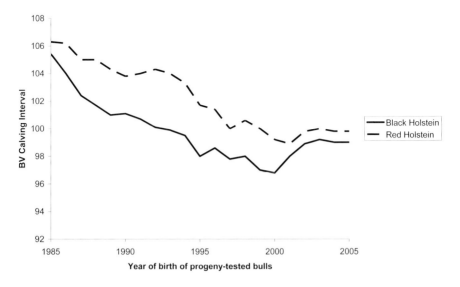

Figure 1. Genetic trend for calving interval for progeny-tested Dutch Black and White and Red and White Holstein bulls. Reproduced from Figure 3 in Van Drie (2007).

Breeding for robustness in cattle

Although multi-trait selection clearly works and is better than single trait selection, there are four reasons why the commonly applied approach multi-trait index might not be satisfactory enough to maintain or improve robustness. Firstly, it is difficult for the many aspects of fitness to clearly define a trait to be measured. For example, mastitis may be caused by a large number of pathogens and may or may not result in clinical symptoms. If all pathogens are combined, the trait may be insufficiently discriminative, resulting in a very low heritability, but if groups of pathogens are distinguished, there may be insufficient observations for each trait to be meaningful. An additional issue is to identify the most relevant traits from the very large number of problem states that are potential indicators of (lack of) robustness.

Secondly, the data for fitness traits is often incomplete or censored. This may take several forms. For example, a cow that was culled after insemination, does not have a record for the success of that insemination. A highly productive cow may get a second and third chance to breed or recover from an infection, whereas a below-average productive cow may be culled after the first appearance of the problem. Further, as the nature of many recorded fitness traits is the presence of a problem, it is often not clear whether the absence of a record should be interpreted as absence of the problem or a missing record. Often it is assumed that herds either do not submit information or submit complete information, but there is no way to verify this assumption. For example, herds without records of clinical mastitis may genuinely not have had clinical cases, but may also not have recorded clinical cases that occurred. Testday SCC is more readily available in practice and provides information on intra-mammary infections

Thirdly, it has been suggested to include traits of the adaptive systems in the multi-trait index, like antibody and cell-mediated immune responsiveness (Groves *et al.*, 1993; Wagter *et al.*, 2000). These systems, however, are highly integrated life systems with a high degree of unpredictability and ambiguity observed (Tada, 1997). There are no mathematically linear cause-effect relationships in many of the important immune phenomena. Although genetic variation exists in immune components, it is unlikely that genetic selection for changes in the adaptive systems will improve robustness in the foreseeable future, except for the rare cases where complete resistance is conveyed through simple inheritance. In practice, it means that breeding for robustness is likely to be more successful when it concerns the result of adaptation (i.e. effective coping), rather than the adaptation process itself.

Fourthly, from the perspective of population biology, fitness traits are expected to exhibit genotype by environment interaction, causing animals to rank differently in different environments. In this way, after any major change in living conditions, there is at least a part of the population that is able to survive in the new environment. For milk production traits, genotype by environment is largely limited to scaling effects, when comparing North-American and Western European production environments, causing the difference between the best and the worse animal to increase with the average level of production, but without affecting the ranking of animals (Mark, 2004). For this reason, breeding value estimation generally includes a correction for heterogeneous variances for milk production traits. However, this is not a guarantee that fitness in all environments is improved (Mark, 2004). When following the definition of robustness given above, robustness is about maintaining homeostasis across environments and environmental challenges, and not only about having a high fitness on average across environments. Clearly environmental sensitivity and genotype by environment interaction play an important role when breeding for robustness.

Environmental sensitivity

Some animals respond to a change in environment for some characteristics in a much stronger way than other animals. Such animals are more environmentally sensitive for these traits. In population biology terms, these animals are called 'specialists' as they have a very high fitness only in specific conditions. Less environmentally sensitive animals are called 'generalists'. These qualifications are not absolute, but relative to the range of environments considered. A cow may be a 'specialist' when considering all possible environments, but a 'generalist' when considering the range of acceptable production systems in a country. A robust cow is more of a 'generalist' as it has a reasonable fitness across relevant production environments (Bryant *et al.*, 2006).

If animals rank differently in different environments, then there is a risk of selecting environmentally sensitive individuals, if selection mainly takes place in the favourable environments. The question is how to identify the 'generalists' for the target range of environments. We distinguish four situations, (1) natural selection, (2) production environments fall into a limited number of categories, (3) production environments differ on a continuous environmental parameter and (4) production environments differ on a large number of aspects with a relatively small effect each.

Natural selection

An implicit way to breed for 'generalist' cows, is to create conditions in which cows that are environmentally sensitive in the target range of environments have a selection disadvantage. In natural populations without artificial selection, genetic fitness is maintained through a self-structuring force, called natural selection. If through a change in the environment, variation in genetic fitness arises, then the increase in fitness in the population is equal to the additive genetic variance of fitness at that time (Falconer and Mackay, 1996). Translated to production environments, genetic variation in fitness at an average level of management may be masked in environments with high-level management. Selecting bull dams in the latter group of herds will increase the dependency on high-level management and hence increase environmental sensitivity, if not accounted for.

Natural selection can not be utilised easily in breeding programmes for practical reasons. Breeding animals are often kept under strict biosecurity control in order to be able to sell semen or breeding stock. However, an option is to ensure that the target range of production environments is properly represented in the environments in which the progeny of the bulls are evaluated to improve general adaptability by selecting on an index based on the EBV of the specific environments (Mulder *et al.*, 2006). Natural selection in livestock production is particularly relevant for the very large number of fitness traits that are not recorded and included in a selection index.

For small, deviating production environments, such as low-input farming or organic farming, it may also be worthwhile to consider a breeding strategy that is radically different from using proven bulls with high-reliability indices and relies on bull dams that have proven to thrive in the specific production environment.

Limited number of environmental categories

For example, the distinction between organic and non-organic dairy herds is discrete. Grouping herds on prevailing soil type may yield five or six categories. Other examples are grouping by geographical region, type of production system or presence or absence of a major disease. If there is evidence of a change in ranking between groups of environments, the most obvious way is to define the trait of interest within groups and estimate genetic correlations between measurements of the trait in different environments. With these genetic parameters it is possible to carry out a multi-trait breeding value analysis for each trait measured in different groups of environments. In fact, this is how Interbull operates with its world wide breeding value estimation. A separate breeding programme for a challenging and a non-challenging environment may be considered if the genetic correlation between the two groups of environments is lower than 0.70 – 0.80 (Mulder, 2007; Mulder *et al.*, 2006).

Continuous environmental parameter

The performance of a genotype across the range of the environmental parameter is called the reaction norm. If a change in a certain environmental parameter affects some genotypes more than others, then there is genetic variation in environmental sensitivity. Conceptually, this could apply to a wide range of environmental parameters, such as temperature, stocking density, bacterial load, concentrates consumption, etc., but in practice, it is very difficult to collect reliable information. Hence, the specific environments are usually quantified by the mean performance of all genotypes, which then becomes the environmental parameter. An example of estimated breeding values for survival being dependent on the herd-year average of fat-protein ratio is shown in Figure 2 (Calus *et al.*, 2005). Each line in this graph represents the reaction norm of one bull. Random regression models have been used successfully in dairy cattle data to estimate reaction norms (Calus, 2006; Kolmodin *et al.*, 2002; Schaeffer and Dekkers, 1994; Windig *et al.*, 2006). The results of these analyses indicate that the genetic correlation between two traits may also be dependent on the level of the environmental parameter (Windig *et al.*, 2006).

Many small differences between environments

In practice, dairy herds differ in many ways and discrete or continuous environmental parameters describe these differences only in part. These unexplained micro-environmental differences may lead to genetic differences in micro-environmental sensitivity, which is observed as differences in residual variance. When bulls have at least 50-100 progeny, breeding values can be estimated for the size of the residual variance (Mulder *et al.*, 2007). Bulls with progeny that exhibit a large residual variation across herds are the ones that are environmentally sensitive. Bulls with progeny that exhibit a small residual variation across herds are the ones that are not environmental sensitive. Figure 3 shows the distribution of residual variance of broiler sires for body weight, showing a large variation between sires in residual variance (H.A. Mulder, unpublished data).

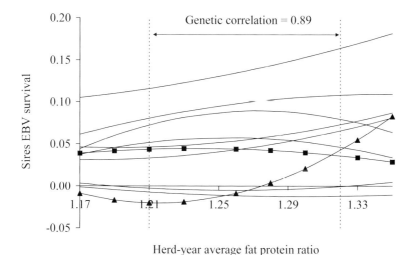

Figure 2. Breeding values for survival of the 10 sires with most daughters in the heifer data, estimated as function of herd-year average fat-to-protein ratio (squares and triangles mark breeding values of 2 particular sires). Tenth and 90th percentiles of the data are shown as dotted lines (reproduced from (Calus et al., *2005)).*

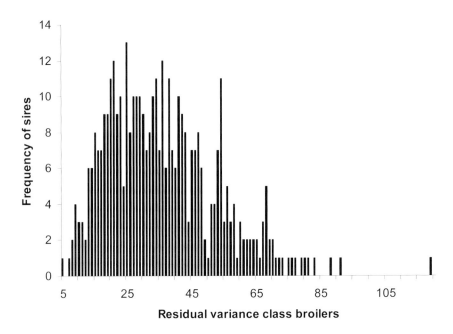

Figure 3. The frequency of sires per class of residual variance for body weight in broilers (H.A. Mulder, unpublished data).

Reduced environmental disturbances through breeding

Virtually all animals in animal production are kept in groups. It means that the genotype of an individual cow not only affects the performance of the cow itself, but it also determines part of the environment of its group mates. Hence, selecting the fittest animals directly, might have deleterious effects on its group mates (Muir, 2005), for example in the form of dominant social behaviour or aggressive behaviour, but also in the form of the amount of pathogens or parasites shed during an infection. The theory of simultaneously estimating direct genetic and associative effects has been developed by Bijma *et al.* (2007a,b). Work in layers showed that these models reveal more genetic variation in survival, which will enhance selection responses in comparison to selection on only direct effects (Ellen *et al.*, 2008). Hence, this work is a good example how selecting more robust animals at the individual level does not always lead to a higher robustness at pen or system level.

Conclusions

A robust dairy cow is able to maintain homeostasis in the commonly accepted and sustainable dairy herds of the near future. Robustness is largely an acquired characteristic through building up experience from exposure to a very large number of minor and major environmental signals. Breeding may give animals an advantage in acquiring robustness. Several methods have been discussed to improve robustness through breeding in practice.

Acknowledgements

This work was financially supported as part of the project 'Innovative and Practical Breeding Tools for Improved Dairy Products from More Robust Dairy Cattle (RobustMilk)' by the Seventh EU Framework Programme.

References

Bijma, P., W.A. Muir and J.A.M. van Arendonk, 2007a. Multilevel selection 1: Quantitative genetics of inheritance and response to selection. Genetics, 175: 277-288.

Bijma, P., W.M. Muir, E.D. Ellen, J.B. Wolf and J.A.M. van Arendonk, 2007b. Multilevel selection 2: Estimating the genetic parameters determining inheritance and response to selection. Genetics, 175: 289-299.

Bryant, J.R., N. Lopez-Villalobos, J.E. Pryce, C.W. Holmes and D.L. Johnson, 2006. Reaction norms used to quantify the responses of New Zealand dairy cattle of mixed breeds to nutritional environment. New Zealand Journal of Agricultural Research, 49: 371-381.

Calus, M.P.L., 2006. Estimation of genotype x environment interaction for yield, health and fertility in dairy cattle. PhD Thesis, Wageningen UR, Wageningen.

Calus, M.P.L., J.J. Windig and R.F. Veerkamp, 2005. Associations among descriptors of herd management and phenotypic and genetic levels of health and fertility. J. Dairy Sci., 88: 2178-2189.

Cannon, W.B., 1932. The wisdom of the Body. W.W. Norton, New York. Accessed through: http://chla.library.cornell.edu/cgi/t/text/text-idx?c=chla;idno=3117174

Dechow, C.D., G.W. Rogers and J.S. Clay, 2002. Heritability and correlations among body condition score loss, body condition score, production and reproductive performance. J. Dairy Sci., 85: 3062-3070.

Dechow, C.D., G.W. Rogers, J.B. Cooper, M.I. Phelps and A.L. Mosholder, 2007. Milk, Fat, Protein, Somatic Cell Score, and Days Open Among Holstein, Brown Swiss, and Their Crosses. J. Dairy Sci., 90: 3542-3549.

Detilleux, J.C., K.J. Koehler, A.E. Freeman, M.E. Kehrli and D.H. Kelley, 1994. Immunological Parameters of Periparturient Holstein Cattle - Genetic-Variation. J. Dairy Sci., 77: 2640-2650.

Ellen, E.D., J. Visscher, J.A.M. van Arendonk and P. Bijma, 2008. Survival of laying hens: Genetic parameters for direct and associative effects in three purebred layer lines. Poultry Science, 87: 233-239.

Falconer, D.S. and T.F.C. Mackay, 1996. Introduction to quantitative genetics. Fourth ed. Longman, Harlow.

Groves, T.C., B.N. Wilkie, B.W. Kennedy and B.A. Mallard, 1993. Effect of selection of swine for high and low immune responsiveness on monocyte superoxide anion production and Class II MHC antigen expression. Veterinary Immunology and Immunopathology, 36: 347-358.

Heins, B.J., L.B. Hansen, A.J. Seykora, D.G. Johnson, J.G. Linn, J.E. Romano and A.R. Hazel, 2008. Crossbreds of Jersey x Holstein Compared with Pure Holsteins for Production, Fertility, and Body and Udder Measurements During First Lactation. J. Dairy Sci., 91: 1270-1278.

Kolmodin, R., E. Strandberg, P. Madsen, J. Jensen and H. Jorjani, 2002. Genotype by environment interaction in Nordic dairy cattle studied using reaction norms. Acta Agriculturae Scandinavica Section A-Animal Science, 52: 11-24.

Mark, T., 2004. Applied genetic evaluations for production and functional traits in dairy cattle. J. Dairy Sci., 87: 2641-2652.

Marshall, D.M., M.D. Monfore and C.A. Dinkel, 1990. Performance of Hereford and 2-Breed Rotational Crosses of Hereford with Angus and Simmental Cattle.1. Calf Production through Weaning. J. Animal Sci., 68: 4051-4059.

McAllister, A.J., 2002. Is crossbreeding the answer to questions of dairy breed utilization? J. Dairy Sci., 85: 2352-2357.

Meuwissen, T.H.E. and A.K. Sonesson, 1998. Maximizing the response of selection with a predefined rate of inbreeding: Overlapping generations. J. Animal Sci., 76: 2575-2583.

Miglior, F., B.L. Muir and B.J. van Doormaal, 2005. Selection indices in Holstein cattle of various countries. J. Dairy Sci., 88: 1255-1263.

Morris, C.A., 2000. Genetics of susceptibility in cattle and sheep. In: Breeding for disease resistance in farm animals, 2nd edition. R.F.E. Axford, S.C. Bishop, F.W. Nicholas and J.B. Owen (eds.) CABI Publishing, Wallingford, pp. 343-356.

Muir, W.M., 2005. Incorporation of competitive effects in forest tree or animal breeding programs. Genetics, 170: 1247-1259.

Mulder, H.A., 2007. Methods to optimize livestock breeding programs with genotype by environment interaction and genetic heterogeneity of environmental variance. PhD Thesis, Wageningen University, Wageningen.

Mulder, H.A., P. Bijma and W.G. Hill, 2007. Prediction of breeding values and selection responses with genetic heterogeneity of environmental variance. Genetics, 175: 1895-1910.

Mulder, H.A., R.F. Veerkamp, B.J. Ducro, J.A.M. van Arendonk and P. Bijma, 2006. Optimization of dairy cattle breeding programs for different environments with genotype by environment interaction. J. Dairy Sci. 89: 1740-1752.

Owen, J.B., R.F.E. Axford and S.C. Bishop, 2000. Mastitis in dairy cattle. In: Breeding for disease resistance in farm animals, 2nd edition. R.F.E. Axford, S.C. Bishop, F.W. Nicholas and J.B. Owen (eds.) CABI Publishing, Wallingford, pp. 243-252.

Pedersen, J. and L.G. Christensen, 1989. Heterosis for Milk-Production Traits by Crossing Red Danish, Finnish Ayrshire and Holstein-Friesian Cattle. Livest. Prod. Sci.,23: 253-266.

Philipsson, J. and B. Lindhé, 2003. Experiences of including reproduction and health traits in Scandinavian dairy cattle breeding programmes. Livest. Prod. Sci., 83: 99-112.

Pryce, J.E., R.F. Veerkamp, R. Thompson, W.G. Hill and G. Simm, 1997. Genetic aspects of common health disorders and measures of fertility in Holstein Friesian dairy cattle. Animal Sci., 65: 353-360.

Rauw, W.M., E. Kanis, E.N. Noordhuizen-Stassen and F.J. Grommers, 1998. Undesirable side effects of selection for high production efficiency in farm animals: a review. Livest. Prod. Sci., 56:15-33.

Royal, M., G.E. Mann and A.P.F. Flint, 2000. Strategies for reversing the trend towards subfertility in dairy cattle. Veterinary Journal, 160: 53-60.

Schaeffer, L.R. and J.C.M. Dekkers, 1994. Random regressions in animal models for test-day production in dairy cattle. Proceedings of the 5th World Congress on Genetics Applied to Livest. Prod., Guelph, 18: 443-446.

Taberlet, P., A. Valentini, H.R. Rezaei, S. Naderi, F. Pompanon, R. Negrini and P. Ajmone-Marsan, 2008. Are cattle, sheep, and goats endangered species? Molecular ecology 17: 275-284.

Tada, T., 1997. The immune system as a supersystem. Annual Review of Immunology, 15: 1-13.

Ten Napel, J., F. Bianchi and M. Bestman, 2006. Utilising intrinsic robustness in agricultural production systems. In: Working Papers no 1: Inventions for a sustainable development of agriculture. TransForum, Zoetermeer, pp. 32-54.

Van Drie, I., 2007. Dieptepunt bereikt? Achteruitgang vruchtbaarheid van Nederlandse veestapel tot stilstand gebracht. [All-time low reached? Deterioration of fertility of the Dutch dairy herd has come to a stand-still]. Veeteelt, issue 2 April 2007: 28-30.

Wagter, L.C., B.A. Mallard, B.N. Wilkie, K.E. Leslie, P.J. Boettcher and J.C.M. Dekkers, 2000. A quantitative approach to classifying Holstein cows based on antibody responsiveness and its relationship to peripartum mastitis occurrence. J. Dairy Sci., 83: 488-498.

Windig, J.J., B. Beerda and R.F. Veerkamp, 2008. Relationship between milk progesterone profiles and genetic merit for milk production, milking frequency, and feeding regime in dairy cattle. J. Dairy Sci. (in press).

Windig, J.J., M.P.L. Calus, B. Beerda and R.F. Veerkamp, 2006. Genetic correlations between milk production and health and fertility depending on herd environment. J. Dairy Sci., 89: 1765-1775.

Windig, J.J., M.P.L. Calus, G. de Jong and R.F. Veerkamp, 2005. The association between Somatic cell count patterns and milk, production prior to mastitis. Livest. Prod. Sci., 96: 291-299.

Zinkernagel, R.M., 2003. On natural and artificial vaccinations. Annual Review of Immunology, 21: 515-546.

Robustness as a breeding goal and its relation with health, welfare and integrity[2]

E.D. Ellen[1], L. Star[2], K. Uitdehaag[1] and F.W.A. Brom[1,3]
[1]Animal Breeding and Genomics Centre, Wageningen University, P.O. Box 338, 6700 AH Wageningen, the Netherlands
[2]Adaptation Physiology Group, Wageningen University, P.O. Box 338, 6700 AH Wageningen, the Netherlands
[3]Rathenau Institute, P.O. Box 95366, 2509 CJ Den Haag, the Netherlands

Abstract

The combination of breeding for increased production and the intensification of housing conditions have resulted in increased occurrence of behavioural, physiological, and immunological disorders in production animals. These disorders affect health and welfare of production animals negatively. For future livestock systems, it is important to consider how to manage and breed production animals. In this paper we develop the concept of robustness as a breeding goal. Improving robustness by selective breeding will increase (or restore) the animals' ability to interact successfully with the environment and thereby to make the animal better able to adapt to an appropriate husbandry system. This, in turn, is likely to improve both welfare and productivity, although this also depends on management and housing conditions. Therefore, in order to breed for sustainable and social acceptable production systems, animal production should accept robustness as a breeding goal.

Keywords: health, integrity, robustness as a breeding goal, welfare, future of livestock production

Introduction

From the 1960's onwards, animal husbandry became more and more intensive. This was supported by agricultural policy and the quest for sufficient, safe and cheap food. Animal breeding – in this context – was directed at an increase in production. The combination of breeding for increased production and the intensification of housing conditions have not been without consequences, especially for agricultural animals. And, despite of the economical and food-policy successes of these new husbandry systems, animal production created societal discussion on the way animals were treated. It was said that animals in animal production systems suffered; not only because of direct injuries, but also because of the confinement that made it impossible for them to behave in certain species-specific ways. In several countries public protests incited inquiries into the welfare of animals kept under intensive husbandry systems, for example the Brambell-report in the UK in 1965.

The traditional strategy to reduce these problems is preventive management. Besides the traditional strategy of preventive management, another possibility is to adapt animals by selective breeding or even genetic modification. Selective breeding can be used to improve health and welfare related traits in production animals. Health can be enhanced by selective breeding for disease resistance. This may be effective in resistance to a wide range of pathogens and can be used to protect animals under different environmental conditions (Lamont, 1998). Welfare can be enhanced by selection against expression of undesirable behaviour.

---

[2] This paper is abbreviated version of L. Star, E.D. Ellen, K. Uitdehaag and F.W.A. Brom (2008). A plea to implement robustness into a breeding goal: poultry as an example. Journal of Agricultural and Environmental Ethics, 21: 109-125.

Improving health and welfare by adapting the animal to the housing system, however, can result in violation of the integrity of the animal; for instance, breeding animals that lack animal specific possibilities that are needed for a good animal life. The breeding of blind laying hens (Ali and Cheng, 1985) is a famous example. Many people, however, intuitively feel that it is a morally wrong approach to improve animal welfare by adapting animals to housing conditions by taking away essential traits (Sandøe et al., 1999).

In this paper, we will focus on selective breeding of production animals. We argue that in future livestock systems it is necessary that breeding goals[3] should not only be defined in terms of production, but that they should also include traits related to animal health and welfare. For this we introduce robustness as a breeding goal.

Robustness is a term that is rapidly becoming a main interest in animal production (Knap, 2005; Ten Napel et al., 2006). We like to explore the discussion on robustness as a breeding goal for animals kept in future livestock systems. The concept of robustness is related to the concepts of health, welfare, and integrity, but in our opinion, robustness is more comprehensive. We expect that robustness as a breeding goal will result in better health and welfare without affecting the integrity of the animal. Based upon this, we argue that it is ethical acceptable to use selective breeding in order to create animals that are able to function better in conventional agricultural systems.

The concept of health, welfare, and integrity

Before going into detail about the concept of robustness, the concepts of health, welfare, and integrity will be explored. For the concept of robustness it is important to have a perception about the definitions and considerations behind the realisation of the concepts of health, welfare, and integrity. The considerations are important for the implementation of the different concepts into a breeding goal for robustness.

The concept of health

Different approaches towards the concept of health can be found in literature. The very basic definition of health is no more than the absence of disease (Gunnarsson, 2006; Nordenfelt, 2007). Boorse (1997 in Nordenfelt; 2007) defined disease as 'a type of internal state that is an impairment of normal functional ability'. This definition indicates that disease (and health) are linked to functional ability, i.e. biological functioning (Nordenfelt, 2007). For Boorse (1997), biological functioning is tied to the individual's survival and reproduction. This is, however, a very narrow concept of biological functioning. The broader concept of biological functioning, as basis for the concept of health, is related to homeostasis, i.e. regulation of the internal environment of living organisms (Gunnarsson, 2006). In addition, an animal may be in pain and disabled by internal bodily causes (failure in regulating homeostasis) without reducing the probability of the animal's survival. This indicates that there are other possible goals than the one of pure survival (Nordenfelt, 2007). One goal related to health, and commonly used in the debate about animal welfare, is *quality of life*, which includes psychological aspects of health (Fraser et al., 1997). Gunnarsson (2006), however, mentioned that if health is defined as physical and psychological well-being, there will be problems associated with applying the definition to all animals, especially production animals. Gunnarsson (2006) stated that a health definition that puts priority to the physical and psychological well-being of a production animal is misleading in relation to the general purpose of livestock production. In livestock production, economical considerations are involved and can be decisive in the judgment of the animals' health. To achieve good health the animal has to be in harmony with itself and its environment, and has to be in a normal physical condition (free of diseases and other physical

---

[3] The definition of breeding goal will be elaborated in the section 'The concept of robustness' in this paper.

disorders) (Rutgers, 1993). Health could than be considered as 'the physical condition required to achieve welfare at an acceptable level' (Brom, 1997 derived from Nordenfelt, 1987).

The concept of welfare

Welfare of farm animals is a major concern, in society, in livestock production, as well as in animal science (Kanis *et al.*, 2004). Animal welfare, however, is a complex concept that is difficult to define operationally, and hence to evaluate empirically (Rowan, 1997). This has led to different welfare definitions.

Fraser *et al.* (1997) suggested that three main ideas are expressed in public discussion concerning animal welfare: feelings, functioning, and natural living. Fraser *et al.* (1997) also argued that a scientific approach to animal welfare has to take into account these ideas expressed in public discussion. Animal feelings are related to experiences of animals, i.e. mental harmony, whereas functioning is related to biological functioning, i.e. physical harmony. The concept of experience is based on the presence of positive experiences and the absence of negative experiences, whereas the concept of functioning is based on 'doing well', so that the animal is functioning as it should do (Stafleu *et al.*, 1999). The idea that animals should live natural lives includes considerations of an animal's nature or *telos* (Appleby and Sandøe, 2002), which is related to the concept of integrity, and will be discussed later.

A definition of animal welfare related to the concept of experience is that 'animals should feel well by being free from prolonged and intense fear, pain, and other negative states, and by experiencing normal pleasures' (Fraser *et al.*, 1997). Kanis *et al.* (2004) considered animal welfare as similar to 'animal happiness', which can be seen as 'the balance between an animal's positive and negative emotions or feelings over a certain time period'. It is, however, impossible to ask an animal directly in which situation it feels comfortable and if its preferences are satisfied. Therefore, making use of the concept of experience in scientific studies is rather difficult. To make animal experiences more applicable, the concept of functioning can be used as a tool. The concept of functioning often involves ideas about evolutionary fitness, including successful breeding. When breeding is strongly affected by human intervention, as for production animals, it might be difficult to apply the concept of functioning (Appleby and Sandøe, 2002). The concept of functioning, however, can still be linked to scientific (biological, physiological, social functioning) animal production theories, or models. Definitions of welfare commonly used are often based on the concept of functioning. For instance, welfare definitions given by Broom (1993) 'welfare of an animal is reflected by the success of its attempt to cope with its environment' and by Siegel (1995) 'welfare depends on physiological ability to respond properly in order to maintain or re-establish homeostatic state or balance'.

For scientific models, the concept of functioning is easier to demonstrate than what an animal experiences (Duncan and Fraser, 1997). Although the concept of functioning is more straightforward to quantify, the link between (biological) functioning and the animal's welfare is not always apparent, e.g. there is little consensus on the baseline that should be used in assessing measures and there is less agreement on which levels necessarily denote a better quality of life for the animal. Therefore, assessment of welfare involves a mixture of scientific knowledge and value judgments.

The concept of integrity

Integrity has been described by Rutgers and Heeger (1999) as the 'wholeness and intactness of the animal and its species-specific balance, as well as the capacity to sustain itself in an environment suitable to the species'. The principle of respect for the integrity of animals leads to considerations and arguments beyond animal health and welfare (Grommers *et al.*, 1995). The integrity theory of King (2004) proposed that the value of animal life is such that animals should not be harmed or destroyed. The loss of life itself is conceived as the ultimate harm to the animal's integrity, i.e. to its 'completeness'.

Integrity gives notion to our own moral position, purposes, and perspectives with regard to animals (Vorstenbosch, 1993; De Vries, 2006). Integrity is not a strictly describing term, but it rather refers to the way we think an animal has to be (Brom, 1997). In the former, we already mentioned the possibility to breed blind laying hens and that many people intuitively feel that this is a morally wrong approach to improve animal welfare. The moral notion that gives voice to this intuition is integrity (Bovenkerk *et al.*, 2002). Another example is non-broody behaviour in laying hens. Selection has resulted in strains of chickens that normally do not incubate eggs or brood chicks (Price, 1999). These laying hens seem to be well adapted to their situation and, probably, are still able to brood. However, they do not have the motivation to express their brooding behaviour; it is just not natural to them. These two examples clearly show that it is important to consider the nature and biological needs of animals.

According to Rollin (1989), the nature and biological needs are related to the *telos* of an animal. He defined *telos* as 'the unique, evolutionarily determined, genetically encoded, environmentally shaped set of needs and interests which characterise the animal in question'. Each animal has a *telos* that is unique to its species, it can be seen as the 'chickenness of the chicken' or the 'pigness of the pig', which are essential to their well-being as speech is to us (Rollin, 1989). He stated that the animal's well-being is determined by the match between its needs and interest and the treatment it receives (Rollin, 1995). Although, the animal's *telos* is unique to its species, Rollin (1995) argued that changing the *telos* of an animal can be justified. He stated that there is no moral problem in making an animal happier or prevent it from suffering by changing its *telos*, unless changes endanger the animal itself, other animals, humans, or the environment. Verhoog (1992), however, insisted that *telos* is of direct moral relevance in itself and should not be violated or changed. He stated that selective breeding is morally questionable, because it represents interference with the natural species integrity and evolutionary development of animals. In our opinion, selective breeding can violate the animals' integrity in extreme cases like breeding blind laying hens. We can use selective breeding to improve animals, but only if the animals' identity is preserved.

The concept of robustness

Introduction to robustness

In the previous paragraph we have explored the concepts of health, welfare, and integrity. All three concepts are related to the quality of life of an animal. To improve the quality of life of an animal in future livestock systems these concepts have to be integrated into a breeding goal. The breeding goal defines which traits have to be improved and how much weight is given to each trait. The breeding goal is the direction in which we want to improve the population (Cameron, 1997). The concepts of health and welfare primarily focus on the state of the animal (mentally and physically) in a specific situation. These concepts do not consider animal related traits and, therefore, could not be implemented into a breeding goal. Integrity considers animal related traits, namely the presence of species specific characters, e.g. it's 'completeness'. It is, however, not possible to optimise the integrity of an animal, and therefore integrity cannot be improved by selective breeding. For this, we would like to introduce the concept of robustness. The concept of robustness includes individual traits of an animal that are relevant for health, welfare, and integrity. Because robustness includes individual traits, it can be integrated into a breeding goal.

The concept of robustness is defined in different fields, e.g. ecology, biological systems, statistics, and animal production. A broad definition of the concept of biological robustness is 'the maintenance of specific functionalities of the system against perturbations, and it often requires the system to change its mode of operation in a flexible way' (Kitano, 2004). This definition can be used as a starting point for definitions of robustness in other fields, like animal production. Knap (2005) defined robust pigs as 'pigs that combine high production potential with resilience to external stressors, allowing for unproblematic expression of high production potential in a wide variety of environmental conditions'. Whereas Ten Napel *et al.* (2006) defined robustness in a broad sense as 'the minimal variation in a target feature following a disturbance, regardless of whether it is due to switching between underlying processes, insensitivity or quickly regaining the balance', and in a narrow sense as 'the ability to switch between underlying processes to maintain balance'. The definitions of Ten Napel *et al.* (2006) are independent of species.

From these definitions, it can be concluded that the main characteristics informative for robustness of production animals are production and adaptation in a wide variety of environmental conditions. Production is important because it is one of the parameters related to the functioning of an animal. Besides, production is important because of its economical value. In the concept of robustness, adaptation can be seen as a mechanism of the animal that enables it to cope with internal or external disturbances, or with changes in the environment. Ideally, we would like to breed a strain of laying hens that can adapt to different environmental conditions. In practice, however, strains of laying hens can perform differently in different environments; this is called genotype by environment interaction (Falconer and Mackay, 1996). As mentioned earlier, there is a limited number of internationally operating poultry breeding companies that provide laying hens worldwide. For these companies, it is favourable to have animals that can function under a wide variety of environmental conditions.

Using the main characteristics informative for robustness, e.g. production, adaptation, and a wide variety of environmental conditions, we define a robust laying hen as 'an animal under a normal physical condition that has the potential to keep functioning and take short periods to recover under varying environmental conditions'. Functioning can be evaluated in terms of physiological, behavioural, and immunological traits. This definition of robustness includes different measurable characteristics and traits that make the concept of robustness applicable for breeding programs.

Implementation of health in the breeding goal for robustness

In the definition of robustness, 'keep functioning' and 'take short periods to recover' are referring primarily to health. The definition of Rutgers (1993), 'the harmony between an animal itself and its environment, where the animal is free of diseases and other physical disorders', primarily focuses on 'functioning'. Whereas the definition of Gunnarsson (2006) 'regulation of the internal environment of living organisms', primarily focuses on 'take short periods to recover'. Robust animals will be less sensitive to disease pressure and are expected to recover more quickly than less robust animals. Therefore, by implementing the concept of robustness as a breeding goal, the health of the animal should improve simultaneously.

Implementation of welfare in the breeding goal for robustness

Together, the welfare definitions given by Broom (1993) and Siegel (1995) 'welfare of an animal is reflected by the success of its attempt to cope with its environment' and 'welfare depends on physiological ability to respond properly in order to maintain or re-establish homeostatic state or balance', respectively, corresponds with the definition of the concept of robustness. The distinction between animal welfare and robustness is that animal welfare is often measured by an animals' response to a current stressor, whereas robustness is based on the possibility to respond adequately to a stressor and is aiming at less disturbed functioning by challenge with a stressor. Implementation of robustness into a breeding goal should result in animals with improved coping abilities for conventional housing systems, and, therefore, should result in improved animal welfare.

Implementation of integrity in the breeding goal for robustness

As described earlier, the concept of integrity indicates how an animal has to be. We have to be aware that selective breeding can have either positive or negative side effects on the ability to function. Sometimes a change in genotype would be an advantage to both animals and humans (Sandøe *et al.*, 1999). But in other cases it could have a negative side effect. These negative side effects are not only morally problematic due to undesired consequences for health and welfare. They are also problematic because two core elements in the concept of integrity, as described by Rutgers and Heeger (1999) are at issue, namely 'the balance in species specifity' and 'to sustain itself in an environment suitable to the species'. According to Rollin (1995), changing the animal by selective breeding does not necessary lead to impoverishment of the *telos*. In line with this, notion of integrity is a requirement for robustness. Therefore, improvement of health and welfare by implementation of the breeding goal of robustness should not be achieved by violation of the integrity or impoverishment of the *telos*.

Robustness as a breeding goal

As mentioned earlier, robustness embraces health, welfare, and integrity. Therefore, different traits can be implemented in the breeding goal of robustness. To utilise robustness as a breeding goal, the traits have to be:
- relevant, i.e. they have to say something about robustness;
- simple, i.e. they have to be understandable for users;
- sensitive, i.e. they have to react to changes in the system;
- reliable, i.e. different measurements must lead to the same outcome;
- it must be possible to establish a target value or trend; and
- data have to be accessible.

Robustness as a breeding goal can be used for different production animals. Each production animal has its species specific characteristics.

In our opinion robustness as a breeding goal can be successful to improve health and welfare of production animals in future livestock systems. Before robustness can be implemented into a breeding goal, large scale genetic research on the different traits has to be done. Large scale genetic research is for most traits labour intensive and expensive. For instance, behavioural measurements and collecting blood samples for immunological parameters have to be done at the individual level.

After determining the most promising traits, the next step will be the implementation of these traits into the breeding goal. Implementation of the traits is difficult and riskful, but the potential of success for robustness as a breeding goal depends on this implementation. One of the difficulties for the implementation is to decide which trait is more important than another, e.g. how much weight is given to each trait. It is, however, important to implement all traits, because the success of selective breeding for robustness depends on all traits and not on a single trait.

Genetic research for robustness traits and the implementation of these traits into the breeding goal have to be established by cooperation between science and breeders. Additionally, successful result of robustness as a breeding goal depends on the opinion and motivation of the farmer. The principle aspects of robustness may be different for each individual farmer (or breeder), but also reference values can change. Besides, in the future, other traits may arise that have to be implemented into the breeding goal of robustness. By implementation of new traits, it is, however, important that these traits concern the animal itself.

Finally, the potential for a successful result of robustness as a breeding goal depends on the economic value. In his decision-making, a farmer has to consider not just animal robustness, but also how to produce efficiently, at competitive cost.

Questions related to robustness

In this paper, we explored the discussion of robustness as a breeding goal for laying hens kept in future livestock systems. Although we think it is possible to implement robustness into a breeding goal, it still raises several ethical questions like: Is it acceptable to adapt animals to the production environment, rather than by changing their environments? Should animals be adapted to all environments, even the worst? And does selection for robustness affect the integrity of the animal?

When looking at the definition of robustness, a robust animal is an animal that has the potential to keep functioning and take short periods to recover under varying environmental conditions. This indicates that the animal has to function under a wide range of circumstances. It is, therefore, preferable to select for robustness traits that are common to different types of production environments. But, are we really aiming at adapting the animal to even the worst environment? No. The aim is to breed animals that can function well in a range of environments and not to breed animals specifically for the worst environments. However, even in the most optimal environments welfare of laying hens can be improved as illustrated by the fact that they show abnormal behaviour. Increasing robustness by selective breeding, therefore, improves welfare by adapting animals to the production environment. This does, however, not take away the need for improvement of housing conditions.

Christiansen and Sandøe (2000) mentioned that breeding for animals that are better suited for intensive farming *instead* of adapting the farming system may be considered violations of animal integrity. This, however, is only the case in those situations where adapting the animal involves diminishing its ability to live a good life or by depriving the animals of natural abilities, such as being able to see. However, improving the ability to cope with stress and improving the ability to recover by using robustness as a breeding goal does not deprive natural abilities, and is, therefore, not a violation of animal integrity. Of course, we have to be aware that when selecting for robust animals it is unknown if problems negatively correlated with the genetic make-up underlying robustness will occur.

Conclusion

The aim of this paper was to develop the concept of robustness as a breeding goal. Improving robustness by selective breeding will increase (or restore) the animals' ability to interact successfully with the environment and thereby to make the animal better able to adapt to an appropriate husbandry system. This, in turn, is likely to improve both welfare and productivity, although this also depends on management and housing conditions.

The implementation and application of robustness as a breeding goal is desirable. We are convinced that this application will result in animals with improved health and welfare without affecting the integrity. Therefore, improving robustness by introducing this concept as a breeding goal is ethically acceptable.

Acknowledgements

This research is part of a joint project, in animal breeding, of Institut de Sélection Animale, a Hendrix Genetics company, and Wageningen University on 'The genetics of robustness in laying hens' which is financially supported by SenterNovem. We would like to thank Franck Meijboom and the six anonymous reviewers of the *Journal of Agricultural and Environmental Ethics* for their valuable comments.

References

Ali, A. and K.M. Cheng, 1985. Early egg production in genetically blind (rc/rc) chickens in comparison with sighted (Rc+/rc) controls. Poultry Science, 64: 789-794.

Appleby, M.C. and P. Sandøe, 2002. Philosophical debate on the nature of well-being: implications for animal welfare. Animal Welfare, 11: 283-294.

Boorse, C., 1997. A rebuttal on health. In: J.M. Humber and R.F. Almeder (eds.), What is disease? (pp. 1-134). Totowa, New Jersey: Biomedical Ethics Reviews, Humana Press.

Bovenkerk, B., W.A. Brom and B.J. van den Bergh, 2002. Brave new birds: the use of 'animal integrity' in animal ethics. Hastings Center Report, 32: 16-22.

Brom, F.W.A., 1997. Onherstelbaar verbeterd: biotechnologie bij dieren als moreel probleem. Assen, The Netherlands: Van Gorcum & Comp. B.V.

Broom, D.M., 1993. Assessing the welfare of modified or treated animals. Livestock Production Science, 36: 39-54.

Cameron, N.D., 1997. Selection indices and production of genetic merit in animal breeding. Oxon, UK: CAB International.

Christiansen, S.B. and P. Sandøe, 2000. Bioethics: limits to the interference with life. Animal Reproduction Science, 60-61: 15-29.

De Vries, R., 2006. Genetic engineering and the integrity of animals. Journal of Agricultural and Environmental Ethics, 19: 469-493.

Duncan, I.J.H. and D. Fraser, 1997. Understanding animal welfare. In: M.C. Appleby and B.O. Hughes (eds.), Animal welfare (pp. 19-31). Oxon, United Kingdom: CAB International.

Falconer, D.S. and T.F.C. Mackay, 1996. Introduction to quantitative genetics; fourth edition. Harlow, England: Pearson Education.

Fraser, D., D.M. Weary, E.A. Pajor and B.N. Milligan, 1997. A scientific conception of animal welfare that reflects ethical concerns. Animal Welfare, 6: 187-205.

Grommers, F.J., L.J.E. Rutgers and J.M. Wijsmuller, 1995. Welzijn - intrinsieke waarde - integriteit: ontwikkeling in de herwaardering van het gedomesticeerde dier (with a summary in English). Tijdschrift voor Diergeneeskunde, 120: 490-494.

Gunnarsson, S., 2006. The conceptualisation of health and disease in veterinary medicine. Acta Veterinaria Scandinavica, 48: 1-6.

Kanis, E., H. van den Belt, A. Groen, J. Schakel and K.H. de Greef, 2004. Breeding for improved welfare in pigs: a conceptual framework and its use in practice. Animal Science, 78: 315-329.

King, L.A., 2004. Ethics and welfare of animals used in education: an overview. Animal Welfare, 13: S221-227.

Kitano, H., 2004. Biological robustness. Nature Reviews Genetics, 5: 826-837.

Knap, P.W., 2005. Breeding robust pigs. Australian Journal of Experimental Agriculture, 45: 763-773.

Lamont, S.J., 1998. Impact of genetics on disease resistance. Poultry Science 77: 1111-1118.

Nordenfelt, L., 1987. On the nature of health: an action-theoretic approach. Dordrecht, The Netherlands: Kluwer/ Reidel.

Nordenfelt, L., 2007. The concept of health and illness revisited. Medicine, Health Care and Philosophy, 10: 5-10.

Price, E.O., 1999. Behavioral development in animals undergoing domestication. Applied Animal Behaviour Science, 65: 245-271.

Rollin, B.E., 1989. The unheeded cry: animal consciousness, animal pain and science. Oxford, UK: Oxford University Press.

Rollin, B.E., 1995. The Frankenstein syndrome: ethical and social issues in the genetic engineering of animals. Cambridge, NY, USA: Press syndicate.

Rowan, A.N., 1997. The concept of animal welfare and animal suffering. In: L.M.F. van Zutphen and M. Balls (eds.), Animal alternatives, welfare and ethics. Amsterdam, the Netherlands: Elsevier, pp. 157-168.

Rutgers, B. and R. Heeger, 1999. Inherent worth and respect for animal integrity. In: M. Dol, M. Fentener van Vlissingen, S. Kasanmoentalib, T. Visser and H. Zwart (eds.), Recognizing the intrinsic value of animals. Assen, The Netherlands: Van Gorcum, pp. 41-51.

Rutgers, L.J.E., 1993. The weal and woe of animals: ethics of veterinary practice. Utrecht, The Netherlands: University Utrecht.

Sandøe, P., B.L. Nielsen, L.G. Christensen and P. Sorensen, 1999. Staying good while playing god - the ethics of breeding farm animals. Animal Welfare, 8: 313-328.

Siegel, H.S., 1995. Stress, strain and resistance. British Poultry Science, 36: 3-22.

Star, L., E.D. Ellen, K. Uitdehaag and F.W.A. Brom, 2008. A plea to implement robustness into a breeding goal: poultry as an example. Journal of Agricultural and Environmental Ethics, 21/2: 109-125.

Stafleu, F.R., F. Grommers and J.M.G. Vorstenbosch, 1999. Animal welfare: a hierarchy of concepts. In: Proceedings of the 1st congress of the European Society for Agricultural and Food Ethics: preprints, Wageningen, The Netherlands.

Ten Napel, J., F.B. Bianchi and M. Bestman, 2006. Utilising intrinsic robustness in agricultural production systems. Zoetermeer, the Netherlands: Transforum.

Verhoog, H., 1992. The concept of intrinsic value and transgenic animals. Journal of Agricultural and Environmental Ethics, 5: 147-160.

Vorstenbosch, J.M.G., 1993. The concept of integrity. Its significance for the ethical discussion on biotechnology and animals. Livestock Production Science, 36: 109-112.

Robustness in dairy cows: experimental studies of reproduction, fertility, behaviour and welfare

A. Lawrence[1], G.E. Pollott[2], J. Gibbons[1], M. Haskell[1], E. Wall[1], S. Brotherstone[3], M.P. Coffey[1], I. White[3] and G. Simm[1]
[1]Sustainable Livestock Systems, SAC, Bush Estate, Penicuik, Midlothian, EH26 0PH, United Kingdom
[2]Royal Veterinary College, Royal College Street, London, NW1 0TU, United Kingdom
[3]Institute of Cell and Population Biology, University of Edinburgh, EH9 3JT, United Kingdom

Abstract

A multi-sponsor project was undertaken to investigate robustness in dairy cows. The project involved both experimental approaches and those involving the analysis of national breeding data. Two of the experimental approaches used are described in this paper. Detailed recording of a 200-cow experimental dairy herd was undertaken to investigate the link between energy balance and fertility. This was facilitated by having two genetic lines and two production systems represented on the same farm. Energy balance was estimated from daily liveweight and weekly condition score data, whilst fertility was monitored using thrice-weekly progesterone assays from milk samples. Both production system and genetic lines were shown to affect several aspects of fertility. The production system effects were shown to be due to differences in energy balance whilst the genetic lines effects on fertility could not be accounted for by energy balance, implying that cows selected for high milk production probably had genes with a negative effect on fertility. The national UK breeding database was used to identify high and low 'robustness' animals. Farms with a reasonable sample of high and low 'robust' heifers were visited and heifers sampled using validated and practical tests for behavioural traits. Significant effects of robustness were found on both aggressive and non-aggressive social behaviour. No effect of robustness was found in both fear behaviour and response to human interaction. The results of these experimental studies, carried out as part of the Robust Cow project, have illustrated that by using more detailed measurements than were commonly available on commercial farms, a greater insight could be made into how energy balance and genetics affected fertility; a key relationship influencing robustness in dairy cows. The second experiment illustrated the potential of applying practical measures to score a wide range of temperament traits on farms. The data could be used to monitor the effects of current breeding on temperament; it could also be used to look at interactions with traits and explore the longer-term possibilities of improving welfare and production on farm through selection on behavioural traits.

Keywords: dairy cattle, environmental sensitivity, reproduction, fertility, behaviour, welfare

Introduction

Traditional concerns about farm animal welfare have been around the impact of intensive environments and management practices on the animal (e.g. battery cages or sow stalls). This emphasis on the physical environment is however changing with greater consideration being given to animal factors and in particular the effects of unbalanced selective breeding of farm animals. This increasing attention on the welfare consequences of genetic change comes partly in response to some well-publicised examples where selective breeding has led to 'undesirable' side effects (e.g. Rauw *et al.*, 1998), and also in part because of the perceived risk to welfare posed by emerging biotechnologies (Christiansen and Sandoe, 2000).

In dairy cattle there is good evidence of unfavourable genetic correlations between selection on milk production and aspects of 'fitness' (where fitness is the overall capacity of the animal to survive, grow and reproduce) including leg and udder health and fertility (e.g. Pryce *et al.,* 1997). With selection pressure in most countries having been largely focused on milk production over the last 50 years the concern is that these unfavourable genetic relationships with milk production have contributed to a reduction of cow health and longevity on farms. This thinking led to the UK's Farm Animal Welfare Council suggesting that 'breeding companies should devote their selection to health traits to reduce lameness, mastitis and fertility' (FAWC, 1997).

These sorts of concerns have generally given rise to the view that there should be a focus on breeding for what are described as 'robust' animals (Star *et al.,* 2008). In response to this The Scottish Agricultural College (SAC) organised a research consortium of government funders, animal breeders and relevant charities to explore the potential for selecting for robust dairy cows. The summary of the completed 'Robust Cow' project is available on the Defra web-site (Defra, 2008).

One of our starting points in Robust Cow was to provide a framework for researching robustness in dairy cows that was seen as relevant to various stakeholders. Our conclusion from discussions with our research partners was that robustness in dairy cows could be viewed in 2 broad ways:
1. The further development of the concept of broadening breeding goals by inclusion of traits such as rate of maturity that may underlie robustness (e.g. Lawrence *et al.,* 2004; Coffey *et al.,* 2006).
2. By exploring the evidence for significant genotype x environment interactions in dairy cows (see Strandberg, 2008).

Our research in Robust Cow consisted of 3 main objectives. The first of these, to develop new robustness traits such as body energy content or rate of maturity has been reported elsewhere (Wall *et al.,* 2007; Brotherstone *et al.,* 2007). Within Robust Cow we have also developed an approach that allowed us to categorise the production environment on a national scale and hence explore the extent of genotype x environment interactions in fitness traits such as longevity (Haskell *et al.,* 2007). In this paper we will describe 2 experimental studies also carried out within Robust Cow that explored:
1. The effect of genetic merit and production system on dairy fertility using data from the Langhill selection experimental herd.
2. The behavioural characteristics associated with robustness.

The separate sections for these studies will provide a brief rationale and description of this research which has either been or is about to be published in detail in other places (e.g. Wall *et al.,* 2008; Pollott and Coffey, 2008).

1. The effect of genetic merit and production system on dairy cow fertility, measured using progesterone profiles and on-farm recording

Introduction

The decline in fertility in dairy herds, both at the phenotypic and genetic level, appears to be a feature of dairy industries in many countries (e.g. Wall *et al.,* 2003). Most studies have used time trends from national or regional dairy cow recording schemes to highlight the growing problem. However, both the type of animals used in such analyses and the production methods may have changed over time, making it difficult to ascribe the decline in fertility to specific causes. When using commercial farm records, as most studies have done, system is usually confounded with both farm (management level, health status, climate, housing environment, husbandry, etc.) and genetics. Many of the measures of fertility used in these studies also rely on farm observation which may not always reflect the real underlying physiological state of the animal. The traditional measurement of fertility on farms serves a useful purpose but is open to missing certain key events through silent heats and/or poor observation. Progesterone profiling provides a more objective method for tracking reproductive events in dairy cows (Lamming and Bulman, 1976). Another limitation of studies carried at the national level is that they cannot directly assess the role of milk yield compared to energy balanced as a factor in the declining fertility (Bulman and Lamming, 1979), because national breeding schemes are not able to collect energy balance data for the animals in question.

The Langhill herd maintained by SAC at its Crichton Royal Farm (CRF) provides the opportunity to study the effects of both system and genetics on dairy traits since it comprises two different levels of genetic merit with two contrasting systems of management. This herd is monitored for feed intake, condition scored and weighed regularly, and feeds sampled and analysed routinely. In addition progesterone profiles have been monitored in this herd and thus provide a more objective view of a cow's reproductive physiology. This unique combination has been used to study the effects of both genetic merit and production system on fertility, in a herd of dairy cows.

Methods

The detailed fertility recording work undertaken at CRF, as part of the RobustCow project, has investigated two genotypes of dairy cow and two contrasting modern production systems. The genotypes were a Selection Line (S), from sires chosen for their maximal levels of milk fat plus protein, and a Control Line (C), representing the average of UK animals at the time of sire selection. The two production systems were a high concentrate (C) diet providing the maximum possible nutrition for the cow and a high forage diet (F) representing a more extensive type of system. Thus four groups of cows were monitored for fertility; SC, SF, CC and CF. Evidence of robustness was sought based on the relative performance of the four groups for the fertility traits measured. Milk progesterone profiles were analysed to include an objective measure of fertility in the early part of lactation as an addition to the more routine farm recording procedures, which would not have detected silent heats.

The cows at CRF were monitored for milk progesterone levels between September 2003 and January 2006. Three milk samples per week were taken from each cow, from calving to 140 days post-calving. A total of 363 lactations from 229 cows provided complete 140 d progesterone profiles, 1114 luteal cycles and 310 calving intervals for analysis. Figure 1 shows parameters which were analysed from the progesterone data.

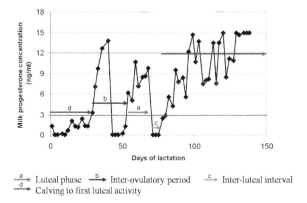

Figure 1. A normal progesterone cycle from a Langhill cow illustrating parameters which were analysed in the comparison of genotype and system effects from the progesterone data.

Results

The progesterone profiles revealed that cycle characteristics were unaffected by the genetic level of the cow, i.e. there was no effect of the Control or Selection Line on cycle length, any of its components or level of circulating progesterone. However, when the first cycle after calving was examined on its own then Selection Line cows were found to have lower levels of milk progesterone than Control cows. Also milk progesterone level was related to cycle length such that short cycles were associated with low levels of progesterone. In addition the Selection Line cows had a higher incidence of delayed ovulation between cycles.

Two aspects of cycling activity were affected by feeding system: inter-luteal interval and length of the luteal phase. However, they cancelled each other out and no effect on overall cycle length was found. The high concentrate feeding system was also associated with a higher incidence of delayed ovulation after calving.

Control cows on both diets and all cows fed on the high-forage system showed evidence of an earlier commencement of luteal activity, or reproductive cycling, after calving (CLA) by 5 to 6 days (Table 1). The day of first heat was 14 days earlier for control cows and they had shorter gestation lengths. However, the experimental policy for rebreeding is to start insemination only after 42 days at CRF and so the advantage of the earlier onset of first heat by the control group was lost such that both the day of first service and the day of successful service were the same for both genetic groups. There was some evidence of cows on the high-forage diet conceiving more quickly than those on the high-concentrate diet but this difference was not observed between the two genetic lines.

Energy balance was calculated from liveweight and body condition score data. The combination of genetic group and feeding system produced four groups of cows with very different energy balance characteristics. The High Forage system provided a less energy-dense diet than the High Concentrate system and so the cows on this system were in negative energy balance for a lot longer than the

Table 1. Least squares means of traits where genetic line and production system were found to contain significantly differences (P<0.05).

Trait	Control	Selection	High-concs.	High-forage
Day of commencement of luteal activity	25.7	30.9	31.2	25.3
Day of first heat	53.7	67.2		
Number of cycles	4.32	3.99	3.98	4.33
Gestation length (d)	283.6	280.9	280.3	284.2
Intervals (d)				
1st heat to 1st AI	11.7	5.9		
1st heat to succ. Service	61.8	49.2		
1st heat to 1st AI (binary)	0.297	0.145		
1st heat to succ. service (binary)	0.771	0.582	0.591	0.762
1st service to succ. service (binary)			0.538	0.664
1st heat >42d to 1st AI	0.93	0.34		
1st heat >42d to 1st AI (binary)			0.003	0.036
1st heat >42d to succ. service (binary)			0.529	0.689
Cycle data)				
Length of luteal phase (d)			12.3	13.2
Inter-luteal period (d)			9.5	8.7
Silent heats (incidence)	0.65	0.71		

cows on the High Concentrate system. A similar situation was found between the two genetic groups such that the return to positive energy balance occurred on days 40 (CC), 46 (SC), 55 (CF) and 68 (SF). Curves were fitted to both the energy balance and condition score data and the characteristics of these curves were used to investigate their link with fertility, as measured by CLA, day of first heat and day of first service.

The effect of system on CLA disappeared when both EB and CS effects were used in the model. Commencement of luteal activity (Energy Balance analysis) was the only trait affected by genetic line; the least-squares means for control and selection line cows were 30.1 and 36.7d respectively. Earlier CLA was associated with higher levels of both condition score (CS) and energy content (EC), lower progesterone levels at first cycle, and less average daily loss in both CS and energy balance (EB) up to the 25th day of lactation. Earlier day of first heat was highly influenced by earlier CLA, higher EC and a higher nadir of EB drop. Day of first service was decreased by an earlier day of first heat, higher progesterone levels at first cycle, and higher average EB loss during the first 25 days of lactation. Day of first service was also associated with an early return to positive condition score.

Discussion and conclusions

There was no evidence of a genotype-by-system interaction for fertility traits in this dataset indicating that the four group means for any trait were the sum of their genetic and system effects. The analysis of the milk progesterone samples provided some addition insight into the reproductive performance of the cows which would not have been available from farm data alone. Significant differences between genetic groups and/or systems were found for CLA, the number of cycles and various cycle characteristics, traits unavailable from farm records.

Differences between the genetic makeup of the two genetic lines did not affect their cycling characteristics, indicating that these reproductive processes were robust to large genetic differences. However, the onset of breeding after calving was heavily influenced by genetic makeup, Control Line cows commencing activity earlier than Selection Line animals. This could be considered as a 'plastic'/adaptive trait that contributed to cows being robust at the level of their fertility and conception (see Strandberg, 2008; this publication). Interestingly, the management regime at CRF nullified this difference resulting in similar overall performance for the two groups. In fact there was some evidence that within the management practised at CRF that Selection Line cows had better conception characteristics than Control Line cows.

Cows in the different systems had different mean CLA values with High-Forage cows starting cycling activity before the High-Concentrate cows. System had more significant effects on aspects of fertility than genetic group. In particular the intervals between first heat/first service and successful service were longer for High-Forage cows. This indicates that cows were less able to show adaptive responses (to be robust) to system differences than genetic differences in these traits.

The energy balance and condition score analyses indicated that the differences in early-lactation characteristics between groups were almost entirely due to these two influences. The only exception was the difference in CLA between genetic groups that could not be entirely attributable to energy balance; a difference of 6 days in favour of the Control cows remained after correction for energy balance. This implies that either genetic and management control of energy balance, or both, are important elements of improving cow fertility. The availability of a UK female fertility index will allow the genetic component to be addressed in the long term but in the short term, farmers will have to rely on improved management practices to minimise the loss of body energy in high yielding Holstein type cattle.

2. Behaviour, temperament, robustness and cow welfare

Introduction

The Robust Cow project has provided a set of options for breeding programmes to use when selecting for more robust dairy cows (see Defra, 2008). However in addition to investigating the feasibility of selecting for robust dairy cows we also wished to address the desirability of this breeding goal (e.g. Jones and Hocking, 1999). We were particularly concerned to understand whether selecting for robust dairy cows would have undesirable outcomes on behaviour and cow temperament.

Temperament is the predisposition to behave in a similar way across time and in different contexts (Bell, 2007; Guitierrez-Gil *et al.,* 2008). There is increasing evidence of the role of temperament in behavioural ecology (Brydges *et al.,* 2008) and that temperament has a genetic basis (Turner *et al.,* 2008) although the genetic contribution will vary depending on the behavioural trait in question (e.g. Turner and Lawrence, 2007; Turner *et al.,* 2008).

In the Robust Cow project we were interested in changes in temperament as a possible (undesirable) consequence of selecting for robustness, but we were also interested in identifying behavioural traits that might contribute to robustness. A major contributor to the dairy cow successfully adapting (being robust) is her ability to obtain sufficient resources (feed, water, clean lying spaces, etc.) whilst negotiating the social and physical environment in a way that does not cause her undue levels of stress. These 'abilities' may be reflected in temperament traits, and as such may be candidates for inclusion in breeding programmes if the traits can be properly defined. We saw four major aspects of temperament as appropriate to study in the context of robustness: aggressive 'style' of the individual cow (or her willingness to compete for feed, or to displace or bully other cows in a competition situation), sociality (ability to cope or thrive in group housing), flexibility (ability to adjust her behaviour to changing conditions) and responsiveness to stimuli (both human and non-human).

In order to determine whether tests are assessing an underlying temperament trait, it is necessary both to estimate repeatability of behavioural responses over time and to compare responses to tests in different contexts. Hence our work was divided into 2 phases:
a. development of behavioural tests to quantify the dairy cow temperament traits we had identified as relevant to robustness;
b. applying the developed tests to assess temperament in daughters of 'high' and 'low' robust bulls on commercial farms.

Development of tests

Methods and results

The tests were developed during an 18 month period at CRF using groups of Holstein-Friesian cows (see Gibbons *et al.,* 2008).

Responsiveness

Two series of tests were carried out, one in which the individual cow's response to humans was assessed, and a second in which the response to novel objects or experiences was assessed. Human approach tests involved recording the response of the cow to being approached in three different contexts:
a. whilst feeding;
b. lying; and
c. standing in the passageway.

Qualitative descriptors of the cow's interaction were also made during the passageway test. The approach in the passageway test showed the highest repeatability (0.72), and a number of the descriptive terms showed good levels of repeatability (0.50 to 0.62). There was a high level of concordance between the tests. As the passageway test had the highest repeatability, this test was used in the on-farm study.

Three different tests were used to assess fear or exploratory responses to novel objects or experiences: a striped board test, in which two yellow and black striped boards were placed on either side of a passageway, a flashing light test and a test in which water was sprayed onto the hindquarters. The level of fear expressed was recorded for all cows in all three tests, and the level of curiosity shown was recorded for the board and light test. The greatest level of fear was shown in the water spray test and the most curiosity was shown in the striped board test. The level of fear shown by individuals was correlated across the novelty tests and the level of fear shown in the human interaction tests was correlated with the fear shown in the novelty tests. As the striped board test allows curiosity and fear to be measured, this test was be used in the on-farm trials.

Aggressiveness

To assess individual aggressiveness, cows were observed in four situations: cows of all ages with unrestricted access to feed, heifers in a separate group with unrestricted access to feed; all cows with length of feed-face available reduced and heifers only with a reduced feed-face. Cows were also observed at peak-, mid- and late-lactation. The results showed that some animals chose to avoid the feed-face at peak feeding times, while others are consistently present, indicating a variation in the level of 'competitiveness' in cows. Cows were consistent across the three stages of lactation in their aggressive style. Cows did not become more aggressive when the length of feedface was reduced. However, heifers show more aggressive behaviour when housed with other heifers, compared to when housed with older cows. These factors were taken into account when designing the on-farm trials in the selection of farms.

Sociability

The social motivation of the same group of cows was tested using a runway test and observations of behaviour within the home pen. In the runway test, the time taken for an individual cow to move down a runway to return to her pen-mates was used as a measure of the individual's social motivation. The time taken for each cow to move down the runway showed the best repeatability (0.54) and there was a significant correlation between trials. When compared to observations in her home pen, cows that showed low social motivation in the runway test were less likely to stand close to other cows in the home pen. There was a high correlation between the number of times the cow was observed at the outside edge of the group (when lying, standing or feeding) and whether she was observed avoiding peak feeding times. These results indicate that sociability can be assessed using observations of the cow's position and behaviour in the home group, relative to the behaviour of other cows.

Flexibility

The side of the parlour that each cow entered was recorded automatically over a 5-day period. Parlour side preference (for either left or right) varied from 60 to 100%. This measure was related to the consistency in temporal organisation, and response to a detour test, which involved placing a barrier in the passageway from the parlour, which forced the cows to move to the right to avoid it. On the final test day, it was moved to the other side of the passageway, and the time taken to move around the barrier in its new location was recorded. However, there was very little variation between cows in the time taken to adjust to the new situation, and subsequently, little relationship with the parlour side preference. Side preferences were also not very stable in heifers, so these measures were not taken forward into the next stage of the trial.

Discussion and conclusions

The development phase, in itself, provided important new evidence on temperamental traits in dairy cows and the potential for measuring these traits on-farm. In most of the four categories of traits studied there was considerable within-cow consistency across time and in different contexts. Given that these studies were of short-term then more evidence will be needed to characterise the full extent of this consistency including the question of whether temperament measured at early stages of the cow's life will predict her behavioural dispositions at later ages. One way of addressing these sorts of questions would be through farm-level studies where easy-to-apply tests of temperament are measured on a sufficiently large sample of cows to estimate genetic parameters and interactions with farming system. This set of studies was also able to identify a set of measures that could be applied under farm conditions to assess the different temperamental categories.

On-farm study

Methods

The aim of the on-farm trial was to compare the behaviour of daughters of 'high robust' and 'low robust' bulls. A 'robustness' index was calculated for each bull comprising the scores for the available functional traits of its daughters. In Figure 2 the robustness index is compared to the PIN value (a UK national production selection index) of each bull; the distribution of the robustness index scores had a normal distribution.

High robust bulls were chosen for high scores and low robust bulls for low scores on our robustness index; two standard deviations separated the scores of the two groups of bulls. There was no difference between the two groups in their PIN value.

To compare the behaviour of daughters of sires of high (H) and low (L) robustness on farms, we used the national databases to identify farms that had at least eight daughters of high robust bulls and eight daughters of low robust bulls. Heifers were used as the focal animals, as young animals are the least likely to have their temperament modified by environment, and the sample would be the least

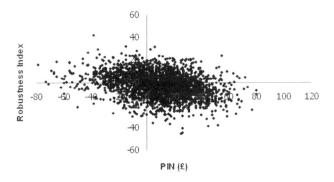

Figure 2. A plot of bulls PIN (a UK national production index) against a robustness index comprising scores for all functional traits measured at a national level in the UK (lameness, mastitis, fertility, body condition score and heifer growth rate. The high and low robust bulls were separated by two standard deviations

biased due to any effects of differential culling on low robust animals. Organic farms, farms using straw courts and farms housing their heifers in a separate group were excluded, as these factors are known to affect behaviour and health. An effort was made to recruit farms across the spectrum of environment types, from low input to high input.

A total of thirty-five farms were visited from October 2005 to May 2006. The following measures of temperament were taken on each of the focal heifers:
- *Aggression*: a video-clip of at least 15 minutes in length of behaviour at the feedface was made at a time when other cows are present.
- *Sociality*: A behavioural scan sample was made every 30 minutes for 3 hours, starting from the time feed was presented in the morning. The behaviour of all cows in the group was tallied. For the focal heifers, the behaviour performed at each time point, their position in the shed (e.g. central position/at the edge) and distance to nearest other cow was scored at each time point. Behavioural synchrony was defined as the proportion of scan points that H and L cows were exhibiting the same activity as the majority of the herd.
- *Responsiveness to humans*: each animal was approached while in the passageway. The observer recorded the nearest distance that can be approached, and scored the animals on a number of qualitative rating scales.
- *Responsiveness to novelty*: two striped boards were attached to either side of a passageway that leaves the milking parlour. Responses were recorded.

Lameness, presence of hock injuries and condition score were recorded on all focal animals and at least 60% of the other milking cows. We also assessed the quality of the housing and the stockhandling, and asked about the management and feeding practices, as these factors will affect welfare at the farm level.

Aggressiveness

Cows from the H group were involved in more aggressive interactions, initiated more aggression and received more aggression than cows from the L group (Table 2). This may indicate that H heifers are intrinsically more competitive. There was a strong influence of management factors influencing aggression such as the quality of stockmanship, feedface design and nutrition.

Sociability

High-robustness heifers showed lower levels of social synchrony, in that they were less likely to be performing the behaviour that the majority of the herd was performing and were more likely to feed at non-peak times (Table 2). However, there were no differences between the groups in how close they chose to stand to other cows or in attempts to isolate themselves by standing close to the edges of the shed.

Responsiveness

A total of 410 heifers were assessed. Significantly more investigative responses towards the novel object were shown by H than L heifers. There was no difference between the groups in the level of fear-related responses shown to the novel object (Table 2). Heifers that displayed highly reactive fear responses showed a low level of investigative response to the novel object. There was no difference in the flight distance between H and L heifers, however, heifers reared in pens as calves were more approachable than those reared in groups. There was no difference between the groups of heifers in their scores for the qualitative terms.

Table 2. Summary of the effects of robustness on temperament trait. Descriptions of how high and low robust animals were identified and the measures taken are given above.

Robustness	High	Low
Agression (counts/15 min.		
Total interactions	5.1	4.5
Initiator	3.8	3.1
Receiver	4.1	3.5
Sociability (%)		
Behavioural synchrony	67.0	74.9
Non-peak feeding	20.5	15.0
Responsiveness (%)		
Exploration of novel object	60.1	53.1

Discussion and conclusions

One of the aims of the project was to provide a behavioural 'audit' for the potential use of an index of 'robustness' and specifically to identify if selection for robustness might be associated with unfavourable effects on behaviour. In particular, we wished to determine whether high robust animals were more aggressive towards other cows and less responsive to the environment.

- In terms of social behaviour we found significant effects of robustness on both aggressive and non-aggressive social behaviour. In terms of aggressive behaviour although the effects were small they appear to signify the need for monitoring effects of selection policies on aggressive behaviour given the potentially adverse effects this may have on welfare and production. In terms of non-aggressive social behaviour our results are equivocal as high robust heifers appear to be less social than low robust heifers in terms of social synchrony, but to show no difference in other measures.
- In tests of responsiveness, there was no difference between high robust and low robust heifers in the level of fear shown towards a novel object. In fact, high robust heifers were more curious and interactive than low robust animals.
- There was no difference between low robust and high robust daughters in their response to human interaction.
- Farm management factors, such as quality of stockhandling and calf rearing systems were found to have significant and potentially important effects on behaviour and temperament
- Overall, it appears that the use of certain types of bulls influences temperament. It may be possible to select for suitable temperament types in the future. Promoting responsiveness and good social behaviour may be useful, although the genetic relationships between the traits and with production would need to be understood.

Final conclusions

In this paper we have addressed the question of robustness in dairy cows by describing two experimental approaches which were undertaken as part of the Robust Cow project; a collaboration between a wide range of industry partners interested in breeding productive and healthy cows that enjoy a high standard of welfare. The first experimental study demonstrated that by using more detailed measurements than were commonly available on commercial farms a greater insight could be made into how energy balance and genetics affected fertility; a key relationship influencing robustness in dairy cows. Both selection for high milk production and the production system employed were shown to affect fertility. Whereas the differences in genetics could not be explained in terms of energy balance, the between system differences were totally accounted for by their different energy balance characteristics.

In the second experimental study, using behavioural monitoring on commercial farms, we have demonstrated the relevance of monitoring the impacts of breeding programmes on temperament in cows. Current scoring of cow temperament relates to the response of cows to handling (e.g. in the milking parlour). The results of the Robust Cow project have illustrated the potential of applying practical measures to score a wider range of temperament traits on farms. The data could be used to monitor the effects of current breeding on temperament; it could also be used to look at interactions with traits and explore the longer-term possibilities of improving welfare and production on farm through selection on temperament traits.

Acknowledgements

We would like to acknowledge the support of our research partners in Robust Cow, who in addition to funding have provided much time and expertise during the course of the project.

References

Bell, A.M., 2007. Future directions in behavioural syndromes research. Proceedings of the Royal Society B-Biological Sciences, 274: 755-761.

Brotherstone, S., M.P. Coffey and G. Banos, 2007. Genetic parameters of growth in dairy cattle and associations between growth and health traits. J. Dairy Sci., 90: 444-450.

Brydges, N.M., N. Colegrave, R.J.P. Heathcote and V.A. Braithwaite, 2008. Habitat stability and predation pressure affect temperament behaviours in populations of three-spined sticklebacks. Journal of Animal Ecology, 77: 229-235.

Bulman, D.C. and G.E. Lamming, 1979. Use of Milk Progesterone Analysis in the Study of Estrus Detection, Herd Fertility and Embryonic Mortality in Dairy-Cows. British Veterinary Journal, 135: 559-567.

Christiansen, S.B. and P. Sandoe, 2000. Bioethics: limits to the interference with life. Animal Reproduction Science, 60: 15-29.

Coffey, M.P., J. Hickey and S. Brotherstone, 2006. Genetic aspects of growth of Holstein-Friesian dairy cows from birth to maturity. J. Dairy Sci., 89: 322-329.

DEFRA, 2008. Summery of completed project 'Robust cow'. Available at: http://www.defra.gov.uk/science/Project_ Data/DocumentLibrary/LK0657/LK0657_5534_FRP.doc

Farm Animal Welfare Council (FAWC), 1997. Report on the welfare of dairy cattle. MAFF Publication, UK.

Gibbons, J., A.B. Lawrence and M.J. Haskell, 2008. Responsiveness of Dairy Cows to Human Approach and Novel Stimuli. Applied Animal Behaviour Science, in press.

Guitierrez-Gil, B., N. Ball, Burton, M.J. Haskell, J.L. William and P. Wiener, 2008. Identification of quantitative trait loci affecting cattle temperament. Journal of Heredity, In press.

Haskell, M.J., S. Brotherstone, A.B. Lawrence and I.M.S. White, 2007. Characterization of the dairy farm environment in Great Britain and the effect of the farm environment on cow life span. J. of Dairy Sci., 90: 5316-5323.

Jones, R.B. and P.M. Hocking, 1999. Genetic selection for poultry behaviour: Big bad wolf or friend in need? Animal Welfare, 8: 343-359.

Lamming, G.E. and D.C. Bulman, 1976. Use of Milk Progesterone Radioimmunoassay in Diagnosis and Treatment of Subfertility in Dairy-Cows. British Veterinary Journal, 132: 507-517.

Lawrence, A.B., J. Conington and G. Simm, 2004. Breeding and animal welfare: practical and theoretical advantages of multi-trait selection. Animal Welfare, 13: S191-S196.

Pollott, G.E.and M.P. Coffey, 2008. The effect of genetic merit and production system on dairy cow fertility, measured using progesterone profiles and on-farm recording. J. Dairy Sci., 91: 3649-3660.

Pryce, J.E., R.F. Veerkamp, R. Thompson, W.G. Hill and G. Simm, 1997. Genetic aspects of common health disorders and measures of fertility in Holstein Friesian dairy cattle. Animal Science, 65: 353-360.

Rauw, W.M., E. Kanis, E.N. Noordhuizen-Stassen and F.J. Grommers, 1998. Undesirable side effects of selection for high production efficiency in farm animals: a review. Livest. Prod. Sci., 56: 15-33.

Star, L., E.D. Ellen, K. Uitdehaag and F.W.A. Brom, 2008. A plea to implement robustness into a breeding goal: Poultry as an example. Journal of Agricultural & Environmental Ethics, 21: 109-125.

Strandberg, E., 2008. The role of environmental sensitivity and plasticity in breeding for robustness: lessons from evolutionary genetics. In: Breeding for robustness in cattle. Klopcic M., R. Reents, J. Philipsson and A. Kuipers (eds.) EAAP Scientific Series No. 126, Wageningen Academic Publishers, Wageningen, the Netherlands, pp. 17-33.

Turner, S.P. and A.B. Lawrence, 2007. Relationship between maternal defensive aggression, fear of handling and other maternal care traits in beef cows. Livest. Sci., 106: 182-188.

Turner, S.P., R. Roehe, W. Mekkawy, M.J. Farnworth, P.W. Knap and A.B. Lawrence, 2008. Bayesian analysis of genetic associations of skin lesions and behavioural traits to identify genetic components of individual aggressiveness in pigs. Behavior Genetics, 38: 67-75.

Wall, E., S. Brotherstone, J.A. Woolliams, G. Banos and M.P. Coffey, 2003. Genetic evaluation of fertility using direct and correlated traits. J. Dairy Sci., 86: 4093-4102.

Wall, E., S. Brotherstone and M.P. Coffey, 2007. The relationship between body energy traits and production and fitness traits in first-lactation dairy cows. J. Dairy Sci., 90: 1527-1537.

Wall, E., M.P. Coffey and P.R. Amer, 2008. A theoretical framework for deriving direct economic values for body tissue mobilization traits in dairy cattle. J. Dairy Sci., 91: 343-353.

An international perspective on breeding for robustness in dairy cattle

H. Jorjani, J.H. Jakobsen, F. Forabosco, E. Hjerpe and W.F. Fikse
Interbull Centre, Dept. of Animal Breeding & Genetics, Swedish University of Agricultural Sciences, Box 7023, S-75007 Uppsala, Sweden

Abstract

Evidence on antagonistic correlated effects of selection for milk production traits in dairy cattle, and the resulting deterioration of functional traits, has been accumulating during the past decade. Consequently, it can be concluded that robustness, which is the well-being and lifelong production and functionality of the cow in terms of superior health, fertility and welfare, should be given a more prominent place in the breeding objectives than before. Unfavourable genetic correlations between production and functional traits are about -0.25 to -0.30. Therefore, combining functional and production traits into a total merit index (TMI) to guarantee breeding of robust cows is necessary. Accordingly, number of countries combining traits in a TMI is increasing. There is, however, much variation among countries in the weights given to different traits. Therefore, international genetic evaluation of TMI is unrealistic. Fortunately, Interbull's service portfolio has been expanded steadily since 1994 and now, in addition to production and conformation traits, includes routine genetic evaluation for udder health, longevity, calving, fertility and evaluation for workability traits is under research. These traits constitute more than 95% of the total weight in any country's TMI. International predicted genetic merits for individual traits can be combined nationally to construct an international robust bull index on each country's scale.

Keywords: multi-trait robustness, lifelong production, functional traits, total merit index, international genetic evaluation

Introduction

'Robust' and 'robustness' can be used in two different ways. In the animal breeding literature, we are more likely to encounter the words robust and robustness in connection with statistical estimators. Those estimators that are not too affected by small departures from the underlying assumptions are considered to be robust.

However, 'Robust animal', a term used by the industry for many years, is a new addition to the scientific literature. What the industry is trying to convey by the word 'robust animal' is an animal in vigorous health, and especially, an animal whose health is not too negatively affected by successful breeding for production traits. The terminology is a bit muddled and may also be used for animals that are resilient to reasonable departures from optimal environmental conditions. In scientific literature the word 'robust' has been used in both of these two new meanings.

An animal can be robust in the sense that it would perform approximately the same under different environmental conditions. In this case, the animal would not be too environment sensitive, i.e. has a flat norm of reaction (Kolmodin *et al.*, 2002; Calus and Veerkamp, 2003). The difference between a flat and a steep norm of reaction is exemplified in the schematic representation of two animals in Figure 1. The animal depicted by the solid line performs approximately the same under a wide range of environmental conditions, while the animal depicted by the dashed line performs worse under poor environmental conditions and better under excellent conditions.

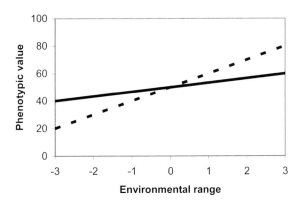

Figure 1. Schematic representation of norm of reaction for two animals.

Alternatively, an animal can be robust in the sense that it would perform relatively good for both production and other traits. In this case the animal is a member of a population simultaneously selected for production and other traits.

One may also simplify the two different meanings of robust and robustness by interpreting the former as the single-trait robustness and the latter as the multi-trait robustness. Here, in this paper, we define robustness as the well-being and lifelong production and functionality of the cow in terms of superior health, fertility and welfare. In other words, we restrict ourselves to the examination of prerequisites of multi-trait robustness, as this seems to be the prominent understanding of robustness.

Livelihood of a farming entrepreneurship

It is very easy to resort to simplification of livelihood of a farming entrepreneurship and reduce it to a few so-called good indicators, let's say, milk yield and calving interval. However, the fact remains that all serious research on the economics of milk production unanimously point towards importance of a large number of traits. Report of an EAAP working group (Groen *et al.*, 1997) indicates the importance of six different trait groups (production, health, fertility, calving, efficiency and milkability traits), comprising tens of potential single biological traits to be included in the breeding goal and breeding index. Economic importance of each of these traits is, of course, dependent on many different factors that vary from country to country and the usual 'average values' are meaningless.

However, economic importance of non-production traits is acknowledged both in research and in practice. Summary of results from one Nordic study on economic values of traits included in the total merit index (TMI) of dairy cattle (Kulak *et al.*, 2004) are included in Table 1.

Comparison of economic weights for the various functional traits with that of milk production is very interesting. Results of Kulak *et al.* (2004) indicate that 1% conception rate is worth 5 kg energy corrected milk (ECM), 1 day of calving to the first heat is worth 6 kg ECM, 1% stillbirth is worth 17 kg ECM and one incidence of mastitis is worth 545 kg ECM (Table 1).

One factor that complicates the decision making processes in a farming entrepreneurship is the genetic correlation structure among the biological traits. Reports of the negative genetic correlations between production traits and functional traits are abundant. For example, genetic correlation between milk production and female fertility traits in the Nordic countries is estimated to be unfavourably

Table 1. Economic values of the traits included the Nordic countries' total merit index in the Holstein breed (from Kulak et al., 2004).

Trait	Unit	Range of economic values in Nordic countries (€)	
Energy corrected milk	kg	0.24 -	0.36
Conception rate	%	0.41 -	2.87
Interval calving-first heat	days	-2.70 -	-0.87
Calving ease	0.1 SD on liability scale	0.33 -	1.05
Stillbirth	%	-8.61 -	-1.63
Mastitis	incidence	-191.00 -	-136.00
Retained placenta	incidence	-167.00 -	-47.00
Milk fever	incidence	-402.00 -	-184.00
Laminitis	incidence	-360.00 -	-96.00
Involuntary culling	%	-7.30 -	-2.45
Body weight	kg	-0.81 -	-0.20
Gain	g	0.85 -	1.27
Carcass classification	0.1 SD on liability scale	0.71 -	1.34
Feed intake capacity	kg	0.04 -	0.10
Milking speed	l/min	5.15 -	16.55
Temperament	0.1 SD on liability scale	0.72 -	0.91

around -0.3 (Roxtröm, 2001; Philipsson and Lindhé, 2003). One interesting aspect of these negative correlations is that the size of genetic correlations shows much variation in different breeds.

Table 2 shows correlation of international breeding values for a number of traits for the Holstein and Guernsey populations. The unfavourable correlation between the three milk production traits (milk, fat and protein yield) and the two functional traits are between -0.17 and -0.32 for the Guernsey population, and between -0.33 and -0.49 for the Holstein population.

The variation in the magnitude of genetic correlations can also be seen for different countries. Table 3 shows just an example of correlations between international breeding values of milk production traits (milk, fat and protein yield) in the Nordic countries and the United States with days open and daughter pregnancy rate in these countries, respectively.

Correlations presented in Table 3 are not, by any means, extreme values. VanRaden (2006) estimated correlations between breeding values for a large number of functional traits and milk production traits in a dozen of countries which indicate that values in Tables 3 should be considered as average values.

International trends

Exact genetic and physiological mechanisms behind the deterioration of functional traits, i.e. poor performance in cows and decreasing breeding values in bulls, are not fully known to us. However, there seems to be a consensus that successful breeding for milk production traits has been associated with poorer robustness. Average international predicted genetic merit shows a steady upward trend for all milk production traits in all countries and breeds. The positive trend in milk production traits is also more pronounced after 1994, which coincides with the start of international genetic evaluations at the Interbull Centre. Figures 2 and 3 show examples of positive trend for protein yield in Red dairy cattle (RDC ~ Ayrshire type breeds) and Holstein populations.

Table 2. Correlation between international breeding values between three milk production traits and two functional traits in the Holstein (HOL) and Guernsey (GUE) populations. For milk, fat, protein and non-return rate higher values are desirable, and for calving interval the smaller values are desirable. Therefore, the negative sign indicates an unfavourably correlation (data from Interbull September 2007 test evaluation).

Trait combination	Correlation of international breeding values	
	GUE	HOL
Milk yield – fat yield	0.85	0.62
Milk yield – protein yield	0.94	0.90
Milk yield – non-return rate	-0.25	-0.48
Milk yield – calving interval	-0.32	-0.49
Fat yield – protein yield	0.92	0.75
Fat yield – non-return rate	-0.17	-0.33
Fat yield – calving interval	-0.26	-0.43
Protein yield – non-return rate	-0.21	-0.49
Protein yield – calving interval	-0.30	-0.49
Non-return rate – calving interval	-0.79	-0.43

Table 3. Correlations of international breeding values between milk production traits (milk and fat yield) and one fertility trait (days open (DO) or daughter pregnancy rate (DPR)) in the Nordic countries and USA for the Red Dairy Cattle (Ayrshire type breed, data from Interbull September 2007 test evaluation).

	DO DFS	DPR USA
Milk	-0.29	-0.16
Fat	-0.29	-0.02
Protein	-0.27	-0.12

At the same time period, an opposite trend can be observed for the functional traits. In Figures 2 and 3, it can be seen that there is a negative trend for some fertility traits, especially in the Holstein populations.

Breeding for multi-trait robustness

Given the economic weights (Table 1) and negative correlations (Table 2), it is no wonder that many countries include a large number of traits in the total merit indices (TMI) used in their countries. Results of two studies (VanRaden, 2004; Miglior *et al.*, 2005), considering the use of 19 and 16 traits/ trait groups, respectively, in the TMIs of different countries are summarised in Table 4.

The use total merit indices in different countries accentuate the old problem of across country choice of the best available bulls. If in the 1980's there was the need to look for bulls with very high predicted genetic merit for milk yield, now one must look for bulls with very high predicted genetic merit for 10-20 different traits. The reason is that across country comparison of total merit indices

Breeding for robustness in cattle

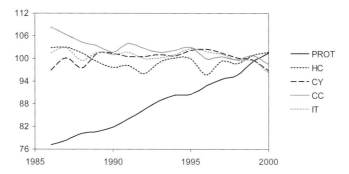

Figure 2. Genetic trends for protein yield (PROT), heifer conception (HC, e.g. interval first to last insemination), cow re-cycling ability (CY, e.g. interval calving to first insemination)), cow conception (CC, e.g. conception rate) and fertility interval traits (IT, e.g calving interval/days open) in the Red Dairy Cattle (Ayrshire type populations). Values shown in graphs are international predicted genetic merits standardised to a mean of 100 and standard deviation of 10 (data from Interbull September 2007 test evaluation).

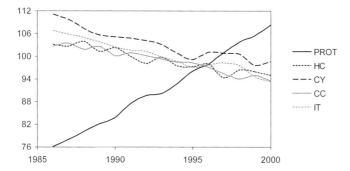

Figure 3. Genetic trends for protein yield (PROT), heifer conception (HC, e.g. interval first to last insemination), cow re-cycling ability (CY, e.g. interval calving to first insemination)), cow conception (CC, e.g. conception rate) and fertility interval traits (IT, e.g calving interval/days open) in the Holstein populations. Values shown in graphs are international predicted genetic merits standardised to a mean of 100 and standard deviation of 10 (data from Interbull September 2007 test evaluation).

is much more complicated than across country comparison of single traits, because index weights and economic weights can vary enormously across countries.

Interbull centre service portfolio

Interbull Centre, the operational unit of Interbull (a permanent sub-committee of International Committee of Animal Recording (ICAR)), started international genetic evaluations in 1994 for the milk production traits in Holstein and Red Dairy Cattle populations from nine populations. Fortunately, Interbull Centre, has progressively increased the size of its service portfolio to include more countries and all major trait groups of economic importance (see Table 5). In the latest routine international genetic evaluation six different trait groups comprising 35 single traits were included in the analyses.

Table 4. Number of constituent traits/trait groups in the total merit indices of some Interbull member countries for the Holstein breed based on the two reports by VanRaden (2004) and Miglior et al. (2005).

Country	Number of traits/trait groups in each country's total merit index	
	VanRaden (2004)	Miglior et al. (2005)
Australia	9	9
Canada	8	7
Switzerland	-	11
Germany	13	11
Denmark	14	13
Spain	9	9
France	11	11
Great Britain	6	6
Great Britain	-	6
Ireland	-	5
Israel	-	5
Italy	9	9
Japan	4	4
The Netherlands	8	7
New Zealand	6	6
Sweden	12	-
United States	9	9
Unites States	-	7

Table 5. Commencement of research and routine international evaluations at the Interbull level for the various trait groups, and size of operations in the latest official international genetic evaluations conducted at the Interbull Centre (August 2007).

Trait group	Research	Routine	Current number of		
			Traits	Breeds	Countries [a]
Production	1991	1994	3	6	9-25
Conformation [b]	1994	1999	19-20	5	4-20
Udder health	1995	2001	2	6	5-23
Longevity	2000	2004	1	6	2-19
Calving	2000	2004	4	3	4-12
Female fertility	2004	2006	5	6	2-16
Beef traits (beef breeds)	2000	-	2	2	-
Workability	2006	-	2	5	-
Locomotion	2007	-	1	1	-
Body condition score	2007	-	1	1	-

[a] Number of populations depends on the breed.

[b] Number of traits for Brown Swiss (20) and other breeds (19) is different. As of September 2008 two new traits (body condition score and locomotion) are added to the list of conformation traits, leading to 22 and 21 traits for Brown Swiss and other breeds, respectively.

Breeding for robustness in cattle

With the current Interbull service portfolio the largest part of traits of any country's total merit index is already included in the international genetic evaluations conducted at the Interbull Centre. Table 6 shows the assembly of index weights, as reported by VanRaden (2004) and Miglior *et al.* (2005), for the traits with international genetic evaluation which are included in the various countries' total merit indices.

Hopefully, with the expansion of Interbull Centre's service portfolio, all traits evaluated in all countries will have an international genetic evaluation very soon. Therefore, it can easily be seen the international tools for conducting selection for multi-trait robustness are already available.

Discussion

Of course, the distinction between the two meanings of robustness is not clear cut and we may in future observe a synthesis of the two meanings. For example, Knap (2005) defines robustness as the combination of 'high production potential with resilience to external stressors, allowing for unproblematic expression of high production potential in a wide variety of environmental conditions'. Two parts of this definition deserve more attention. First, 'resilience to external stressors' not only covers environmental stressors, but also can be interpreted as stresses exerted on the animal because of the shortcomings in the functional traits. In other words, constraints caused by functional traits act as external stressors for high production. Second, 'unproblematic expression of high production potential in a wide variety of environmental conditions' is dependent on environmental conditions under which animal's well-being and lifelong functionality in terms of superior health, fertility and welfare is ensured.

Table 6. Proportion of total merit index weight covered by current Interbull Centre service portfolio. Data are adopted from VanRaden (2004) and Miglior et al. *(2005) and matched to the Interbull Centre's portfolio.*

Country name	Total merit Index (TMI)	Weight of TMI covered by Interbull service portfolio
Australia	APR	93
Canada	LPI	96
Switzerland	ISEL	95
Germany	RZG	95
Denmark	S-index	87
Spain	ICO	97
France	ISU	96
Great Britain	PLI	100
Great Britain	TOP	83
Ireland	EBI	100
Israel	PD01	100
Italy	PFT	95
Japan	NTP	100
The Netherlands	DPS	100
New Zealand	BW	100
Sweden	TMI	90
United States	NM$	97
Unites States	TPI	91

The generally held idea, put forward by Falconer (Falconer and Mackay, 1996: 343), that successful selection for any metric character (e.g. milk production traits) eventually leads to the deterioration of fitness (e.g. health and fertility) gains more support from laboratory and field research. The time of denial seems to be over. Instead, the holistic view of the animal is being acknowledged more and more. The positive trend in the adoption of total merit indices by more countries and inclusion of many more functional traits in these indices is very commendable.

International trade of frozen semen and embryos or live animals should not fall behind by pursuing single-trait marketing of elite bulls. The use of international genetic evaluation results for more traits should get higher priority in advertisements so that construction of total merit indices appropriate for the use in a specific country or a specific farming condition becomes easier.

In summary, it can be concluded that international genetic evaluation results of any trait of economic importance for dairy cattle are available and should be used in the national, as well as, international comparisons. The above statement is true irrespective of the reason for inclusion of more traits in the national genetic evaluations and the TMIs or the weights given to the constituent traits of TMIs.

References

Calus, M.P.L. and R.F. Veerkamp, 2003. Estimation of environmental sensitivity of genetic merit for milk production traits using a random regression model. J. Dairy Sci., 86: 3756-3764.

Falconer, D.S. and T.F.C. Mackay, 1996. Introduction to Quantitative Genetics. 4th edition. Pearson Education Ltd., Essex, England.

Groen, A.B., T. Steine, J.J. Colleau, J. Pedersen, J. Pribyl and N. Reinsch, 1997. Economic values in dairy cattle breeding, with special reference to functional traits. Report of an EAAP-working group. Livest. Prod. Sci., 49: 1-21.

Knap, P.W., 2005. Breeding robust pigs. Australian Journal of Experimental Agriculture, 45: 763-773.

Kolmodin, R., E. Strandberg, P. Madsen, J. Jensen and H. Jorjani, 2002. Genotype by environment interaction in Nordic dairy cattle studied by use of reaction norms. Acta Agric. Scand., Sect. A, Anim. Sci., 52: 11-24.

Kulak, K., H.M. Nielsen and E. Strandberg, 2004. Economic values for production and non-production traits in Nordic dairy cattle populations calculated by stochastic simulation. Acta Agric. Scand., Sect. A, Animal Sci., 54: 127-138.

Miglior, F., B.L. Muir and B.J. van Doormaal, 2005. Selection indices in Holstein cattle of various countries. J. Dairy Sci., 88: 1255-1263.

Philipsson, J. and B. Lindhé, 2003. Experiences of including reproduction and health traits in Scandinavian dairy cattle breeding programmes. Livest. Prod. Sci. 83: 99-112.

Roxström, A., 2001. Genetic aspects of fertility and longevity in dairy cattle. Doctor´s dissertation. Agraria 276. Swedish University of Agricultural Sciences, Uppsala, Sweden,

VanRaden, P.M., 2004. Selection on Net Merit to improve lifetime profit. J. Dairy Sci., 87: 3125-3131.

VanRaden, P.M., 2006. Fertility traits economics and correlations with other traits. Interbull Bulletin, 34: 53-56.

Co-operation between Nordic and US Holstein alters the ratio between the traits that contribute to total merit

L.H. Buch, A.C. Sørensen, J. Lassen, P. Berg and M.K. Sørensen
Department of Genetics and Biotechnology, University of Aarhus, Blichers Allé 20, P.O. Box 50, DK-8830 Tjele, Denmark

Abstract

Co-operation between populations can result in higher selection intensity and thus higher genetic gain. However, co-operation may also change the ratio between the traits contributing to total merit. The objective of this study was to test how the composition of genetic progress in total merit is influenced by co-operation between dairy cattle populations. Two dairy cattle populations were simulated stochastically over a 25 year period with 15 replicates. One population mimicked the Nordic Holstein population and the other mimicked the US Holstein population. Both breeding goals consisted of production, udder health, and female fertility. Two scenarios were simulated: (1) all breeding animals were selected within their own population, and (2) proven bulls could be selected across populations. The results showed that co-operation increased genetic progress in total merit in the Nordic population. The genetic progress was obtained by higher genetic progress in protein yield at the expense of lower genetic progress and larger decline in the functional traits. Selection across populations did not change genetic progress in total merit in the US population. However, the genetic progress in protein yield decreased whereas the genetic progress in the functional traits was either maintained or it increased. Thus, co-operation with Nordic Holstein may limit the deterioration of animal health and reproduction in US Holstein.

Keywords: co-operation between populations, selection, breeding scheme simulation

Introduction

Co-operation between dairy populations offers the potential to select the best breeding animals across populations. Selection across populations should therefore always result in at least as much gain as selection within populations (Smith and Banos, 1991). Co-operation can be beneficial because the required number of animals can be selected from a larger population. This results in higher selection intensity and thus in higher genetic gain (Mulder and Bijma, 2006). The possibility of profitable co-operation depends among other things on population size and similarity of the breeding goals (Smith and Banos, 1991). Former studies have focused on the effect of co-operation on genetic progress in total merit for single trait breeding goals. To our knowledge, there are no studies on whether co-operation affects the composition of genetic progress in total merit for breeding goals containing more traits.

Dairy cattle breeding has been an international enterprise for many years. The US Holstein breed had for instance a strong influence on the genetic make-up of the Nordic black and white strains through exportation of semen and embryos. This co-operation continues to take place, and in recent years the exportation of semen from Nordic Holstein bulls to the US has started. The latter is due to greater international attention on functional traits. Because co-operation between the Nordic and the US Holstein breeds already has a long history it is interesting to mimic these two populations. The objective of this study was to test how the composition of genetic progress in total merit is influenced by co-operation between Nordic and US Holstein. The objective was tested in a stochastic simulation study of the Swedish, Finnish, and Danish (here the Nordic) Holstein population and the US Holstein population.

Materials and methods

The Nordic and the US Holstein populations, each with an individual breeding program and environment, were simulated. The populations were simulated using a modified version of the stochastic simulation program Dairysim (Sørensen *et al.*, 1999). Fifteen replicates of two scenarios were investigated, and the simulations covered a 25 year period. Genetic progress in total merit and in the traits contributing to the total merit index (TMI) were quantified and compared across the scenarios.

The breeding goal of both populations was production, udder health, and female fertility. Due to genotype by environment interactions these traits were seen as different traits in each environment. Partly different index traits were chosen within each population as indicators of the breeding goal traits. For the Nordic population, these traits were protein yield, mastitis resistance, somatic cell score (SCS), number of inseminations, days from calving to first insemination, and days from first to last insemination. For the US population, the traits were protein yield, SCS, and daughter pregnancy rate. All traits are defined so that higher values are desirable. All traits were simulated as normal distributed traits, and the initial phenotypic variances were set to 1. The economic values given to the Nordic and the US breeding goal traits corresponded to the economic values used at present in the Danish S-index and the US Lifetime Net Merit index, respectively (Danish Cattle Federation, 2006; VanRaden, 2006). In this study, the relative emphasis on production, udder health, and female fertility were 0.29, 0.44, and 0.27 in the Nordic breeding goal and 0.55, 0.16, and 0.29 in the US breeding goal.

The breeding goals were identical with the TMI in both scenarios. Animals were selected within their own population in scenario I, i.e. the populations operated individually. In scenario II, the populations could co-operate as up to 9 foreign bulls could be selected. Scenario II was supposed to reflect the current situation in the Nordic countries and the US.

Genetic and phenotypic parameters within populations and genetic correlations between the same traits across populations were obtained from a literature study.

The two populations of females were each divided in two groups, a breeding population and a test population. Also the two populations of males were each divided in two groups, a group of test bulls and a group of proven bulls. In the breeding populations, only semen from proven bulls was used, and selection of proven bulls was across populations in scenario II. In the test populations, semen from test bulls was used, and bulls were solely selected within their own population. Each group of test bulls was chosen among the bull calves from the breeding population within their own population. Bulls were progeny tested within their own environment, and only selected test bulls became proven bulls. Females were not moved from one population to another.

The Nordic and the US breeding populations consisted of 32,000 cows and 100,000 cows, respectively. Females in the breeding populations were simulated separately and distributed equally on a number of herds. Herd size was 100 cows in the Nordic population and 200 cows in the US population. The number of test bulls per year was 355 and 1,100 in the Nordic and the US populations. The progeny tests were based on daughter group sizes of 125 and 70 in the Nordic and the US populations, respectively. Hence, the Nordic test population and the US test population consisted of 44,400 and 77,000 cows, respectively. All females in the progeny group had observations for protein yield and SCS. Ninety percent and eighty percent of the females in the progeny group were measured for mastitis incidence and the fertility traits, respectively.

Selection of test bulls was performed when the bull calves were 11 months old, and it was based on the average breeding value of the parents. Test bulls were used for test inseminations at the age of 15 months, and they were not used in the time period between the test inseminations and the estimation of breeding values based on daughter performances. Selection of proven bulls was by truncation. Proven bulls were available until the age of eight years, and the maximum number of life-born calves sired by a single bull within one year was set to 2,000 and 5,000 in the Nordic population and the US population.

Breeding values were calculated in a linear animal model by means of the following model:

$$y_{i,m} = hys_i + a_{i,m} + e_{i,m} \qquad (1)$$

where $y_{i,m}$ is the record on animal m for trait i, hys_i is the herd-year season effect for each trait i, and $a_{i,m}$ and $e_{i,m}$ represent the additive genetic and residual term for trait i of animal m. This is done using the following (co)variance structure to predict breeding values:

$$\begin{pmatrix} a \\ e \end{pmatrix} \sim N\left(0; \begin{bmatrix} G_0 \otimes A & 0 \\ 0 & R_0 \otimes I \end{bmatrix}\right) \qquad (2)$$

G_0 and R_0 were 2×2 matrices for the analysis of protein yield in the two populations, 3×3 matrices for the analysis of mastitis resistance and SCS in the Nordic population and SCS in the US population, and 4×4 matrices for the analysis of number of inseminations, days from calving to first insemination, and days from first to last insemination in the Nordic population and daughter pregnancy rate in the US population. The matrix A was the numerator relationship matrix among animals, and the matrix I was an identity matrix. The use of covariances across populations was necessary if each animal should be assigned breeding values for both populations. The breeding value estimation was carried out using the DMU package (Madsen and Jensen, 2008). The variance components were those used in the simulations. Each animal was assigned a TMI for both populations.

Results

Average genetic change per year for the six traits registered in the Nordic population and the three traits registered in the US population was estimated to study how the contribution of the traits to progress in total merit changed when co-operation was possible.

For the Nordic population, scenario II gave significantly higher genetic progress in protein yield than scenario I (0.217 and 0.242 for scenario I and II). Scenario I gave 39% and 27% higher increase in mastitis resistance and SCS than scenario II; however the scenarios were not significantly different from each other. Both scenarios led to declines in genetic merit for female fertility, but scenario I gave smaller declines in all fertility traits than scenario II.

For the US population, scenario I led to significantly higher genetic progress in protein yield than scenario II (0.298 and 0.284, respectively). Scenario II gave, on the other hand, higher genetic progress in SCS than scenario I. Both scenarios led to declines in genetic merit for female fertility, but scenario II was significantly better than scenario I for daughter pregnancy rate (-0.070 for scenario I and -0.048 for scenario II).

Average genetic progress in total merit per year were 0.244 and 0.258 genetic standard deviation units of the true breeding goal in the Nordic population when bulls were selected within and across populations, respectively. Thus, scenario II was significantly better than scenario I for the Nordic population. In the US population average genetic progress in total merit per year were 0.301 and 0.294 genetic standard deviation units of the true breeding goal when the population operated individually and when it co-operated, respectively. Scenario I and II were not significantly different from each other for the US population. The Nordic population had significantly lower genetic progress in total merit than the US population across both scenarios.

Discussion

The possibility of co-operation led to significant higher genetic progress in total merit in the Nordic population. Hence, it is recommended that Nordic Holstein co-operates with US Holstein. Co-operation with the US population changed, however, the ratio between the traits that contribute to total merit. Thus, the scenario where co-operation was possible led to significantly higher genetic progress in protein yield than the scenario where the Nordic population operated individually, but the scenario where the Nordic population operated individually tended to be better in regard to functional traits. These results show that co-operation with populations that give more emphasis to high heritability traits, test more bulls, and use lower progeny group sizes has influence on the direction dairy cattle breeding takes in small populations. Therefore, great care should be exercised in the choice of collaborators if the Nordic breeding organisations wish genetic progress in the functional traits and wish to market Nordic Holstein as a breed that is characterised by a high genetic level for the functional traits.

The possibility of co-operation did not change genetic progress in total merit in the US population significantly but the ratio between the traits that contribute to progress in total merit changed when co-operation was possible. The scenario where the US population operated individually led to higher genetic progress in protein yield than the scenario where co-operation was possible. On the other hand, the scenario where co-operation was possible tended to give genetic progress or smaller decline in the functional traits. Deterioration of animal health and reproduction concerns the producers and the consumers (Miglior *et al.*, 2005) as it decreases animal welfare (Sandøe *et al.*, 1999). Therefore, if genetic progress in total merit remains the same whether co-operation is possible or not it must be in common interest if genetic progress in total merit is due to gain in functional traits such as disease resistance and fertility.

Selection across populations is characterised by always being at least as good as selection within populations (Smith and Banos, 1991). The effect of co-operation between Nordic Holstein and US Holstein on genetic progress in total merit was investigated under the current circumstances. The results showed that the Nordic population benefited from co-operation with the US population, but the US population did not benefit from co-operation with the Nordic population. The possibility of profitable co-operation depends among other things on population size. Smith (1981) found that genetic gain increases with the number of bulls tested but at a declining rate. Consequently, the small population, in this case the Nordic population, gained more from combining with the large population than the large population gained from combining with the small population. The US population obtained higher genetic progress in total merit than the Nordic population across both scenarios both because of the large population size and because relative more emphasis were given to high heritability traits in the US population than in the Nordic population. It is not possible to quantify the individual effect of these two causes in this study.

The economic values given to the simulated breeding goals corresponded to those used in the Danish TMI and the US TMI. These values are primarily based on economic models of the production systems (Danish Cattle Federation, 2006; VanRaden, 2006). However, it is difficult to derive economic values for traits associated with animal health and reproduction from profit functions (Groen *et al.*, 1997). Furthermore, using a breeding goal based on objective economic values may result in deterioration of animal health and female fertility even though these traits are included in the TMI. Farmers and/or consumers may find that this response to selection is unacceptable, although it results in the highest profit. A solution could be to add a non-market value to the objective economic value or to express the breeding goal as a set of desired gains (Nielsen *et al.*, 2005).

All scenarios led to declines in genetic merit for female fertility. A decline in genetic merit for female fertility has also been found in the Danish Holstein population until the turn of the millennium (Danish Cattle Federation, 2008). Within the last years a slight increase has been observed. The decline was due to an unfavourable genetic correlation between milk production traits and female fertility traits, a relatively low emphasis on female fertility in the breeding goal and to the fact that sires of sons were not selected on the basis of the Danish TMI. A continuing decline in genetic merit for female fertility is not acceptable in sustainable breeding schemes. To turn the tide, the economic value on female fertility is higher in the joint Nordic breeding goal, which will be applied from the autumn 2008, than in the present Danish breeding goal.

Conclusion

The possibility of selection across populations increased genetic progress in total merit in the Nordic population. Hence, co-operation with US Holstein is recommended. The genetic progress in total merit was obtained by higher genetic progress in protein yield at the expense of lower genetic progress in the functional traits. The possibility of selection across populations did not change genetic progress in total merit in the US population. However, the genetic progress in protein yield decreased whereas the genetic progress in the functional traits was either maintained or it increased. Thus, co-operation with Nordic Holstein may limit the deterioration of animal health and reproduction in US Holstein. However, all results are sensitive to the assumptions made in the simulation study. The results showed, in addition, that the composition of genetic progress is changed in the direction of the foreign breeding goal when populations co-operate. This happens even though the imported bulls are selected in accordance with the breeding goal of the domestic population.

References

Danish Cattle Federation, 2006. Principles of Danish cattle breeding. Eighth edition. Online. Available http://www.lr.dk/kvaeg/diverse/principles.pdf. Accessed July 4, 2008. 55 p.

Danish Cattle Federation, 2008. Genetic trend in female fertility - Danish Holstein. Online. Available at: http://www.lr.dk/kvaeg/informationsserier/abonnement/genetisk_trend_hunfrgt.pdf. Accessed July 4, 2008.

Groen, A.F., T. Steine, J.J. Colleau, J. Pedersen, J. Pribyl and N. Reinsch, 1997. Economic values in dairy cattle breeding, with special reference to functional traits. Report of an EAAP-working group Livest. Prod. Sci., 49: 1-21.

Madsen, P. and J. Jensen, 2008. DMU: a user's guide. A package for analysing multivariate mixed models. Version 6, release 4.7, Danish Institute of Agricultural Sciences, Foulum, Denmark. http://dmu.agrsci.dk.

Miglior, F., B.L. Muir and B.J. van Doormaal, 2005. Selection indices in Holstein cattle of various countries. J. Dairy Sci., 88: 1255-1263.

Mulder, H.A. and P. Bijma, 2006. Benefits of cooperation between breeding programs in the presence of genotype by environment interaction. J. Dairy Sci., 89: 1727-1739.

Nielsen, H.M., L.G. Christensen and A.F. Groen, 2005. Derivation of sustainable breeding goals for dairy cattle using selection index theory. J. Dairy Sci., 88: 1882-1890.

Sandøe, P., B.L. Nielsen, L.G. Christensen and P. Sørensen, 1999. Staying good while playing God – The ethics of breeding farm animals. Anim. Welfare, 8: 313-328.

Smith, C., 1981. Levels of investment in testing the genetic improvement of livestock. Livest. Prod. Sci., 8: 193-201.

Smith, C. and G. Banos, 1991. Selection within and across populations in livestock improvement. J. Animal Sci., 69: 2387-2394.

Sørensen, M.K., P. Berg, J. Jensen and L.G. Christensen, 1999. Stochastic simulation of breeding schemes for total merit in dairy cattle. GIFT Seminar on Genetic Improvement of Functional Traits in Cattle, Wageningen, the Netherlands, November 7-9, 1999. Interbull Bulletin, 23: 183-192.

VanRaden, P.M., 2006. Net merit as a measure of lifetime profit: proposed 2006 revision. Online. Available http://www.aipl.arsusda.gov/reference/nmcalc.htm. Accessed July 4, 2008. 15 p.

Predictive ability of different models for clinical mastitis in joint genetic evaluation for Sweden, Denmark and Finland

K. Johansson[1,5], S. Eriksson[1], W.F. Fikse[1], J. Pösö[2], U. Sander Nielsen[3] and G. Pedersen Aamand[4]
[1]*Swedish University of Agricultural Sciences, P.O. Box 7023, 75007 Uppsala, Sweden*
[2]*FABA Breeding, P.O. Box 40, FI-01301 Vantaa, Finland*
[3]*Danish Agricultural Advisory Service, Udkaersvej 15, 8200 Aarhus, Denmark*
[4]*Nordic Cattle Genetic Evaluation, Udkaersvej 15, 8200 Aarhus, Denmark*
[5]*Swedish Dairy Association, P.O. Box 7023, 75007 Uppsala, Sweden*

Abstract

Records on clinical mastitis (CM) and somatic cell count (SCC) in three lactations, and first lactation udder conformation traits (UC), are included in the joint genetic evaluation of Red Dairy Cattle in Sweden, Denmark and Finland since 2006. The aim of this study was to compare predictive ability of different multiple-trait models for udder health on a Nordic level. Linear sire models including different number of udder health traits (CM, SCC, UC) were used to estimate breeding values (EBVs), based on data comprising 2.7 million Red Dairy Cattle cows recorded until 2002. Correlations were estimated between these EBVs and daughter group means for clinical mastitis recorded from 2003 and onwards. The comparison involved 99 proven bulls born 1992-1995 and 486 young bulls born 1997-2000. Genetic evaluation models utilising CM data and data for at least one of the correlated traits SCC or UC gave the highest correlations between EBVs in early data and daughter group means in recent data. Compared with the full model including all three traits, a genetic evaluation utilising CM data only gave 9% weaker correlation between EBVs and daughter group means for CM (for proven bulls). The corresponding difference was 17% weaker using a genetic evaluation utilising SCC and UC data, and 39% weaker for a genetic evaluation utilising only UC data.

Keywords: udder health, multiple-trait models, prediction, somatic cell count, udder conformation

Introduction

Clinical mastitis (CM) is one of the most common diseases in dairy cattle, causing financial losses to the farmers due to reduced milk yield, discarded milk, veterinary treatments, labour and replacement costs, see Heringstad *et al.* (2000) for a review. It also reduces animal welfare considerably (Milne, 2005). In the Nordic countries, CM has long been recorded and included in the breeding goal and in the genetic evaluation (Aamand, 2006). The heritability of CM is low (e.g. Pösö and Mäntysaari, 1996; Carlén *et al.*, 2004); hence based on knowledge from general animal breeding theory, indirect selection through correlated traits with higher heritabilities, such as somatic cell count (SCC) or udder conformation traits (UC) is used.

The average daughter group size is large but variable in the Nordic countries (e.g. Heringstad *et al.*, 2000), and the merit of including data on SCC and UC over and above data on CM in a genetic evaluation of CM has been debated (Philipsson *et al.*, 1994). Reliabilities depend clearly on progeny group sizes when estimating breeding values. However, while the reliability of CM breeding values based on only CM data is relatively high for proven bulls (85-95%, depending on breed and country), this is not the case for selection candidates and utilising data on correlated traits may increase the accuracy of selection (e.g. Philipsson *et al.* 1994; Nielsen *et al.*, 1996).

Ultimately, the purpose of genetic evaluations is to predict the phenotypic performance of offspring of selected animals. Thus, the quality of a genetic evaluation can be measured by the predictive ability of the breeding values. An appealing approach to assess predictive ability is to compare breeding values predicted from a subset of the data with the phenotypic observations that were excluded from the complete data set (e.g. Thompson, 2001). Along these lines, Ødegård et al. (2003) reported that CM observations for second crop daughters were best predicted in a genetic evaluation utilising records on both CM and SCC from first crop daughters compared to utilising records from just one of these traits.

A routine genetic evaluation of udder health traits jointly for Sweden, Denmark and Finland was launched in 2006 (Johansson et al., 2006). The data utilised in this genetic evaluation are records on CM and SCC in lactations 1 to 3, and first lactation records of fore udder attachment (FU) and udder depth (UD). The initial development of the joint Nordic udder health evaluation focused on selection of a model for genetic evaluation yielding unbiased genetic trend estimates (Johansson et al., 2006). However, the predictive ability of this genetic evaluation deserved further attention, in particular the importance of indicator traits to increase efficiency of selection.

The aim of this study was to compare predictive ability of different multiple-trait models for clinical mastitis derived from the routine joint Nordic genetic evaluation of Red Dairy Cattle. Predictive ability was measured as the correlation between breeding values predicted from a subset of the data, where the 4 most recent years of data were excluded, with the sire daughter group means or breeding values computed from the 4 most recent years of data.

Materials and methods

The routine joint Nordic genetic evaluation model

The model for estimation of breeding values currently used in joint Nordic udder health routine evaluations is a multiple-trait, multiple-lactation sire model (Johansson et al., 2006). In total there are 9 traits (Table 1): 4 clinical mastitis traits, 3 somatic cell count traits and 2 udder conformation traits (udder depth [UD] and fore udder attachment [FU]).

The effect of herd×year is included as random in the model for CM and SCC, which includes fixed class effects of herd×5-year-period, calving age×country, and year×month of calving×country. Effects of Original Red Danes (RDM), Danish Friesian (SDM), Finnish Ayrshire (FAY), Norwegian Red (NRF), American Brown Swiss (ABK), American Holstein (HOL), Swedish Red Cattle (SRB), Canadian Ayrshire (CAY) and Finn cattle (FIC), are accounted for by regressions on population proportions. Heterosis is accounted for using a regression on expected total heterosis of all included populations.

The model for udder conformation traits in the joint Nordic udder health evaluation is almost the same as the one used in joint Nordic type evaluation (Fogh et al., 2004). The only difference is that a sire model is used in the udder health evaluation whereas an animal model is used in the type evaluation.

All traits are pre-corrected for heterogeneous variance due to year of calving and country using a standard linear transformation. The genetic parameters used for the 9 traits in the evaluation were presented by Johansson et al. (2006); heritabilities are in Table 1. The genetic correlations used between CM traits and SCC traits were in the range of 0.55 to 0.66, and between CM traits and UC traits in the range of -0.34 to -0.54. For computational reasons residual correlations between lactations were set to zero.

Table 1. Trait abbreviations, heritabilities (h^2), definitions of traits included in the joint Nordic udder health evaluations.

Trait abbrev.	h^2	Definition
CM11	0.032	Clinical mastitis (1) or not (0) between -15 and 50 days after 1st calving
CM12	0.024	Clinical mastitis (1) or not (0) between 51 and 300 days after 1st calving
CM2	0.032	Clinical mastitis (1) or not (0) between -15 and 150 days after 2nd calving
CM3	0.034	Clinical mastitis (1) or not (0) between -15 and 150 days after 3rd calving
SCC1	0.140	Log. somatic cell count average 5-170 days after 1st calving
SCC2	0.133	Log. somatic cell count average 5-170 days after 2nd calving
SCC3	0.115	Log. somatic cell count average 5-170 days after 3rd calving
FU	0.240	Fore udder attachment score in 1st lactation
UD	0.360	Udder depth score in 1st lactation

The four CM traits and the three SCC traits are combined into a CM and SCC index, respectively, for publication purposes. The actual weights are given in Table 2 and these were also used in the present study.

Data

Data on Red Dairy Cattle from Denmark, Finland and Sweden were used in this study. CM data was available for Sweden and Finland since 1984 and for Denmark since 1990. SCC has been available since 1984 for Finland, 1988 for Sweden and 1990 for Denmark. Udder conformation data since 1992 in Sweden and Finland and since 1990 in Denmark were utilised. In total, data on 3,281,094 cows were available. Three generations of pedigree were traced for all bulls with daughters in the data set.

The dataset was split into two smaller datasets; one with records until 2002 ('early' data) comprising 2,568,297 cows, the other with 563,815 cows having records from 2003 to 2006 ('recent' data).

Analysis

The model for the joint Nordic udder health evaluation was used to predict breeding values in the 'early' data set. The 'early' data was also analysed with several other sub-models, in which data for one or several of the udder health traits were discarded. The sub-models considered were: CM+SCC, CM+UC, SCC, UC and SCC+UC (Table 2).

The CM and SCC indices, as used in the joint Nordic routine evaluation, were considered in this study. In addition, another index was computed that combined the SCC and UC traits. The weights

Table 2. Definition of indices.

Abbreviation	Index definition
CM	Clinical mastitis: 0.25*CM11 + 0.25*CM12 + 0.3*CM2 + 0.2*CM3
SCC	Log somatic cell count: 0.5*SCC1 + 0.3*SCC2 + 0.2*SCC3
UC	Udder conformation: -1*(0.5*FU + 0.5*UD)
SCC+UC	Both indicator traits: b1*SCC+b2*UC, where b1 = 0.195 and b2 = 0.020

were calculated from data using a regression of the CM index from the CM evaluation (including records on CM only) on SCC and UC indices from the SCC+UC evaluation (including records on SCC and UC only).

Information on mastitis from the 'recent' dataset was used, summarised as either sire daughter group means, or as CM-index of breeding values, to study the predictions from the 'early' dataset. Calculation of the daughter group means considered only daughters in the country where the sire had most records. Daughters with records in the 'early' dataset were omitted in the 'recent' dataset.

Estimation of predictive ability

Different sets of EBVs (actually: indices) from the 'early' data were correlated with clinical mastitis information (daughter group means or CM index) in the 'recent' data. Production levels as well as genetic levels varied between countries. Therefore, correlations were estimated using the residuals from a model that contained fixed effects of sire birth year and country of recording.

When CM data were included in the analysis of 'early' data the CM index (Table 2) was considered as predictor of the daughter group means or CM index based on the 'recent' data. When CM data were not included, the respective indices (SCC, UC or SCC+UC; Table 2), were correlated with the daughter group means or CM index in the 'recent' data.

Two different categories of bulls were studied: bulls born 1992-1995 and bulls born 1997-2000. Bulls born 1992-1995 were required to have at least 200 daughters with first lactation records for CM or SCC in the 'early' dataset, and 99 bulls met these criteria. Bulls born 1997-2000 were required to have at least 20 first lactation daughters in total and at least 10 daughters with third lactation CM records in the 'recent' dataset.

Different amounts of information on udder health of daughters were available for these two categories of bulls in the two datasets. The bulls born 1992-1995 were proven bulls with EBVs mainly based on records on first crop daughters in the 'early' dataset, and with CM records on second crop daughters in the 'recent' dataset. The proven bulls had on average 1,533 first lactation daughters.

The bulls born 1997-2000 were young bulls, and EBVs estimated from the 'early' dataset were based on pedigree index for these bulls, whereas the daughter group means and CM index in the 'recent' dataset were calculated based on first crop daughter records. The young bulls had on average 142 first lactation daughters.

The number of records considered for the calculation of the daughter group means is shown for each trait in Table 3.

Results and discussion

Estimated correlations between EBVs (indices) obtained from the different genetic evaluation models of the 'early' dataset, and CM index or daughter group means in the 'recent' dataset, are in Table 4. These correlations reflect the accuracy of the models in the prediction of clinical mastitis occurrence.

Predictive ability of different multiple-trait models

Multiple-trait models utilising CM data gave generally higher correlations than models ignoring CM data. For example, the correlation for sire daughter group means for proven bulls ranged from 0.39 to 0.53 when the genetic evaluation of the 'early' data was based on SCC and/or UC data, but

Table 3. Number of bulls and number of daughters with records for clinical mastitis in first (CM11, CM12), second and third lactations (CM2, CM3) included in the calculation.

Country	Trait	No. of bulls[1]		Average no. of daughters 2003-2006[1]	
		Proven bulls (1992-1995)[2]	Young bulls (1997-2000)[2]	Proven bulls (1992-1995)[2]	Young bulls (1997-2000)[2]
Sweden	CM11	28	198	2,537	139
	CM12	28	198	2,359	134
	CM2	28	198	1,465	87
	CM3	28	198	561	42
Denmark	CM11	13	91	1,158	71
	CM12	13	91	1,103	69
	CM2	13	91	715	43
	CM3	13	91	320	20
Finland	CM11	58	197	1,134	178
	CM12	58	197	1,107	174
	CM2	58	197	833	128
	CM3	58	197	394	61

[1] Bulls with at least 10 daughters with 3rd lactation records.

[2] Birth year of bulls.

Table 4. Estimated correlations between EBVs (indices) from different models in the 'early' data (records until 2002) and clinical mastitis (from means or estimated breeding values) for proven and young bulls in the 'recent' data (records 2003-2006).[1]

Data included in genetic evaluation of 'early' data	Sire daughter group means for CM; 'recent' data		CM index from estimated breeding values; 'recent' data	
	Proven bulls (1992-1995)[2]	Young bulls (1997-2000)[2]	Proven bulls (1992-1995)[2]	Young bulls (1997-2000)[2]
CM+SCC+UC	0.64	0.36	0.75	0.44
CM+SCC	0.64	0.31	0.75	0.42
CM+UC	0.62	0.36	0.73	0.44
CM	0.58	0.25	0.71	0.38
SCC+UC	0.53	0.28	0.58	0.34
SCC	0.45	0.23	0.53	0.32
UC	0.39	0.12	0.36	0.12

[1] Abbreviations as in Table 1.

[2] Birth year of bulls.

increased to 0.62-0.64 when CM data was also included. For young bulls, however, utilising just CM data gave a slightly lower correlation than a combination of the indirect traits SCC and UC (0.25 vs. 0.28; Table 4).

For proven bulls all multiple-trait models where CM data were utilised resulted in a much better prediction of CM in future daughters, compared with models ignoring data on this target trait. This is in agreement with the results from Ødegård *et al.* (2003) who estimated that predictive ability for CM alone in first lactation was 43% higher than for a model with only SCC. The corresponding figure in this study, utilising data on 3 lactations, was somewhat lower (29%). Including SCC or UC data in the multiple-trait model further increased the correlations between EBVs (indices) in 'early' data and daughter group means or CM index in the 'recent' data (from 0.58 to 0.62-0.64). The advantage of using data in all three traits together seemed small judging from the estimated correlations for proven bulls. The three-trait alternative was not better than either of the two two-trait alternatives. One reason could be that CM data and either SCC or UC data provide sufficient information, and that the value of including additional information from another correlated trait is low.

For young bulls the scenario is different. In this case, the predictions in the 'early' dataset were based on pedigree information only. The more information available, the better predictions can be made. Consequently, including UC data in the multiple-trait models appeared to give better predictions especially for young bulls. Philipsson *et al.* (1994) found that it was more efficient to select for a correlated trait with higher heritability (SCC) than directly on resistance to clinical mastitis if progeny groups were small (less than 100 daughters). With large daughter groups, direct selection on resistance to clinical mastitis was more efficient, however. They also found that a combination of both measures was the most efficient in both cases.

Nielsen *et al.* (1996) compared the efficiency of including information on different udder health traits in the Danish evaluation of clinical mastitis resistance. Similarly to our results, they found that all indices including clinical mastitis gave higher correlation between the index and aggregate genotype for mastitis resistance than indices based on SCC, UC or SCC and UC traits together.

Predictive value of UC

For both young and proven bulls, the value of UC data alone to predict CM seemed low in our study. The UC data comprised first lactations only, whereas CM and SCC data were available from three lactations. Moreover, records on UC were just available for a subset of cows that had observations for SCC. In addition, the genetic correlations between UC and CM were lower than between SCC and CM. Nielsen *et al.* (1996) found that udder depth and fore udder attachment together were as effective as SCC for evaluation of mastitis resistance. Different progeny group sizes in the different countries may at least partly explain the difference between their result and our observations.

Including UC data provides valuable information as udder conformation data is available at an early stage. Looking at the results for young bulls, there was a significant value of utilising UC data, even when SCC data were already included. This is important in the practical situation when selecting bulls. The main part of the selection will take place when most of the daughters only have first lactation records.

It would be interesting to pursue this issue further and do comparisons between genetic evaluation utilising only first lactation records for both SCC and UC, or records for all three lactations for both indicator traits.

Proven vs. young bulls

Correlations were higher for proven bulls than for young bulls (0.39-0.64 vs. 0.12-0.36 for daughter group means; Table 4). There are several factors affecting the comparison of results for both groups of bulls. In this dataset the number of proven sires was small. This means that the influence of an individual sire on the results could be high. Proven bulls had randomly sampled daughter groups in the 'early' dataset but had daughters with records in the 'recent' dataset after being selected based on the first crop daughter information. Young bull daughter groups, found in the 'recent' dataset, were randomly sampled, but information on these bulls in 'early' dataset was based on pedigree only.

EBVs or daughter group means

Correlations between EBVs (indices) in 'early' and 'recent' dataset were stronger than correlations between EBVs (indices) in 'early' dataset and index from daughter group means in the 'recent' dataset (0.36-0.75 vs. 0.39-0.64 for proven bulls and 0.12-0.44 vs. 0.12-0.36 for young bulls). All alternatives that utilised CM data in the genetic evaluation of the 'early' dataset gave somewhat higher correlations when correlating EBVs (indices) in the 'early' dataset with CM index rather than daughter group means in the 'recent' dataset. To calculate the daughter group means for CM from lactation means in the 'recent' dataset the same formula was used as for the official NAV index (Table 2). The index weights are not optimal for calculating daughter group means. Thus correlations to daughter group means may be a little too low for alternatives that utilised CM data in the genetic evaluation of 'early' data. A more detailed study on the (part-) lactation specific EBVs (CM11, CM12, CM2 and CM3 instead of CM index) could shed light on this.

The setup of the current study is somewhat similar to the setup used in Interbull's validation Method 3. One obvious comparison would be that between breeding values from the reduced dataset, here called the 'early' data, and the breeding values from the complete data ('early' and 'recent' data). The part-whole relationship resulted in unrealistically high correlations in a preliminary analysis relating 'early' predictions to predictions from the whole dataset, which therefore was not pursued further.

Conclusions

Including direct information on clinical mastitis in the genetic evaluation improved the prediction of clinical mastitis in future daughters compared to only use of somatic cell count and udder conformation.

Including information on somatic cell count or udder conformation together with clinical mastitis increased the predictive ability of the estimated multiple-trait model breeding values for clinical mastitis. The combination of clinical mastitis and udder conformation information gave a comparatively high predictive value, especially for young bulls.

Results of predictive ability for both proven bull and young bull breeding values proved to be informative when discriminating between models for prediction of breeding values.

References

Aamand, G.P., 2006. Use of health data in genetic evaluation and breeding. In: Breeding, production recording, health and the evaluation of farm animals. Proc. of the 35th Biennial Session of ICAR, EAAP publication No. 121: 275-282.

Carlén, E., E. Strandberg A. Roth, 2004. Genetic parameters for clinical mastitis, somatic cell score, and production in the first three lactations of Swedish Holstein cows. J. Dairy Sci., 87: 3062-3070.

Fogh, A., J.-Å. Eriksson, J. Juga, M. Toivonen, J. Pösö, M. Simpanen, U.S. Nielsen and G.P. Aamand, 2004. A joint Nordic model for type traits. Interbull Bulletin 32: 13-17.

Heringstad, B., G. Klemetsdal and J. Ruane, 2000. Selection for mastitis resistance in dairy cattle: a review with focus on the situation in the Nordic countries. Livest. Prod. Sci. 64: 95-106.

Johansson, K, S. Eriksson, J. Pösö, M. Toivonen, U. Sander Nielsen, J.-Å. Eriksson and G.P. Aamand, 2006. Genetic evaluation of udder health traits for Denmark, Finland and Sweden. Interbull Bulletin 35: 92-96.

Milne, H.M., 2005. Mastitis is a welfare problem. Proc. of the British Mastitis Conference 2005. Available at: http://www.iah.bbsrc.ac.uk/bmc/2005/Papers 2005/Mastitis_is_a_welfare_problem.doc.

Nielsen, U.S., G.A. Pedersen and J. Pedersen, 1996. Indices for resistance against diseases. Interbull Bulletin 14: 161-166.

Ødegård, A., G. Klementsdal and B. Heringstad, 2003. Genetic improvement of mastitis resistance: Validation of somatic cell score and clinical mastitis as selection criteria. J. Dairy Sci. 86: 4129-4136.

Philipsson, J., G. Ral and B. Berglund, 1994. Somatic cell count as a selection criterion for mastitis resistance in dairy cattle. Livest. Prod. Sci. 41: 195-200.

Pösö, J. and E.A. Mäntysaari, 1996. Relationships between clinical mastitis, somatic cell score and production for the first three lactations of Finnish Ayrshire. J. Dairy Sci. 79: 1284-1291.

Thompson, R., 2001. Statistical validation of genetic models. Livest. Prod. Sci. 72: 129-134.

Breeding for improved disease resistance in ruminants

S.C. Bishop
The Roslin Institute and Royal (Dick) School of Veterinary Studies, University of Edinburgh, Midlothian EH25 9PS, United Kingdom

Abstract

Breeding for enhanced resistance to infectious disease is an effective means of improving the health, fitness and robustness of ruminant livestock. The most amenable endemic diseases to genetic selection are likely to be mastitis, bovine leukaemia, gastrointestinal (GI) parasitism, tuberculosis (TB) and paratuberculosis in cattle; and mastitis, GI parasitism and footrot in sheep. For bovine mastitis, selection on clinical signs or somatic cell count (SCC) is well established; however the longterm wisdom of decreasing SCC is often questioned. A solution may be to decompose SCC into baseline and response variables, along with a liability to become infected, and select for reduced liability. Genetic markers are often sought for mastitis resistance, but the complexity of the host-pathogen interactions may mean that individual markers are of insufficient value, requiring whole-genome approaches. Bovine TB is an emerging zoonotic threat, and a disease of major importance in some geographical regions. Studies are currently underway to assess options for breeding cattle for increased resistance. In particular, identification of cases and controls from herds with disease outbreaks may allow efficient genome scans. Paratuberculosis is a similar bacterial disease, but its prevalence is currently unknown and improved diagnostic tests are required before effective genetic approaches can be contemplated. For nematode resistance and footrot in sheep, readily measured indicators of relative resistance are available, and observed genetic gains may be larger than predicted by genetic theory, due to decreased contamination from selected animals. Although it has yet to be explored, this result is also likely to apply for several dairy cattle diseases. In summary, selection for increased resistance to specific diseases may have considerable benefits for cow health, welfare and robustness. However, this conclusion is dependent upon the disease prevalence, the higher the prevalence the greater the benefits of selection and the greater the impact on robustness.

Keywords: disease resistance, genetics, selection, epidemiology, ruminant

Introduction

This paper considers issues related to breeding cattle for improved disease resistance, as a means of improving robustness. General issues related to the importance of diseases and disease resistance, the benefits of genetically improving resistance and prioritising diseases are discussed. The paper will then explore options for breeding cattle for resistance to various important infectious diseases and draw parallels from experiences in sheep.

Background

The competitiveness of livestock industries in Europe is determined by productivity and quality values, with quality values including food quality, food safety and animal welfare. Sustainable and socially acceptable animal production methods and the prevention of human disease are also becoming more important. Infectious disease has impacts on animal production at all of these levels, as well as having a major impact on animal welfare and robustness. Infectious disease costs are estimated to account for 10-20% of turnover in *developed countries*. This figure does not take account of the devastating impact that outbreaks of exotic diseases, such as foot and mouth disease or bluetongue, have on the livestock industry and the wider national economy. On a global scale, inadequacies in animal production are a significant contributory factor to food shortages and poverty in the *developing world*. With demand for animal products predicted to increase by 50% by 2020, mostly as a consequence of human population expansion in developing countries, this situation is likely to deteriorate. Infectious disease is one of the most important constraints to animal production in these regions, with financial losses due to infectious disease amounting to 35-50% of turnover. This situation largely reflects absence or unsustainability of disease control measures.

Disease-related suffering of animals is increasingly recognised as an important welfare issue, and food-borne and other zoonotic pathogens pose a major risk to human health. Our ability to control many infectious diseases of livestock is also threatened by the emergence of pathogens resistant to antimicrobial and anti-parasitic drugs, as well as the stricter regulations governing chemical residues in animal products and the prophylactic use of antibiotics. More effective and sustainable methods of disease control are required to deal with these problems, particularly with the emergence of drug-resistant pathogens. Recent advances in animal genomics and immunology and in the understanding of host-pathogen interactions provide new opportunities to develop more effective control strategies. The focus of this paper is the selection of animals for increased genetic resistance to disease, leading to healthier, more productive and more robust animals.

Benefits of selection for resistance

Benefits of selection for resistance depend upon the epidemiological context of the disease. A framework within which the benefits may be calculated was presented by Bishop and Stear (2003) and further elaborated by Gibson and Bishop (2005). Essentially the distinction must be made between endemic diseases, i.e. those whose presence is easily predicted, and epidemic diseases, i.e. those that can spread rapidly through the population. Epidemic diseases will be generally absent from livestock populations, therefore genetic selection for resistance would aim to reduce risks of epidemics, or their severity should an epidemic occur. For endemic diseases, the aim of a selection program is to breed animals that are more resistant to a disease that is predictable in its occurrence. As well as improved productivity, welfare and robustness, impacts may include a reduced abundance of the pathogen or parasite in the population or environment, leading to further benefits. This has been quantified for nematode resistance (Bishop and Stear, 2003) and footrot (Nieuwhof *et al.*, 2009) in sheep.

Independent of how an improvement in disease resistance is achieved, i.e. either by genetic or environmental means, the economic benefits of the improvement require careful consideration. A method to calculate these benefits was presented by Bennett *et al.* (1999), and illustrated for several endemic sheep diseases by Nieuwhof and Bishop (2005). Essentially, total annual disease costs may be estimated as the sum of annual losses in expected output and wasted inputs, annual treatment costs and annual prevention costs. Improving disease resistance doesn't necessarily reduce all cost categories proportionately, as the necessary prevention costs may remain unchanged. For example, in Great Britain, annual costs for nematode infections and footrot were estimated to be £84 million and £24 million, respectively. Whilst nematode costs are linear to the severity of infection, approximately half of the footrot costs are due to preventive measures which are not influenced by marginal improvements in resistance. Therefore, the relative marginal benefit of improvements in resistance is greater for nematode infections than for footrot. Similarly for dairy cattle, economic benefits may not be directly proportional to the total disease costs.

Opportunities to select for resistance

To perform effective selection for disease resistance it is necessary to prioritise diseases; unless an attempt is being made to select for general disease resistance it is unlikely that it will be possible to simultaneously select for resistance to more than a small number of diseases. A method of prioritising diseases, based on aspects of disease importance, has been outlined by us (Davies *et al.*, 2009). Six disease-based assessment criteria were defined: industry concern, economic impact, public concern, threat to food safety or zoonotic potential, impact on animal welfare and threat to international trade barriers, and subjective scores were assigned to each category for each disease according to the relative strength of available evidence. Evidence for host genetic variation in resistance was determined from available published data, including breed comparisons, heritability studies, QTL studies, evidence for candidate genes with significant effects and host gene expression analyses. Combined, these enable a qualitative but robust ranking of the amenability of diseases for host genetic studies and eventually for genetic improvement.

From these analyses, the top ranking dairy cattle diseases were, in order, mastitis, bovine leukaemia, gastrointestinal (nematode) parasitism, paratuberculosis and bovine tuberculosis. Many important epidemic diseases, such as foot and mouth disease, were lower ranked because of the difficulty of obtaining host genetic information and hence the associated lack of evidence for host genetic variation. It is interesting to note that these diseases are already the focus of much research. When compared across livestock species, mastitis was the second highest ranked disease overall, only beaten by *Salmonella* infections in chickens.

Opportunities in cattle

Three examples are illustrated and discussed for dairy cattle, where genetic selection has been, or potentially could be, used to breed animals for increased disease resistance.

Mastitis

Mastitis, inflammation of the mammary gland, is usually caused by bacterial organisms such as *Staphylococcus* spp., *Streptococcus* spp., *Pseudomonas* spp., *Mycoplasma* spp. and various coliforms such as *E. coli*. Mastitis incidence in the dairy industry has been estimated at 30% of cows per year, and each case has been estimated to cost between 150 to 300 euros per diseased cow. Therefore, it is a disease of considerable importance.

Selection for increased milk yield will generally worsen the incidence of mastitis, due to the unfavourable genetic correlation between milk yield and mastitis susceptibility. Therefore, efforts to reduce mastitis, or prevent its incidence from rising, are a part of most dairy cattle evaluations. Currently, selection to reduce the incidence of mastitis is based on udder conformation, somatic cell count (SCC) and mastitis infection history. SCC and clinical mastitis generally have low heritabilities, usually in the range 0.05 to 0.15. Mastitis resistance is probably due to structural attributes of the udder or teat, as well as immune responses. This is suggested by the observation that mastitis incidence is correlated with aspects of udder conformation (up to 0.37; Van Dorp *et al.*, 1998) as well as SCC (ca. 0.7; e.g. Carlén *et al.*, 2004, Heringstad *et al.*, 2006).

QTL associated with mastitis resistance traits have been reported on almost all of the 29 bovine chromosomes, in a variety of populations and breeds including US, German and Dutch Holsteins, Finnish Ayrshire, Swedish Red and White, Danish Red and Norwegian Cattle (see reviews: Khatkar *et al.*, 2004; Rupp and Boichard, 2003). This large number of QTL suggests that gene-assisted selection, using causative mutations underlying these QTL, may be inefficient if each of these mutations explains only a small proportion of the observed variation in SCC or clinical mastitis. Possibly a whole genome approach using a dense single nucleotide polymorphism array ('SNP chip'), i.e. genome-wide selection (Meuwissen *et al.*, 2001), may be advantageous in this case.

However, there is also likely to be considerable benefit in redefining traits describing mastitis resistance, and this may also address concerns as to whether continued selection for reduced SCC is a long-term solution to mastitis. SCC is used as an indicator of mastitis, with high SCC values indicating that an animal is likely to be infected. However, SCC measurements on a group of animals comprise a mixture distribution trait describing baseline SCC in unaffected animals as well as elevated SCC in infected animals. The concern is that reducing SCC too far may reduce baseline SCC levels, hence an animals ability to respond to infection. In actual fact, the trait that is of interest to the breeder is liability to mastitis. Therefore, a rational aim when considering SCC data is to decompose it into baseline SCC values for uninfected animals, response SCC values for animals that are infected, along with the probability that a particular animal falls into one distribution and not the other. This concept was introduced by Odegard *et al.* (2005), and the quantitative genetic properties of such mixture distribution traits were formalised by Gianola *et al.* (2006). The primary selection criterion arising from this data decomposition is the liability of an animal to be affected by mastitis. The secondary question is whether to increase or decrease SCC; however, this question must be asked separately for baseline and response SCC. To answer this, it is necessary to calculate genetic correlations between SCC and mastitis liability, separately for baseline and response SCC.

The fact that mastitis is caused by different species of bacteria raises further issues of potential importance. Can these separate categories of infection be teased apart and is it beneficial to do so? Further, for the infectious (as opposed to 'environmental') sources of mastitis, further insight may be gained by assessing cow liability to mastitis in relation to the force of infection. For example, are there genetic influences on the order in which animals become infected, and do genetic effects alter as disease prevalence changes? If these effects exist, they may point to additional epidemiological benefits from selection for increased resistance. However, it may require considerable quantities of detailed data to assess these effects.

Bovine tuberculosis

Bovine tuberculosis (TB) is an infectious and contagious disease of cattle caused by *Mycobacterium bovis*, characterised by the development of tubercles in any organ of the body. Aerosol exposure is the main route of infection, and wildlife reservoirs of infection are often implicated in the disease epidemiology. This is particularly the case in the UK, where badgers are a source of infection, and New Zealand where opossums are similarly implicated. Significant economic loss occurs due to the loss of stock. Bovine TB is zoonotic and may be transmitted to humans through unpasteurised milk and dairy products. Worldwide, annual costs to agriculture due to bovine TB are estimated around $3 billion (Garnier *et al.*, 2003). In the UK, as an example, the costs incurred in attempting to eradicate TB in 2005 were £90 million and it is estimated that could be as much as £1 billion between 2008 and 2013 (The Veterinary Record, 2008).

Although published evidence of genetic variation in the host resistance to TB in cattle is currently weak (i.e. little has been published), substantial evidence exists in deer (Mackintosh *et al.*, 2000), mice and humans (Hill, 2006; Fortin *et al.*, 2007). It would be biologically surprising if similar host genetic variation did not exist for cattle. However, current control measures, along with the associated data collection procedures, provide an opportunity to explore host genetic variation. At present, control is through routine skin testing of all cattle and immediate culling of any animal exhibiting a positive reaction. This information is recorded, at least in the UK. Therefore, combining information from databases containing TB test results and databases containing pedigree information provides the opportunity to quantify host genetic variation, at least for skin test reactions if not for TB resistance itself. This process has been successful in the UK (unpublished data), potentially enabling the ranking of bulls for TB resistance. Further, linking these data to milk recording data further allows relationships between TB resistance and performance to be explored.

SNP chip technology can also be used to assist in the identification of animals with increased resistance to TB. Similar to the identification of animals responding to routine skin testing, cases and controls can also be identified from high prevalence TB herds, defining cases on either skin test results or clinical signs of disease. These cases and controls can then be genotyped using a high density SNP chip, and overall genetic merit for TB resistance can be predicted, akin to the genomic-wide selection strategy outlined by Meuwissen *et al.* (2001). Currently (2008), this approach is being implemented using data recorded in Northern Ireland. Using these approaches, it is technically possible to provide both phenotype and genotype-derived estimated breeding values for TB resistance for dairy cattle bred in high TB risk areas.

Paratuberculosis

Paratuberculosis, or Johne's disease, is a bacterial infection of the gastrointestinal tract caused by *M. avium sub. paratuberculosis*. It is characterised by chronic diarrhoea, persistent weight loss, decreased milk production and eventually death. The disease is not treatable and vaccinations do not prevent infection. Therefore, economic losses are substantial in both the dairy and beef industries. Conflicting opinions have been published which indicate a potential link between the causative agent (i.e. *M. avium sub. paratuberculosis*) and Crohn's disease in humans (a severe and incurable inflammatory bowel disease), via the consumption of infected dairy products (Chiodini and Rossiter, 1996; Bakker *et al.*, 2000).

Studies of infection status of cattle have indicated that susceptibility to paratuberculosis appears to be heritable, with heritability estimates ranging from 0.06 to 0.18 (Koets *et al.*, 2000; Mortensen *et al.*, 2004; Gonda *et al.*, 2006). A QTL affecting susceptibility has been mapped recently in US Holsteins to BTA20 (Gonda *et al.*, 2007). However, application of genetic approaches to paratuberculosis, similar to those used for TB, is hindered by difficulties in diagnosing infected or diseased animals. Diagnosis is a major difficulty, being time-consuming, expensive and prone to error, and is the rate-limiting step in any study of host genetics using field data. Development of new diagnostic tests remains a critical area requiring further research investment; if this is successful then genetic approaches may well be successful for this disease as well.

Lessons from sheep

Nematode resistance

Nematode infections constitute a major disease problem to domestic livestock worldwide (Perry *et al.*, 2002), with most grazing livestock being at risk from nematode parasite infections. Growing lambs with immature immune responses are particularly vulnerable to nematode infections, with even sub-clinical infections causing marked decreases in productivity (Coop *et al.*, 1982; 1985). Effective control of gastrointestinal nematode parasites is becoming difficult, mainly due to the well-documented evolution of drug resistance in nematode parasite populations (e.g. Jackson and Coop, 2000; Waller, 1997), which threatens sustainable sheep production throughout the world. Although nematode problems are generally less severe in cattle, they nevertheless constitute a problem in calves and lactating cows.

Genetic selection is increasingly viewed as a means of helping to control nematode infections in sheep. Many studies have quantified within-breed heritabilities, usually using faecal egg count (FEC) as the indicator of relative nematode resistance (see summary by Bishop and Morris, 2007). In almost all cases FEC, once appropriately transformed, is a moderately heritable trait and one which responds to selection. Genome scans to detect QTL are now well advanced in many countries, again summarised by Bishop and Morris (2007). With the exception of a QTL near the interferon gamma locus on chromosome 3, a feature of these studies is the difficulty in detecting QTL that are consistent between studies. As with mastitis, selection based on either phenotypic data or whole genome results obtained using a dense SNP chip would appear to be the most promising ways of achieving genetic progress.

An important feature of selection for nematode resistance is the interaction between host genotype and disease epidemiology. As described in the Introduction, altering host genotype can also change the force of infection faced by the population. In this case, by creating a population of animals that is more resistant to infection the larval contamination on pasture will tend to decrease. This, in turn, will lead to reduced parasite challenge to all animals, furthering the benefits of selection. This phenomenon was quantified *in silico* by Bishop and Stear (1997 and 1999), with experimental verification provided by Gruner *et al.* (2002) and Leathwick *et al.* (2002).

The key lesson from nematode resistance is that the total benefits from selection can be larger than those arising directly from genetic change in the host, i.e. there may well be additional environmental or epidemiological benefits as well. This benefit would clearly arise for cattle selected for nematode resistance, but in principle it could also arise for other diseases as well, including TB and contagious mastitis.

Footrot

Footrot is a common cause of lameness in both lambs and mature sheep, and it is considered to be one of the major welfare problems in sheep. Footrot is a highly contagious bacterial disease caused by *Dichelobacter* (*Bacteroides*) *nodosus*. In addition to the welfare concerns, it is also a major cause of economic loss and currently it is estimated to have economic costs to the UK sheep industry of £24 million per annum (Nieuwhof and Bishop, 2005). Lameness in dairy cattle has some similarities as a disease, with severity being a function of both structural properties of the hoof and infectious agents.

Substantial genetic variation in resistance to footrot has been demonstrated by Raadsma *et al.* (1994), from deliberate challenges, and Nieuwhof *et al.* (2008), from field data. In particular, Nieuwhof *et al.* (2008) found that data describing 'infected or not' was at least as heritable as data giving more detailed descriptions of the severity of infection, possibly because these data only describe the ~10% of animals that have clinical signs of disease and not the ~90% of animals that do not. A further finding from Nieuwhof *et al.* (2008) was that heritability of disease risk appeared to increase with flock disease prevalence, even when corrections were made for disease prevalence effects. This suggests that the greater likely force of infection in high prevalence flocks has allowed genetic variation in disease resistance to be more strongly expressed. Lastly, using modelling techniques, Nieuwhof *et al.*, 2009) demonstrated that total benefits arising from selection, i.e. genetic plus epidemiological, are usually larger than expected from genetic change alone, making selection for increased resistance an attractive option.

A key lesson from footrot is that hoof disorders, even when assessed using simple scoring systems are heritable traits which should respond to selection. When the disorder has an infectious component, responses to selection will be greater in higher prevalence herds or flocks and, once again, total gains may exceed those predicted by quantitative genetic theory alone.

Discussion and conclusions

Infectious diseases are likely to affect the functional fitness and robustness of dairy cattle, with the impact depending upon the prevalence of the disease. Therefore, endemic diseases are likely to play a much greater role in robustness than sporadic diseases that pose epidemic risks. Of the major endemic diseases, mastitis is the most predictable and common disease with the greatest impact on cow health, welfare and robustness. Consequently, it has had the greatest research focus. However, previous methods of selecting for mastitis resistance have been somewhat clumsy and have ignored important aspects of the underlying disease biology. Therefore, it is likely that greater genetic progress can be made in terms of increasing resistance than has been achieved hitherto, due to a combination of more strategic data interpretation and use of new genomic technologies. Strategic data interpretation, i.e. separating concepts of baseline SCC, response SCC and liability to infection, in principle should enable genetic progress to be increased whilst at the same time addressing concerns about deleterious changes in animals' ability to respond to infection. In other words, decreasing mastitis liability whilst at the same time maintaining the ability of an animal to respond to infections (in general) is likely to be a key component of breeding more robust cows.

The other two dairy cattle diseases discussed, TB and paratuberculosis, are important for contrasting reasons. TB is a high risk disease with major impacts on animal welfare, robustness, etc., but only in specific and defined geographical regions. Therefore, breeding decisions can be specifically targeted to herds in those regions, for example avoiding the use of sires whose daughters are highly susceptible. In contrast, paratuberculosis is of unknown prevalence, with unknown geographical 'hot-spots', yet it potentially poses major constraints on the health, welfare and robustness of animals in affected herds. This makes it a high priority disease, however the initial research priorities should be focused towards diagnosis; only once satisfactory diagnostic tests are available can genetic solutions be effectively addressed.

Sheep provide valuable lessons for cattle in terms of breeding for resistance, particularly as sheep have a number of advantages over cattle in terms of obtaining data and testing breeding strategies. Their smaller body size, but larger flock size, means that data (apart from milk-related trait data) are more easily acquired than for dairy cattle, often in sufficiently large quantities to enable within-flock selection. Further, breeders have greater opportunity to select for non-standard traits than in dairy cattle, hence flock-customised selection is feasible. A critical finding from the sheep studies is that there are often epidemiological benefits from selection for increased disease resistance that are realised in addition to the direct benefits predicted from genetic gain. These additional benefits make selection for increased disease resistance an attractive option under some circumstances. However, the corollary is that if the disease prevalence is low, then selection for resistance in that environment becomes inefficient with limited benefits, other than as an insurance against future disease outbreaks.

In conclusion, several opportunities exist to breed dairy cattle for increased resistance to endemic diseases. The opportunities are currently under-exploited and somewhat inefficiently implemented. A combination of techniques, i.e. interpreting existing data in more sophisticated ways, using new phenotypic information that is potentially available and applying new genomic tools, should enable considerably greater genetic progress in key traits. Further, under some circumstances, the benefits from genetic progress may be greater than anticipated from genetic theory alone. Decisions have to be made on the most appropriate diseases for genetic selection, however mastitis should remain a focus of attention and in some environments TB and paratuberculosis are likely to be important diseases as well. Increased resistance to these diseases will have multiple benefits on productivity, welfare and cow robustness. In particular, it should be possible to increase resistance to these diseases whilst simultaneously improving overall dairy cow robustness.

Acknowledgements

I wish to thank Defra, BBSRC and EU for funding, particularly the European Animal Disease Network of Excellence for Animal Health and Food Safety (EADGENE).

References

Bakker, D., P.T. Willemsen and F.G. van Zijderveld, 2000. Paratuberculosis recognized as a problem at last: a review. The Veterinary Quarterly, 22: 200-203.

Bennett, R.M., K. Christansen and R.S. Clifton-Hadley, 1999. Modelling the impact of livestock disease on production: case studies of non-notifiable diseases of farm animals in Great Britain. Animal Science, 68: 681-689.

Bishop, S.C. and C.A. Morris, 2007. Genetics of disease resistance in sheep and goats. Small Ruminant Research, 70: 48-59.

Bishop, S.C. and M.J. Stear, 1997. Modelling responses to selection for resistance to gastrointestinal parasites in sheep. Animal Science, 64: 469-478.

Bishop, S.C. and M.J. Stear, 1999. Genetic and epidemiological relationships between productivity and disease resistance: gastrointestinal parasite infection in growing lambs. Animal Science, 69: 515-525.

Bishop, S.C. and M.J. Stear, 2003. Modeling of host genetics and resistance to infectious diseases: understanding and controlling nematode infections. Veterinary Parasitology, 115: 147-166.

Carlén, E., E. Strandberg A. Roth, 2004. Genetic parameters for clinical mastitis, somatic cell score, and production in the first three lactations of Swedish Holstein cows. J. Dairy Sci., 87: 3062-3070.

Chiodini, R.J. and C.A. Rossiter, 1996. Paratuberculosis: a potential zoonosis. The Veterinary clinics of North America. Food Animal Practice, 12: 457-467.

Coop, R.L., A.R. Sykes and K.W. Angus, 1982. The effect of three levels of *Ostertagia circumcincta* larvae on growth rate, food intake and body composition of growing lambs. J. Agric. Sci., Cambridge, 98: 247-255.

Coop, R.L., R.B. Graham, F. Jackson, S.E. Wright and K.W. Angus, 1985. Effect of experimental *Ostertagia circumcincta* infection on the performance of grazing lambs. Research in Veterinary Science, 38: 282-287.

Davies, G., S. Genini, S.C. Bishop and E Giuffra, 2009. An assessment of the opportunities to dissect host genetic variation in resistance to infectious diseases in livestock. Animal, in press.

Fortin, A., L. Abel, J.L. Casanova and P. Gros, 2007. Host genetics of mycobacterial diseases in mice and men: forward genetic studies of BCG-osis and tuberculosis. Annul Review of Genomics and Human Genetics, 8: 163-192

Garnier, T., K. Eiglmeier, J.C. Camus, N. Medina, H. Mansoor, M. Pryor, S. Duthoy, S. Grondin, C. Lacroix, C. Monsempe, S. Simon, B. Harris, R. Atkin, J. Doggett, R. Mayes, L. Keating, P.R. Wheeler, J. Parkhill, B.G. Barrell, S.T. Cole, S.V. Gordon and R.G. Hewinson, 2003. The complete genome sequence of *Mycobacterium bovis*. Proceedings of the National Academy of Sciences, U.S.A., 100: 7877-7882.

Gianola, D., B. Heringstad and J. Odegaard, 2006. On the quantitative genetics of mixture characters. Genetics, 173: 2247-2255.

Gibson, J.P. and S.C. Bishop, 2005. Use of molecular markers to enhance resistance of livestock to disease: A global approach. OIE Scientific and Technical Review, 24: 343-353.

Gonda, M.G., Y.M. Chang, G.E. Shook, M.T. Collins and B.W. Kirkpatrick, 2006. Genetic variation of *Mycobacterium avium ssp paratuberculosis* infection in US Holsteins. J. Dairy Sci., 89: 1804–1812.

Gonda, M.G., B.W. Kirkpatrick, G.E. Shook and M.T. Collins, 2007. Identification of a QTL on BTA20 affecting susceptibility to *Mycobacterium avium ssp. paratuberculosis* infection in US Holsteins. Animal Genetics, 38: 389–396.

Gruner, L., J. Cortet, C. Sauve, C. Limouzin and J.C. Brunel, 2002. Evolution of nematode community in grazing sheep selected for resistance and susceptibility to *Teladorsagia circumcincta* and *Trichostrongylus colubriformis*: a 4-year experiment. Veterinary Parasitology, 109: 277-291.

Heringstad, B., D. Gianola, Y.M. Chang, J. Odegard and G. Klemetsdal, 2006. Genetic associations between clinical mastitis and somatic cell score in early first-lactation cows. J. Dairy Sci., 89: 2236-2244.

Hill, A.V., 2006. Aspects of genetic susceptibility to human infectious diseases. Annual Review of Genetics, 40: 469-486

Jackson, F. and R.L. Coop, 2000. The development of anthelmintic resistance in sheep nematodes. Parasitology, 120: S95-S107.

Khatkar, M.S., P.C. Thomson, I., Tammen and H.W. Raadsma, 2004. Quantitative trait loci mapping in dairy cattle: review and meta-analysis. Genetics, Selection, Evolution, 36: 163-190.

Koets, A.P., G. Adugna, L.L. Janss, H.J. van Weering, C.H. Kalis, G.H. Wentink, V.P. Rutten and Y.H. Schukken, 2000. Genetic variation of susceptibility to *Mycobacterium avium subsp. paratuberculosis* infection in dairy cattle. J. Dairy Sci., 83: 2702-2708.

Leathwick, D.M., D.S. Atkinson, C.M. Miller, A.E. Brown and I.A. Sutherland, 2002. Benefits of reduced larval challenge through breeding for low faecal egg count in sheep. Novel Approaches III. A Workshop meeting on helminth control in livestock in the new millennium. Moredun Research Institute, July 2-5, 2002.

Mackintosh, C.G., T. Qureshi, K. Waldrup, R.E. Labes, K.G. Dodds and J.F.T. Griffin, 2000. Genetic resistance to experimental infection with *Mycobacterium bovis* in red deer (*Cervus elaphus*). Infection and Immunity, 68: 1620-1625.

Meuwissen, T.H.E., B.J. Hayes and M.E. Goddard, 2001. Prediction of total genetic value using genome-wide dense marker maps. Genetics, 157: 1819-1829.

Mortensen, H., S.S. Nielsen and P. Berg, 2004. Genetic variation and heritability of the antibody response to *Mycobacterium avium subsp. paratuberculosis* in Danish Holstein cows. J. Dairy Sci., 87: 2108-2113.

Nieuwhof, G.J. and S.C. Bishop, 2005. Costs of the major endemic diseases of sheep in Great Britain and the potential benefits of reduction in disease impact. Animal Science, 81: 23-29.

Nieuwhof, G.J., J. Conington, L. Bűnger and S.C. Bishop, 2008. Genetic and phenotypic aspects of foot lesion scores in sheep of different breeds and ages. Animal, 2: 1289-1296.

Nieuwhof, G.J., J. Conington and S.C. Bishop, 2009. A genetic epidemiological model to describe resistance to an endemic bacterial disease in livestock: application to footrot in sheep. Genetics, Selection, Evolution, in press.

Odegard, J., P., Madsen, D. Gianola, G. Klemetsdal, J. Jensen, B. Heringstad and I.R. Korsgaard, 2005. A Bayesian threshold-normal mixture model for analysis of a continuous mastitis-related trait. J. Dairy Sci., 88: 2652-2659.

Perry, B.D., J.J. McDermott, T.F. Randolph, K.R. Sones and P.K. Thornton, 2002. Investing in animal health research to alleviate poverty. International Livestock Research Institute (ILRI), Nairobi, Kenya, 138 p.

Raadsma, H.W., J.R. Egerton, D. Wood, C. Kristo and F.W. Nicholas, 1994. Disease resistance in Merino sheep. III. Genetic variation in resistance to footrot following challenge and subsequent vaccination with an homologous rDNA pilus vaccine under both induced and natural conditions. Journal of Animal Breeding and Genetics, 111: 367-390.

Rupp, R. and D. Boichard, 2003. Genetics of resistance to mastitis in dairy cattle. Veterinary Research, 34: 671-688.

The Veterinary Record, 2008. Bovine TB: EFRACom calls for a multifaceted approach using all available methods. The Veterinary Record, 162: 258-259.

Van Dorp, T.E., J.C. Dekkers, S.W. Martin and J.P. Noordhuizen, 1998. Genetic parameters of health disorders, and relationships with 305-day milk yield and conformation traits of registered Holstein cows. J. Dairy Sci., 81: 2264-2270.

Waller, P.J., 1997. Anthelmintic resistance. Veterinary Parasitology, 72: 391-405.

Do robust dairy cows already exist?

J.E. Pryce[1,3], B.L. Harris[1], J.R. Bryant[2,4] and W.A. Montgomerie[1]
[1]LIC, Private Bag 3016, Hamilton, New Zealand
[2]AgResearch Ltd, Grasslands Research Centre, Tennent Drive, Private Bag 11008, Palmerston North, New Zealand
[3]Farmax Limited, Waikato Innovation Park, Ruakura Lane, P.O. Box 1036, Hamilton, New Zealand
[4]Current address: Department of Primary Industries Victoria, Bundoora, Victoria, 3083, Australia

Abstract

Robustness is generally considered to be the ability of an animal to withstand environmental disturbances. The environments on New Zealand dairy farms are characterised by the herds' exposure to fluctuations in the quantity and quality of the pasture supply, so that environmental disturbances feature as part of the milk production system. These disturbances cannot be so readily mitigated by herd management as they are in production systems characterised by high levels of supplementary feeding or provision of total mixed rations. Selection of diverse types of animals suited to this pasture based production system is aided by the introduction of multi-breed genetic evaluation, which allows breeds and crosses to be compared. Good fertility is a fundamental part of the way in which farm systems operate, as most farmers aim to have a single concentrated seasonal calving pattern so that feed usage is optimised. The production environment also influences body tissue mobilisation patterns. Genetic correlations between body condition score and milk production traits change from being positive in early lactation to negative in late lactation, which could be because there is an advantage to cows having body tissue still available to mobilise in the later stages of lactation, which is late summer, when pasture availability is sometimes limited. The New Zealand environment has indirectly selected for robust cows, as only these animals perform well in the systems used. In practice, Holstein-Friesian cross Jersey cows are particularly well suited. This cross now forms the largest proportion of replacements being born per year.

Keywords: dairy, crossbreeding, heterosis, New Zealand

Introduction

Dairy cow production systems in New Zealand are characterised by high reliance on pasture, approximately 60% of which is grown in the first five months of the spring and early summer. The efficient conversion of pasture into milk is fundamental to the success of the system and requires a close match between herd demand for feed and pasture supply. Farming systems have therefore evolved to capture the seasonal patterns of pasture growth by adopting compact calving periods in the spring. Stocking rate can be changed to increase or decrease the feed demand. Calving date is set earlier or later to shift the feed demand earlier or later in the spring, with supplements used strategically to fill any feed deficits. Purchasing and culling policies are also important for matching feed demand with pasture supply.

New Zealand dairy farmers sell their milk at current world market milk prices (without subsidies). This imposes tight economic constraints on the production systems and methods that can be used to profitably produce milk. Pasture feeding appears to be the most profitable strategy, as grain prices are relatively expensive in comparison to pasture (Holmes *et al.*, 2002). The climate lends itself well to pastoral systems.

The aim of this paper is to describe firstly the current method of genetic evaluation in New Zealand, which allows breeds and crosses to be compared fairly using a multi-breed test-day-model and a selection index that accounts for liveweight. Then, the drivers of robustness are described, including fertility, health and body condition score.

Selection for economic efficiency

Dairy cow production is commonly expressed as some form of measure of milk yield with no reference to the feed consumed (e.g. energy requirements for maintenance and production). This rationale is supported by several studies that have shown that the highest yielding strains are generally the most efficient converters of feed energy into milk energy (Oldenbroek, 1988; Veerkamp *et al.*, 1995). The higher gross efficiency of higher yielding breeds is thought to be largely because of a dilution of maintenance costs over a higher total output of milk, rather than any intrinsic difference in net efficiency of food use. However, in the divergent body size lines established by the University of Minnesota in 1964, it was observed that small line cows were more efficient producers of milk and had fewer health and fertility problems and therefore greater ability to stay in the herd than animals selected for large size (Hansen, 2000). Furthermore, in experiments run in New Zealand, (Ahlborn and Bryant, 1992) demonstrated that while Holstein-Friesians yielded more than Jerseys, Jersey farmlets were more efficient in terms of yield of milk solids produced per hectare (MS/ha). Penno (1998) went on to report data for the completed trial. The higher stocking rate of the Jerseys (by 1.2 extra cows/ha) was not balanced by the value of the extra milk/ha from these cows. Pasture in the main feed source in New Zealand and an obvious constraint is the feed available per hectare. For the majority of New Zealand systems there are two predominant breeds worth considering in breeding decisions (Jerseys and Holstein-Friesians). As these two breeds differ in size, then to make a fair and relevant comparison (e.g. on a per hectare basis) some account needs to be taken of liveweight.

In 1996, an Animal Model was implemented that accounted for different breeds and crossbreeding, allowing a direct comparison of all breeds and crosses (Harris *et al.*, 1996). The index adopted at that time is known as Breeding Worth (BW), which is an economic index that measures net farm income per 4.5 tonnes of dry matter consumed. The original version of the index included yield of milk, fat and protein weighted by their respective economic weights. Liveweight was included to account for the fact that heavy cows have larger energy costs for maintenance than small cows. Since the introduction of the Breeding Worth index, three other traits have been added: longevity, fertility and somatic cell score (SCS). The longevity trait, known as residual longevity, estimates survival independent of genetic effects of production, liveweight, fertility and SCS on herdlife. The motivation to include these traits in BW is similar to other countries (Miglior *et al.,* 2005), in that reducing costs of production in addition to halting the genetic decline in these traits through selection for production has led to broadening of breeding goals.

Robustness

The interpretation of robustness varies; however recently there appears to be a consensus that robustness is 'the capacity to handle disturbances in common, sustainable and economical farming systems' e.g. (Veerkamp *et al.*, 2007). The environments on New Zealand dairy farms are characterised by the herds' exposure to fluctuations in the quantity and quality of the pasture supply, so that environmental disturbances feature as part of the milk production system. These disturbances cannot be so readily mitigated by herd management as compared to production systems characterised by high levels of supplementary feeding or provision of total mixed rations. In this respect dairy cattle in New Zealand have to be robust to survive the heterogeneity of pasture-based systems.

The ideal robust cow in New Zealand traditionally has the following characteristics:
1. An ability to get in calf every 365 days, even when exposed to shortages of grazed feed (fertility).
2. To efficiently graze and forage for pasture in both hot and cold conditions and achieve moderate levels of milk production (foraging and resilience to climate).
3. To maintain an acceptable level of body condition to aid reproductive performance and to support milk production at times of nutritional stress (body condition score).
4. Average body size to minimise maintenance costs per cow and to prevent excessive pasture damage to pastures and soils (body size).
5. Easy calving, minimal health disorders and excellent feet traits to support long distances walked to and from the milking shed and longevity to ensure as many cows survive in the herd when they are most productive at 4-8 years of age (easy calving, healthy and longevity).

It is along these lines that the New Zealand dairy cow has evolved both through selection imposed by farmers and by breeding companies. Selection indices have changed to reflect the need for these characteristics. Each component of the robust cow will now be discussed in more detail.

Fertility

For the past century, typical New Zealand farm management practices have featured retention of calves only from cows calving in the first third of the herd's calving period. Also cows failing to conceive within 170 days from the commencement of the herd's calving period are generally culled. These attributes of the seasonal production system impose a form of natural selection for fertility on New Zealand cattle populations, which has been necessary in order to match feed requirements to pasture supply. Therefore, good fertility is a fundamental part of the way in which farm systems operate to achieve a single concentrated seasonal calving pattern so that feed usage is optimised. Consequently New Zealand dairy farmers have managed calving intervals to an average of 368 days, which compares favourably to other countries (Figure 1).

Although dairy cow fertility in New Zealand is good by international standards, selection for production could erode this advantage, as unfavourable genetic correlations exist with milk production (e.g. Harris *et al.*, 2005). As a consequence of the genetic correlation between production and fertility there has already been a genetic decline in fertility, which can clearly be seen in the trend of fertility breeding values of dairy cows born over the last 20 years (Figure 2). The unfavourable genetic trend is particularly apparent in the Holstein-Friesian breed, which has seen a decline of around 1% in 42

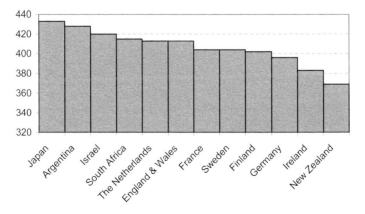

Figure 1. Average calving interval in ICAR countries (http://www.icar.org/).

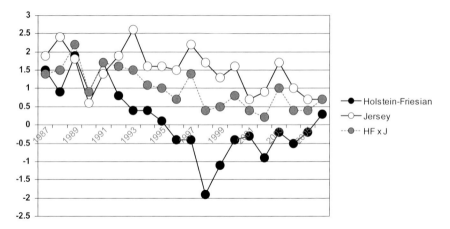

Figure 2. Trend in cow fertility breeding values by year of birth.

day calving rate over 10 years. The decline in Jersey cow fertility has not been so dramatic. This can partly be attributed to 'Holsteinisation'. Harris *et al.* (2005) reported that sires of New Zealand origin have around a 10% advantage in BVs for 42-day calving rate subsequent to first lactation over sires originating in North America and Europe (Harris *et al.,* 2005). A similar pattern is seen even at the heifer level, with heifer pregnancy rates being around 5% more in NZ strain heifers than those with overseas origin (Pryce *et al.,* 2008). Additionally, two strain comparison trials were established in New Zealand (Macdonald *et al.,* 2007) and Ireland (Horan *et al.,* 2005) to address farmer concerns of appropriate strains of Holstein-Friesian cattle for the major farming systems in each country. Both New Zealand and Ireland are reliant on pasture as a major feed source. Differences between a modern New Zealand Holstein-Friesian strain (NZ-HF) and a modern strain of North American descent (NA-HF) were found for many measures of fertility. For example, the NA-HF was approximately 20 days older and 20 kgs heavier at puberty than the NZ-HF strain. Furthermore, in the NZ trial, only 54% of the NA-HF cattle were pregnant after 6 weeks of mating, compared to 69% of the NZ-HF strain, with a remarkably similar pattern in the Irish trial.

Many countries have seen a gradual breed replacement to the modern Holstein. In New Zealand, most of the recent importation of Holstein genetics has been from the Netherlands and USA, with many of the sires from these countries being related due to the high use of popular sires. The increased use of North American genetics occurred to broaden the genetic base of New Zealand Holsteins and because this breed has comparably high productivity (Harris and Kolver, 2001). The North-American Holstein percentage has increased from 1.7% in animals born in 1970 to 43.8% in animals born in 2006.

Foraging and resilience to climate

New Zealand dairy cows have traditionally had to forage for pasture (ryegrass and clover), which is lower in energy, and higher in protein and fibre than the diets offered to animals managed overseas. Pastoral raised animals are selected for characteristics that promote fast intakes of standing ryegrass/ clover, quick degradation of fibre and an ability to process and excrete high levels of protein (Bryant *et al.,* 2007a). Evidence suggests breeds, crosses and individuals within breeds can differ in their ability to efficiently process forage and respond to supplementation (Oldenbroek, 1988; Kolver *et al.,* 2002; Bryant *et al.,* 2006). These differences can be partly attributed to eating behaviour (Thorne *et al.,* 2003; Linnane *et al.,* 2004; Rossi *et al.,* 2005). However, differences in rumen function such as the factors mentioned above are likely to be a contributing factor to the greater suitability of some breeds and individual animals to specific systems. More recently the wider use of supplements has slightly reduced the need to maximise foraging ability, with the ability to process highly digestible, energy dense feeds becoming more important in an increasing proportion of farms. Under these conditions, robustness may become an ability to quickly adapt to different feeds and achieve high levels of performance on either feed or feed combination.

Whilst the climate that a cow exists in is not a trait *per se*, the ability to produce well and forage under a range of environments is of importance, especially for cattle that are not housed at any point during the year. Cattle housing in New Zealand is still unusual, although some large herds in the South Island are starting to build cubicle sheds similar to those widely used in Europe. The average daily atmospheric temperature difference between Northland in the far North of New Zealand and Southland at the bottom of the South Island is around 5.0 °C (National Institute of Weather and Atmospheric Research, 2005), the average yield of milksolids (MS) between these two regions differs by 108 kg, with Southland being higher (the overall mean yield of MS in NZ is 330 kg/cow; LIC, 2007). There are other differences, such as herd-size and breed composition, yet differences in climate contribute as well.

Temperature-humidity or heat load (combination of temperature, solar radiation, humidity and wind speed) indices are sometimes used to compare thermal environments to which cattle are exposed (Bryant *et al.*, 2007a,b). In a hot environment, cattle are susceptible to heat stress due to a combination of heat from the environment and heat produced by the cow itself; for example Aharoni *et al.* (2002) showed reductions in milk yield at temperature humidity index levels greater than 72.

Recent research by Bryant *et al.* (2007a,b) showed that there was little re-ranking among sires when comparing sire breeding values estimated in different summer average HLI environments. In a corresponding study, hot conditions significantly reduced daily milk solids (fat plus protein) yield by more than 10 g per 1-unit increase in 3-day average temperature-humidity index at temperature-humidity indices of 68 in Holstein-Friesians, 69 in crossbred, and 75 in Jersey cattle (Bryant *et al.*, 2007c). There was also evidence that high merit Holstein-Friesian animals had a greater reduction in milk yield at high temperature-humidity values than low genetic merit animal.

Body condition score

Genetic correlation estimates between milk yield and dry matter intake are less than unity, so genetic improvement in yield has not been accompanied by a parallel increase in intake (Veerkamp *et al.*, 1995). Consequently there is a greater reliance on mobilisation of body tissue reserves to sustain lactation. The genetic correlation between production and BCS has been reported to be negative in European and North American studies (e.g. Dechow *et al.*, 2001; Berry *et al.*, 2003), yet the same relationship appears to be close to zero (on average) in New Zealand, changing from being positive in early lactation to negative in late lactation (Pryce and Harris, 2006). The hypothesis put forward for this observation, is that in production systems typical of New Zealand, there is an advantage to cows having body tissue still available to mobilise in the later stages of lactation, which coincides with late summer, when pasture availability is sometimes limited.

The genetic correlation between BCS and fertility appears to be reasonably consistent across countries, with many studies reporting that BCS is a good indicator of fertility (e.g. Pryce and Harris, 2006). BCS and milk volume are both included as predictors of fertility in the NZ fertility index (Harris *et al.*, 2005), as they are available relatively early in bull proofs and substantially improve the reliability.

In its role as a predictor of fertility and as a potential indicator of robustness through timely replenishment of body reserves, BCS is undoubtedly an important trait in the genetic improvement of modern dairy cattle. However, recent research by Pryce *et al.* (2006a) in New Zealand and Wall *et al.* (2008) in the UK both concluded that there was limited value in adding BCS as an additional trait in the breeding goal, implying that the economic drivers for including BCS in national breeding objectives are not currently warranted. Despite this there may still be value in genetically improving BCS by including aspects related to non-market values, such as animal welfare (e.g. Olsen *et al.*, 2000). This approach is difficult to quantify within an economic framework, but may be justified if genetic improvement in traits related to robustness is desired.

Body size

As outlined earlier, cows are selected for moderate body size with a negative economic weighting on body size. Under the low per cow feeding levels (by international standards) in New Zealand, moderate sized cows have less maintenance requirements than large size animals. This allows more of the feed to be directed toward milk production in the lighter cows. The dual benefit is that smaller cows cause less damage to pasture and soils in wet conditions.

Easy calving, healthy and longevity

Over the last 15 years, the national herd has expanded by 1.5 million cows (approximately 60%) to around four million dairy cows (LIC, 2007). The cows available have enabled farmers to expand their businesses rapidly. In fact the average number of lactations per cow has increased from 4.7 in the mid-1990s to 4.9 currently. However, during the rapid expansion of the national herd many cows have been retained that would otherwise have been culled for reasons such as elevated milk somatic cell counts or late calving dates. Consequently, more current cows are likely to be carrying non-genetic effects, such as tissue damage from previous infections, than ten years ago. In addition, replacement heifer calves may be subsequently retained from cows with inferior fertility. Farmers also face a greater challenge in operating effective herd reproduction and animal health programs, with much larger numbers of cows being managed per staff member. These factors together have led to an increased focus on functional traits in breeding programs.

Amongst disease and health problems, mastitis accounts for around 47% of dairy health problems recorded in New Zealand, feet and leg problems for 20%, and calving difficulty for 13%. Retained placenta and uterine infections at 8%, and metabolic diseases at 7%, are the next most important categories (Xu and Burton, 2003). In order to include individual health-related traits in the BW index, it is important to have a cost-effective method for collection of data. Increased use of on-farm computers to record health treatments, and increased requirements for such recording from the milk-quality audit perspective can be expected. Consequently, the availability of health treatment records may improve in the near future. Currently, somatic cell score (SCS) is used as a proxy for the occurrence of mastitis and SCS BVs are estimated and included in the BW index (Harris *et al.*, 2005). Selection for mastitis resistance could be improved by collecting incidences of clinical and subclinical mastitis in individual cows.

Analysis of lameness data collected from LIC sire-proving-scheme herds suggests a heritability of 0.02 (D.L. Johnson, personal communication). The limitation with these data is that the actual type of lameness was not defined. Other indicators can potentially provide valuable additional information, for example New Zealand research has shown that cows with intermediate curvature of their rear legs survive longer than cows with either sickled legs or straight legs (Berry *et al.*, 2005). However, this conformation characteristic is not a strong predictor of survivability of New Zealand cows. Further work is needed to find the most useful characteristics of cows to record in progeny testing schemes, in order to provide better genetic information to assist in reducing feet and leg problems.

Total longevity is a genetic measure of survival in dairy herds. In New Zealand it is expressed as days of life. The genetic trend of total longevity is positive (Figure 3) in all breeds, demonstrating

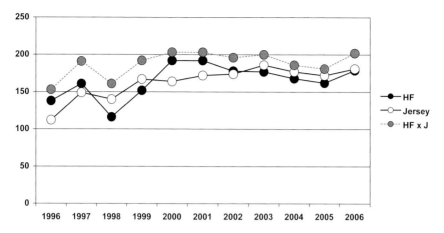

Figure 3. Trend in cow total longevity breeding values (days of life) by year of birth.

that there has been genetic improvement in survival rates over the last 10 years. This result indicates that there has been an improvement in workability and conformation attributes of dairy cows. The same genetic trend is seen in functional longevity. Functional longevity is the ability of the cow to delay culling for reasons other than low production. It is calculated by correcting total longevity for the effect that higher production has on increasing the length of time cows are retained in herds. In New Zealand, this concept is extended to calculate a breeding value known as 'Residual Survival'. Residual Survival is defined as herd-life after accounting for the genetic effects of production, liveweight, milk somatic cells and fertility on herd-life and therefore avoids double counting of effects in the BW index – and measures the expected ability of a cow to resist culling for reasons other than traits included directly elsewhere in the index.

Studies from Sweden and Denmark have noted an increase in the proportion of calvings that result in stillborn calves in recent years (Steinbock *et al.*, 2003; Hansen *et al.*, 2004). In both studies the increased use of North-American Holstein genetics has been implicated as a possible reason for this trend, somewhat supported by the fact that the incidence of stillbirth in primiparous cows has risen from 9.5% in 1985 to 13.2% in 1996 in USA Holsteins (Meyer *et al.*, 2001).

The prevalence of calf mortality in New Zealand was reported to be 7.2% (Pryce *et al.*, 2006b). Trauma at calving (calving difficulty) is one of the reasons why calves are born dead, the age of the dam, gestation length and sex of the calf are other important factors some of which affect calving difficulty themselves (Pryce *et al.*, 2006b). Compared to no assistance at calving, the two categories of greatest calving difficulty resulted in 22% and 41% mortality.

Calf survival was favourably affected by heterosis, with 1% fewer stillbirths arising in first cross Holstein-Friesian and Jerseys (Table 1 and Pryce *et al.*, 2006b). As the heritability estimate of calf mortality is very low in New Zealand (<0.02), crossbreeding and devising strategies to manage calving assistance and gestation length appear to be more worthwhile in reducing the prevalence of calf mortality.

Breeds, strains and crosses

The adoption of the across breed genetic evaluation model has enabled farmers to compare breeds and crosses directly. As the Holstein-Friesian and Jersey are of similar genetic merit when compared using the BW index, there has been growing trend in crossing these two breeds to make use of heterosis.

Table 1. Heterosis estimates for first cross Holstein-Friesian × Jersey.

Trait	Phenotypic SD	Heterosis (units of measurement)	Heterosis in SD
Liveweight	47.8 kg	9.4 kg[1]	0.20
Cow fertility	36%	3.4%[1]	0.09
SCS	1.1 (log SCC)	-0.06[1]	-0.06
Total longevity	893.5 days	227 days[1]	0.25
BCS	0.39	0.06[2]	0.16
Heifer fertility	44%	2.2 to 2.9%[3]	0.05 to 0.07
Calf survival	25%	1%[4]	0.04

[1] Harris (2005).
[2] Pryce and Harris (2006).
[3] Pryce *et al.* (2007), heterosis estimate of New Zealand and North American HF crossed with Jersey respectively.
[4] Pryce *et al.* (2006b).

A second index, the production worth (PW) index, is designed to compare cows for their expected ability to convert feed into farm profit over a typical lifetime. This index is based on milkfat, protein and milk volume production, and liveweight. It includes additive and non-additive genetic effects, and permanent environmental effects that remain with the cow for her lifetime. Farmers use the PW index to guide them on culling decisions, whilst the BW index is used for breeding decisions. When compared on PW, the Holstein-Friesian crossed with the Jersey (HFxJ) is the most profitable strain for NZ pasture conditions. As a consequence, the number of crossbred replacements being born in New Zealand now exceeds the number of single breed replacements (Figure 4). Other breeds and crosses also exist in New Zealand, for example the Ayrshire breed. However, the representation of these breeds is small in comparison to Jerseys and Holstein-Friesians.

Estimates of heterosis in first cross HFxJ are presented in Table 1. When the heterosis estimates are standardised, the largest estimates were seen for longevity (0.25 standard deviation units) and liveweight (0.20 standard deviation units). Fertility exhibited heterosis between 0.05 SDs (heifers) and 0.09 SDs (cows). Incidentally, heterosis is also observed in New Zealand Holstein-Friesians

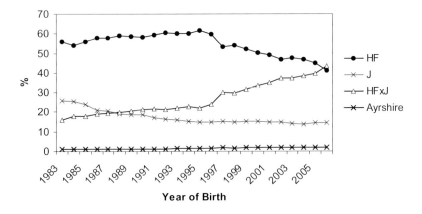

Figure 4. Percentage of major breed dairy replacements born in New Zealand.

Breeding for robustness in cattle

crossed with 'overseas' Holstein. For example, the heterosis estimate for this cross has been estimated to be +0.9% (Pryce *et al.,* 2007).

For both fertility and longevity, the Jersey breed is genetically superior to the Holstein-Friesian breed (Figure 2 and Figure 3). In terms of additive genetic merit, the HFxJ may be intermediate to the two purebreds, but it obviously has the added benefit of heterosis.

Farmers who use crossbreeding generally use two main strategies. Firstly, crossing each generation back to a different breed of sire, i.e. mating the first cross to Jersey and the backcross progeny to Holstein-Friesian, and so on (criss-crossing). Secondly, maintaining uniformity in the size of cows within the herd, larger cows are mated to Jersey sires and smaller cows are mated to HF cows. Recently, around 20-25% of bulls progeny-tested by AI companies in New Zealand are themselves crossbred, predominantly HFxJ first-cross sires. Motivation for progeny testing crossbred sires includes maintaining a high selection intensity in the selection pathway of cows used to breed bulls. Consequently, many farmers are now using crossbred bulls rather than the first two strategies described here.

Discussion

Robustness is an abstract term that may be interpreted in different ways. Our interpretation is that it encompasses traits required by dairy cows to survive and cope with environmental disturbances typical of the New Zealand production environment. In a seasonal production system, fertility is a key factor important for robustness, as cows that fail to get in calf early are culled. Also, the pattern of BCS mobilisation in New Zealand is interesting, as there appears to be an advantage for cows to conserve body reserves for later in the lactation, when drought conditions often arise. BCS and fertility are not independent, as high BCS has a favourable genetic correlation with fertility in early lactation and as a consequence is a good selection criterion for genetically improving fertility (Harris *et al.,* 2005). BCS therefore appears to be strongly implicated in 'robustness' of New Zealand dairy cows. Health and longevity are obviously also important factors, as they also contribute to cows that could be described as easy-care, requiring little veterinary attention throughout their lives. However, the New Zealand dairy farmer has recognised and used crossbreeding to great advantage. The HFxJ is the most profitable 'strain' and also the most robust in terms of longevity, health and fertility (due to the beneficial effects of heterosis). Consequently, it is not surprising that the number of crossbreds being born currently exceeds other breeds (Figure 4).

In terms of the most robust strain or animal, Bryant *et al.* (2007a,b) demonstrated that Jerseys are able to sustain high levels of milk production over a wider range of thermal environments than Holstein-Friesian. Furthermore in both fertility and longevity the Jersey is superior to the Holstein-Friesian (Figures 2 and 3). Obviously, the HFxJ has the additional benefit of heterosis, especially in health and fertility traits. The crossbred is the favoured milking cow in New Zealand currently, as they have been found to be the most profitable breed or strain under New Zealand conditions (e.g. Lopez-Villalobos *et al.,* 2000).

Conclusions

The New Zealand pasture environment creates a challenging environment for dairy cattle. To fit a seasonal pasture system, it is essential that cows calve on an annual basis. Cows that do not fit the calving pattern are generally culled. The New Zealand dairy industry has gone through a phase of massive expansion over the last 15 years, so that cows that would not normally be kept are being given a second chance. Thus, there are concerns that cows are not as 'robust' as they used to be. However, this does not appear to be entirely true, as genetic trends in longevity show an improvement over the last 10 years, although fertility is decreasing. Poor fertility in US/European Holstein animals in comparison to NZ HF animals has been recognised in NZ and Ireland. The Jersey has better fertility than the HF animal, the HFxJ is even better due to heterosis. Fertility and overall profitability are key drivers towards an increased number of crossbred (HFxJ) replacements being born as farmers recognise and use the beneficial effects of heterosis to improve the profitability of their farms.

References

Aharoni, Y., O. Ravagnolo and I. Misztal, 2002. Comparison of lactation responses of dairy cows in Georgia and Israel to heat load and photoperiod. Animal Science, 75: 469-476.

Ahlborn, G. and A.M. Bryant, 1992. Production, economic performance and optimum stocking rates of Holstein-Friesian and Jersey cows. Proceedings of the New Zealand Society of Animal Production, 52: 7-9.

Berry, D.P., F. Buckley, P. Dillon, R.D. Evans, M. Rath and R.F. Veerkamp, 2003. Genetic relationships among body condition score, body weight, milk yield, and fertility in dairy cows. J. Dairy Sci., 86: 2193-2204.

Berry, D.P., B.L. Harris, A.M.Winkelman and W.A. Montgomerie, 2005. Phenotypic associations between traits other than production and longevity in New Zealand dairy cattle. J. Dairy Sci., 88: 2962-2974.

Bryant, J.R., N. López-Villalobos, J.E. Pryce, C.W. Holmes and D.L. Johnson, 2006. Reaction norms used to quantify the responses of New Zealand dairy cattle of mixed breeds to nutritional environment. New Zealand Journal of Agricultural Research, 49: 371-381.

Bryant, J.R., N. Lopez-Villalobos, J.E. Pryce and C.W. Holmes, 2007a. A model for simulating environmental sensitivity differences between dairy cattle breeds in New Zealand. Meeting the challenges for pasture-based dairying: Proceedings of the Australasian Dairy Science Symposium, pp. 520-525

Bryant, J.R., N. López-Villalobos, J.E. Pryce, C.W. Holmes, D.L. Johnson and D.J. Garrick, 2007b. Environmental sensitivity in New Zealand dairy cattle of mixed breeds. J. Dairy Sci., 90: 1538-1547

Bryant, J.R., N. López-Villalobos, J.E. Pryce, C.W. Holmes and D.L. Johnson, 2007c. Quantifying the effect of thermal environment on three breeds of dairy cattle in New Zealand. New Zealand Journal of Agricultural Research, 50: 327-338.

Dechow, C.D., G.W. Rogers and J.S. Clay, 2001. Heritabilities and correlations among body condition scores, production traits, and reproductive performance. J. Dairy Sci., 84: 266-275.

Hansen, L.B., 2000. Consequences of selection for milk yield from a geneticist's viewpoint. J. Dairy Sci., 83:1145-1150.

Hansen, M., M.S. Lund, J. Pedersen and L.G. Christensen, 2004. Genetic parameters for stillbirth in Danish Holstein Cows using Bayesian threshold model. J. Dairy Sci., 87: 706-716.

Harris, B.L., 2005. Breeding dairy cows for the future in New Zealand. New Zealand Veterinary Journal, 53: 384-389.

Harris B.L. and E.S. Kolver, 2001. Review of Holsteinization on intensive pastoral dairy farming in New Zealand. J. Dairy Sci., 84: E56-E61.

Harris, B.L., J.M. Clark and R.G. Jackson, 1996. Across breed evaluation of dairy cattle. Proceedings of the New Zealand Society of Animal Production, 56: 12-15.

Harris, B.L., J.E. Pryce, Z.Z. Xu and W.A. Montgomerie, 2005. Fertility breeding values in New Zealand, the next generation. Interbull Bulletin, 33: 47-51.

Holmes C.W., I.M. Brookes, D.J. Garrick, D.D.S. MacKenzie, T.J. Parkinson and G.F. Wilson, 2002. Milk Production from Pasture. Butterworths, Palmerston North, New Zealand, 602 pp.

Horan, B., P. Dillon, P. Faverdin, L. Delaby, F. Buckley and M. Rath, 2005. The interaction of strain of Holstein-Friesian cows and pasture-based feed systems on milk yield, body weight, and body condition score. J. Dairy Sci., 88: 1231-1243.

Kolver, E.S., J.R. Roche, M.J. de Veth, P.L. Thorne and A.R. Napper, 2002. Total mixed rations versus pasture diets: Evidence for a genotype x diet interaction in dairy cow performance. Proceedings of the New Zealand Society of Animal Production, 62: 246-251.

LIC, 2007. Dairy Statistics 2006-07. Hamilton, New Zealand, 44 pp.

Linnane, M., B. Horan, J. Connolly, P. O'Connor, F. Buckley and P. Dillon, 2004. The effect of strain of Holstein-Friesian and feeding system on grazing behaviour, herbage intake and productivity in the first lactation. Animal Science, 78: 169-178.

Lopez-Villalobos, N., D.J. Garrick, C.W. Holmes, H.T. Blair and R.J. Spelman, 2000. Profitabilities of some mating systems for dairy herds in New Zealand. J. Dairy Sci., 83: 144-153.

Macdonald, K.A., G.A. Verkerk, B.S. Thorrold, J.E. Pryce, J.W. Penno, L.R. McNaughton, L.J. Burton, J.A.S. Lancaster, J.H. Williamson and C.W. Holmes, 2008. A comparison of three strains of Holstein-Friesian grazed on pasture. Part 2: Milk production, reproduction, bodyweight, body condition and mastitis when managed under different feed allowances. J. Dairy Sci., 91: 1693-1707.

Meyer, C.L., P.J. Berger, J.R. Thompson and C.G. Sattler, 2001. Phenotypic trends in the incidence of stillbirth for Holsteins in the United States. J. Dairy Sci., 84: 515-523.

Miglior, F., B.L. Muir and B.J. van Doormaal, 2005. Overview of different breeding objectives in various countries. J. Dairy Sci., 88: 1255-1263.

National Institute of Weather and Atmospheric Research. 2005: The climate of New Zealand. Available at: http://www.niwa.co.nz/edu/resources/climate/overview/ Accessed: 22 March 2005.

Oldenbroek, J.K., 1988. The performance of Jersey cows and cows of larger dairy breeds on two complete diets with different roughage contents. Livest. Prod. Sci., 18: 1-17.

Olsen, I., A.F. Groen and B. Gjerd, 2000. Definition of animal breeding goals for sustainable production systems. J. Dairy Sci., 78: 570-582.

Penno, J.W., 1998. Principles of profitable dairying. Proceedings of the Ruakura Dairy Farmers' Conference, pp. 1-14.

Pryce, J.E. and B.L. Harris, 2006. Genetics of body condition score in New Zealand dairy cattle. J. Dairy Sci., 89: 4424-4432.

Pryce, J.E., B.L. Harris and L.R. McNaughton, 2007. The genetic relationship between heifer and cow fertility. Proceedings of the New Zealand Society of Animal Production, 67: 388-391.

Pryce, J.E., B.L. Harris, D.L. Johnson and W.A. Montgomerie, 2006a. Body condition score as a trait in the breeding worth dairy index. Proceedings of the New Zealand Society of Animal Production, 66: 103-106.

Pryce, J.E., B.L. Harris, S. Sim and A.W. McPherson, 2006b. Genetics of stillbirth in dairy calves. Proceedings of the New Zealand Society of Animal Production, 66: 98-102.

Rossi, J.L., K.A. Macdonald, B.S. Thorrold, J. Hodgson and C.W. Holmes, 2005. Differences in grazing behaviour and herbage intake between genotypes of Holstein-Friesian dairy cows grazing short or long swards. Proceedings of the New Zealand Society of Animal Production, 65: 236-240.

Steinbock, L.; A. Nasholm, B. Berglund, K. Johansson and J. Philipsson, 2003. Genetic effects on stillbirth and calving difficulty in Swedish Holsteins at first and second calving. J. Dairy Sci., 86: 2228-2235.

Thorne, P.L., J.G. Jago, E.S. Kolver and J.R. Roche, 2003. Diet and genotype affect feeding behaviour of Holstein-Friesian dairy cows during late lactation. Proceedings of the New Zealand Society of Animal Production, 63: 124-127.

Veerkamp, R.F., G.C. Emmans, A.R. Cromie and G. Simm, 1995. Variance components for residual feed intake in dairy cows. Livest. Prod. Sci., 41: 111-120.

Veerkamp, R.F., H.A. Mulder, P. Bijma and M.P.L. Calus, 2007. Genetic concepts to improve robustness of dairy cows. Book of abstracts of the 58th annual meeting of the European Association for Animal Production, Dublin 26-29 August 2007. Wageningen Academic Publishers, Wageningen, the Netherlands, p. 1.

Wall, E., M.P. Coffey and P.R. Amer, 2008. Derivation of direct economic values for body tissue mobilisation in dairy cows. In: Breeding for robustness in cattle. Klopcic M., R. Reents, J. Philipsson and A. Kuipers (eds.) EAAP Scientific Series No. 126, Wageningen Academic Publishers, Wageningen, the Netherlands, pp. 201-205.

Xu, Z.Z. and L.J. Burton, 2003. Reproductive performance of dairy cows in New Zealand: final report of the monitoring fertility project. Hamilton: Livestock Improvement. http://www.aeu.org.nz/page.cfm?id=58&nid=26

Part 2
Longevity

A review on breeding for functional longevity of dairy cow

F. Miglior[1,2] and A. Sewalem[1,2]
[1]*Agriculture and Agri-Food Canada - Dairy and Swine Research and Development Centre, Sherbrooke, QC, J1M1Z3, Canada*
[2]*Canadian Dairy Network, Guelph, ON, N1K1E5, Canada*

Abstract

Dairy cattle breeding programs have led to substantial genetic changes in production traits which represent only one component contributing to overall efficiency and profitability of the dairy industry. Functional traits, such as reproduction, longevity, and health traits, are of increased interest to producers in order to improve herd profitability. However, despite their economic significance only recently some attention has been given in genetic selection programs, with the exception of Scandinavian countries. This has resulted in a shift towards a balanced breeding approach and demands dairy cow robustness. A robust cow can be defined as one that adapts well to a wide range of environmental conditions or in genetic terms expresses a reduced genotype by environment interaction. Genetic improvement of longevity involves breeding of dairy cows that can produce a live calf, cycle normally, show observable heat, conceive when inseminated, sustain adequate body condition, avoid udder injuries, resist to infection diseases, walk and stand comfortably and produce milk of desirable composition. Selection for longevity is achieved by combining past information of productive life with early indicators of future longevity, namely conformation, reproduction, health and management traits. As countries expand genetic evaluations for reproduction and health traits, enhanced indicators will be available to better predict and improve functional longevity of our dairy herd.

Keywords: longevity, direct and indirect herd life, selection

Introduction

Dairy cattle breeding programs have led to substantial genetic changes in production traits which represent only one component contributing to overall efficiency and profitability of the dairy industry. Functional traits, such as reproduction, longevity, and health traits, are of increased interest to producers in order to improve herd profitability. However, despite their economic significance only recently some attention has been given in genetic selection programs, with the exception of Scandinavian countries. This has resulted in a shift towards a balanced breeding approach and demands dairy cow robustness. A robust cow can be defined as one that adapts well to a wide range of environmental conditions or in genetic terms expresses a reduced genotype by environment interaction. Genetic improvement of longevity involves breeding of dairy cows that can produce a live calf, cycle normally, show observable heat, conceive when inseminated, sustain adequate body condition, avoid udder injuries, resist to infection diseases, walk and stand comfortably and produce milk of desirable composition. Therefore, longevity is a trait of interest for animal breeders in general and dairy breeders in particular because of its effect on economic performance. With increased longevity, the mean production of the herd increases because a greater proportion of the culling decisions are based on production and the proportion of mature cows, which produce more milk than young cows is increased. Furthermore, the economic importance of herd life compared to milk production is considered higher than other non-production traits (Rogers and McDaniel, 1989; Van Arendonk, 1991; Allaire and Gibson, 1992; Dekkers, 1993). However, genetic improvement of herd life is very hard to achieve because of its low heritability. Heritability estimates range from 0.03 to 0.05 (Van Doormaal *et al.*, 1985; Jairath *et al.*, 1998) using a linear model, while estimates from survival analysis using Weibull proportional hazards models range from 0.10 to 0.20 (Ducrocq, 2002; Roxstrom *et al.*, 2003; Sewalem *et al.*, 2005). Breeding for this trait resulted in only small improvements over time. Actual longevity measurements can be obtained only when a cow is culled or disposed of, after selection decisions have been made. In order to obtain reliable information for sires on the longevity of their daughters, it is necessary to wait until a minimum number of daughters have been culled or have died. These evaluations may be available too late to be useful in breeding programs. Instead, genetic evaluations for direct longevity information based on the number of culled daughters is combined with indirect information based on early predictors. Therefore, selection for longevity is achieved by combining past information of productive life with early indicators of future longevity, namely conformation, reproduction, health and management traits. Type traits have been used as indirect selection criteria for herd life (Short and Lawlor, 1992; Dekkers *et al.*, 1994; VanRaden and Wiggans, 1995; Weigel *et al.*, 1998; Cruickshank *et al.*, 2002; Sewalem *et al.*, 2004). Type traits are recorded relatively early in life, most often in the first lactation and are more heritable than longevity which makes selection relatively more efficient. Several reports showed that udder health problems are a major reason for culling dairy cattle (Beaudeau *et al.*, 1995; Neerhof *et al.*, 2000; Samoré *et al.*, 2001). Bascom and Young (1998) reported mastitis was the primary culling reason for 15% of the cows that were culled. Sewalem *et al.* (2004) showed that udder traits were the second most important traits that influenced the culling decision of Canadian dairy breeds. In Finland, Pösö and Mäntysaari (1996) noted that 34.8% of culling was due to udder problems. Furthermore, in Sweden udder diseases together with high somatic cell count (SCC) was the second most important reason for culling in the year 2001, accounting for nearly 24% of culled cows (Carlèn *et al.*, 2004). De Vliegher *et al.* (2005) reported udder health problems accounted 10% of the culling reasons for culled heifers in their study. Various reports also indicated that udder health and longevity were related traits (Beaudeau *et al.*, 1995; Neerhof *et al.*, 2000; Caraviello *et al.*, 2005; Sewalem *et al.*, 2006). Reproduction traits have been also linked to cow survival. Sewalem *et al.* (2008) have shown significant associations between calving performance and female traits with longevity in all Canadian dairy breeds. In particular, they found that increased risk of culling was observed for cows that:

- required hard pull at calving;
- calved small calves or dead calves;
- required more services per conception;
- required a longer interval between first service to conception; and
- required a longer interval between calving to first service.

As countries expand genetic evaluations for reproduction and health traits, enhanced indicators will be available to better predict and improve functional longevity of our dairy herd. The objective of this investigation was to illustrate how Canada selects for longevity and how it compares with other countries.

The Canadian case

Longevity is a trait of interest for the Canadian dairy industry because of its effect on economic performance. In Canada, balanced breeding has long been the philosophy of the Canadian dairy cattle breeding industry and considerable improvement on type traits has been made compared to other international countries. Because type classification data is already recorded its use as an indirect predictor for longevity would also be very cost effective. Several strategies have been suggested to estimate the breeding value of an animal still in the herd. Longevity can be measured in several ways (Brotherstone *et al.*, 1998, VanRaden and Klaaskate, 1993; Vollema and Groen, 1998) and genetic evaluation systems are not standardised across countries making the comparison of sire rankings difficult. In Canada, for instance, the survival of cows in each of the first three lactations was recorded as a binary trait and evaluated with a multiple trait linear animal model in which survival in each lactation was considered as a separate trait (Jairath *et al.*, 1998). An international genetic evaluation for longevity using direct herd life carried out by the Interbull centre (February, 2005) showed that Canada had the highest average correlation across countries compared to other countries. The report also showed that compared to other countries the current Canadian system appears more stable and results are in favourable proof correlation with other countries. However, with this genetic evaluation system for herd life, a two-year opportunity window was used to determine if a subsequent calving occurred, thus delaying when Canadian bulls were able to get their first genetic evaluations for herd life. This extended time period, before newly proven sires had direct survival data in their proof, meant that those bulls were not receiving an international evaluation for direct herd life as quickly as newly proven bulls in other countries. Sewalem *et al.* (2007) initiated a study to reduce the two-year lag time by breaking the first lactation into three time periods and to include direct survival data into bull proofs quickly. This project was based on the findings of Sewalem *et al.* (2005) using a Weibull proportional hazards model. Results from that study showed that there were two critical time periods during the first lactation where the relative risk of a cow being culled was very high, at 80 and 235 DIM. The same study concluded that the risk of culling in first lactation cows was lowest at 120 and 240 DIM. Thus, a genetic evaluation was developed in Canada using the following five-trait animal model, in which the five survival traits were considered as different traits. The five traits were:

1. survival from first calving to 120 DIM;
2. survival from 120 DIM to 240 DIM;
3. survival from 240 DIM to 2nd calving;
4. survival from 2nd calving to 3rd calving; and
5. survival from 3rd calving to 4th calving.

The model used is the following:

$$y_{ijkm} = hy_{im} + rhs_{jm} + (age_k)_m + (prot_k \times rhs_{jl})_{jm} + (prot_k)_m + (fat_k)_m + animal_{km} + e_{ijkm}$$

where:

y_{ijkm} = observation for survival (0 or 1) in trait m (m = 1 to 5) on cow k that calved in herd year i;

hy_{im} = fixed effect of herd year i for trait m;

rhs_{jm} = fixed effect of subclass j for registry status × herd size change × season of calving for trait m;

$(age_k)_m$ = linear and quadratic regressions of survival in trait m on age at first calving k; trait m on $prot_{km}$ by rhs_{jm} subclasses,

$prot_k$ is the normal rank of the cow for protein yield in first lactation within herd year of first calving;

$(prot_k)_m$ and $(fat_k)_m$ = linear, quadratic, and cubic regressions of survival in trait m on the cow's normal ranks for protein and fat yield in lactation one within herd year of first calving;

$animal_{km}$ = random additive genetic effect of animal k for survival in trait m; and

e_{ijkm} = random residual.

Contemporary groups of herd year of calving were formed in relation to the dairy quota year (July to June) rather than a calendar year because of the impact of the quota system in Canada on culling. Genetic parameters for the Holstein breed are shown in Table 1. Heritabilities were lower for first lactation than later lactations and ranged between 0.01 and 0.05. Genetic correlations were very high among the 3 survival traits in first lactation (0.94 to 0.98), and lower among the other traits (0.73 to 0.85). The five individual EBV are then combined by considering the average survival rate of daughters for each bull (Sewalem *et al.*, 2007).

Let S_i = the mean survival rate for trait i + Sire ETA for trait i and cumulative survival

$$C_i = \prod_{l=1}^{6} S_i \text{ and } C_6 = 0.$$

Table 1. Estimates of heritability and genetic correlation for the five herd life traits of Canadian Holstein (heritabilities are on diagonal, genetic correlations are above the diagonal). Source: Sewalem et al. *(2007).*

Trait	Traits				
Survival	$S1_a$	$S1_b$	$S1_c$	S2	S3
From 1st calving to 120 DIM ($S1_a$)	0.01	0.97	0.94	0.85	0.78
From 120 DIM to 240 DIM ($S1_b$)		0.01	0.98	0.85	0.78
From 240 DIM to 2nd calving ($S1_c$)			0.02	0.81	0.73
Through 2nd lactation (S2)				0.04	0.76
Through 3rd lactation (S3)					0.05

$$DHL = \frac{(\sum_{i=1}^{6}(C_{i-1} - C_i) \times (N_i + D_i)) + K}{365}$$

where:
DHL is the direct herd life,
N_i = mean calving interval in days trait i
D_i are the means days of production for cows culled in trait i,
K is a constant, i.e. the expected the mean calving intervals and average days of production for cows culled after the fourth lactation.

Published Herd Life proofs are then a combination of Direct Herd Life (DHL) and Indirect Herd Life (IHL). DHL reflects true daughter survival while IHL is an indirect measure of longevity based on proofs for non-production traits. The prediction of DHL using early predictors has been improved over time as new genetic evaluations for relevant traits became available in Canada. Until 2001, the prediction was based only on conformation traits [Mammary System (57%), Feet and Legs (29%), Overall Rump (7%), Overall Capacity (7%)]. The R-square of that prediction was 0.16. From 2001 to 2004, the prediction included also SCS and Milking Speed and that increased twofold the R-square to 0.32 [Mammary System (22%), Feet & Legs (16%), Rump Angle (10%), Overall Capacity (-15%), SCS (-24%), Milking Speed (13%)]. In November 2004, a new 4-trait evaluation system for female fertility was released in Canada. The inclusion of 2 fertility traits in the prediction helped to increase the R-square to 0.49 [Mammary System (16%), Feet & Legs (3%), Rump Angle (11%), Overall Capacity (-6%), SCS (-14%), Milking Speed (12%), Non-Return Rate for cows (13%), Calving to First Service (-22%)]. Finally in January 2008, a new genetic evaluation system for reproductive performance was released in Canada. The evaluation system uses a 16-trait animal model, which includes both female fertility and calving performance traits. The inclusion in the prediction for direct herd life of some additional traits from this new evaluations system yielded an R-square of 0.57. The new prediction includes 6 type traits [Mammary System (9.9%), Feet & Legs (7.4%), Overall Rump (7.7%), Angularity (-6.4), Stature (-5%), Chest Width (-4.7%)], 4 reproduction traits [Calving to First Service (10.9%), First Service to Conception (10.7%), Calving Ease (3.9%), Calf Survival (6.9%)], and 4 other functional traits [SCS (-12.9%), Milking Speed (5.8%), Milking Temperament (3.5%), Lactation Persistency (4.9%)]. The genetic evaluations for DHL and IHL are then combined into an overall genetic evaluation for herd life (HL) using a MACE procedure as previously described by Jaraith *et al.* (1998).

Genetic evaluations for Herd Life have been available in Canada since 1996. Even if Herd Life has been included in the Canadian selection index (Lifetime Profit Index, LPI) only in 2001, there has been consistent genetic improvement of longevity in the cow population before then. Chesnais (2006) has shown that in the last 15 years there has been a slow but steady genetic progress of longevity in Canadian Holsteins, approximately 1 unit of SD. The beginning of this period studied by Chesnais (2006) also coincides with the implementation of the LPI in Canada in 1991, when conformation traits (indicators of herd life) had 40% emphasis (see Figure 1) of the overall index. Later on in 2001, as stated before, Herd Life was included in the LPI with a weight of 8%, and, at the same time, udder health was also included in the LPI at 5% emphasis. Thus the total of longevity traits (direct and indirect) became first 43% in 2001, then 46% in 2005 and 49% in 2008 with the inclusion of reproductive traits in the LPI.

Evaluation and selection for longevity worldwide

Interbull (2007) reports that most European countries (10) use the survival kit to evaluate longevity (Denmark, Finland, France, Germany, Hungary, Italy, the Netherlands, Spain and Switzerland). Seven countries (Australia, Belgium, Canada, Great Britain, Ireland, New Zealand and Sweden) use linear models (single or multiple-trait) where the longevity trait is defined as survival per lactation. Finally, Israel and United States use also linear models, but the dependent variable is productive life. Table 2 reports the genetic correlations among all the countries that participated to Interbull evaluations in August 2007. The average genetic correlation among the countries that use the survival kit for the genetic evaluation of longevity was the same (0.72) as the average genetic correlation among the countries that use linear models and survival per lactation for the genetic evaluation of longevity. However, if two countries that have very different production systems (New Zealand and Australia) are excluded from the group of countries that use linear models and survival per lactation, then the average genetic correlation within this cluster raises to 0.81. The genetic correlation between the two countries that use productive life evaluation (US and Israel) was much lower (0.56). This low value is also related by the low number of bulls in common between the two countries, and by possible differences in production management and culling decisions. Israel had low correlations with most countries. The average genetic correlations between the 3 clusters show that the countries that use survival analysis are higher correlated to the countries that use productive life than the countries that use linear models (0.67 vs. 0.63), even higher when US alone was considered (0.80 vs. 0.76).

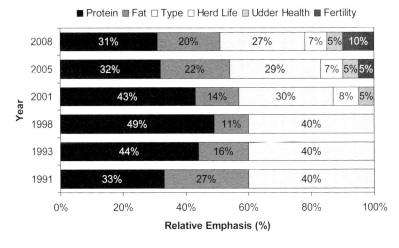

Figure 1. Relative emphasis of various traits on the Canadian Lifetime Profit Index (LPI) over time.

Table 2. Genetic correlations among countries for direct longevity (Interbull, 2007).

	Countries that use linear models and survival per lactation						Countries that use Survival Kit										Productive life	
	BEL	CAN	GBR	IRL	NZL	SWE	CHE	DEU	DNK	DNR	ESP	FIN	FRA	ITA	NLD	HUN	ISR	USA
Australia	0.73	0.61	0.68	0.62	0.70	0.65	0.55	0.57	0.67	0.60	0.36	0.55	0.69	0.36	0.64	0.31	0.53	0.66
Belgium, BEL		0.77	0.86	0.81	0.65	0.76	0.72	0.82	0.75	0.70	0.53	0.58	0.67	0.51	0.76	0.43	0.59	0.75
Canada, CAN			0.86	0.84	0.55	0.77	0.83	0.89	0.79	0.77	0.83	0.65	0.69	0.81	0.83	0.76	0.55	0.91
Great Britain, GBR				0.87	0.63	0.76	0.81	0.86	0.79	0.79	0.69	0.64	0.65	0.68	0.73	0.54	0.51	0.85
Ireland, IRL					0.68	0.79	0.76	0.89	0.68	0.76	0.71	0.68	0.63	0.63	0.64	0.51	0.45	0.81
New Zealand, NZL						0.55	0.43	0.52	0.50	0.52	0.48	0.45	0.41	0.33	0.38	0.30	0.33	0.56
Sweden, SWE							0.62	0.76	0.74	0.68	0.53	0.64	0.65	0.54	0.68	0.46	0.50	0.77
Switzerland, CHE								0.86	0.81	0.85	0.78	0.74	0.70	0.71	0.89	0.58	0.66	0.83
Germany, DEU									0.82	0.85	0.78	0.72	0.66	0.74	0.82	0.64	0.49	0.88
Denmark, DNK										0.81	0.72	0.85	0.77	0.68	0.89	0.57	0.61	0.86
Red Danish, DNR											0.71	0.71	0.66	0.60	0.78	0.59	0.44	0.84
Spain, ESP												0.68	0.58	0.81	0.72	0.71	0.46	0.85
Finland, FIN													0.76	0.51	0.76	0.38	0.61	0.71
France, FRA														0.64	0.81	0.54	0.64	0.73
Italy, ITA															0.73	0.83	0.45	0.75
The Netherlands, NLD																0.63	0.73	0.82
Hungary, HUN																	0.37	0.75
Israel, ISR																		0.56

Breeding for robustness in cattle

Direct selection for longevity worldwide has been implemented only in the last 5-10 years. An investigation by Leitch (1994) showed that until 1994 only the US Net Merit index included some emphasis (27%) on Productive Life. Since then, most countries now have some degree of emphasis on direct longevity in their national selection indices. Table 3 shows a summary of selection indices from around the world. Table 3 was compiled using previous work from VanRaden (2004), Miglior *et al.* (2005) and extension information provided by each country on their websites. Germany (RZG) has the highest emphasis on direct longevity (25%), followed by Great Britain (PLI; 20%) and United States (NM; 17%). Japan (NTP), Ireland (EBI) and South Africa (BVI) do not have any emphasis on direct longevity. However, Ireland has 37% emphasis on fertility and 21% on other traits, likely correlated to longevity. In fact, if we assume that any traits other than production included in the national selection indices are related to longevity, then the proportion allotted to longevity can be as high as 69% both in Sweden (TMI) and in Denmark (S-Index) and as low as 25% in Japan (NTP), as shown in the last two columns of Table 3. It should be noted that those relative emphases are in general an indication of the approximate weight of each EBV with the national selection index. Correlations among traits are likely not accounted for. However, in a separate analysis of the Canadian LPI where correlations among EBV were accounted for, the resulting relative emphases on each trait were in general similar to those emphases before accounting for the correlations among traits.

In a separate investigation, results from the January 2008 Interbull run were analysed. Bull EBV were standardised within each country scale and averaged by bull's year of birth for each country scale by country of origin for a series of traits. Country of origin was assumed as the country where the bull had the highest number of daughters. Nine major dairy countries were identified as those countries that sample at least 200 Holstein bulls per year. Genetic progress was then computed within each country scale as the slope of proofs for bulls born between 1996 and 2002. Slopes were then averaged across country scales. Table 4 reports the result of this part of the investigation. The average genetic progress for longevity was 5%/year of proof SD unit. This value is lower than correspondent values for production and conformation, but higher than SCS. Thus, even if the heritability of longevity is generally low, and direction selection for longevity is quite recent, a significant genetic progress for longevity has been achieved worldwide.

Table 3. Relative emphasis of national selection indices on various traits.

	Protein	Fat	Milk	Type	Longevity	Udder health	Fertility	Other	Production	'Longevity'*
Japan - NTP	0.55	0.20		0.25					0.75	0.25
Australia - APR	0.38	0.13	0.19		0.09	0.06	0.05	0.11	0.70	0.31
New Zealand - BW	0.37	0.10	0.15		0.06	0.07	0.07	0.20	0.62	0.40
Italy - PFT	0.45	0.14		0.23	0.08	0.10			0.59	0.41
Spain - ICO	0.35	0.12	0.12	0.35	0.03	0.03			0.59	0.41
Israel - PD07	0.42	0.15			0.08	0.13	0.16	0.06	0.57	0.43
Switzerland - ISEL	0.36	0.17		0.24	0.07	0.10	0.06		0.53	0.47
South Africa - BVI	0.26	0.26		0.45		0.03			0.52	0.48
Canada - LPI	0.31	0.20		0.27	0.07	0.05	0.10	0.04	0.51	0.49
Germany - RZG	0.40	0.10		0.15	0.25	0.05	0.01		0.50	0.50
France - ISU	0.37	0.13		0.13	0.13	0.13	0.13		0.50	0.52
United States - TPI	0.30	0.17		0.32	0.09	0.05	0.05	0.02	0.47	0.53
United States - NM	0.23	0.23		0.13	0.17	0.09	0.09	0.06	0.46	0.54
Great Britain - PLI	0.22	0.12	0.11	0.10	0.20	0.06	0.19		0.45	0.55
Ireland - EBI	0.24	0.05	0.13				0.37	0.21	0.42	0.58
The Netherlands - NVI	0.23	0.05	0.12	0.26	0.09	0.09	0.16		0.40	0.60
Denmark - S-Index	0.21	0.05	0.05	0.15	0.07	0.14	0.08	0.25	0.31	0.69
Sweden - TMI	0.15	0.08	0.08	0.18	0.04	0.14	0.15	0.18	0.31	0.69

*Traits other than production.

Breeding for robustness in cattle

Table 4. Genetic progress (SD unit/year) of various traits for Holstein bulls born in 1996-2002 by country of origin (Interbull, January 2008).

Country	Milk yield	Fat yield	Protein yield	Overall mammary	Feet and legs	Longevity	SCS*
A	0.131	0.121	0.139	0.091	0.084	0.053	-0.007
B	0.104	0.127	0.121	0.103	0.081	-0.008	0.011
C	0.101	0.130	0.103	0.111	0.107	0.026	-0.017
D	-0.005	0.134	0.050	0.000	-0.038	0.124	-0.019
E	0.122	0.149	0.130	0.120	0.126	0.035	-0.008
F	0.117	0.083	0.102	0.153	0.118	0.061	-0.046
G	0.064	0.069	0.100	0.023	0.023	0.081	
H	0.103	0.113	0.099	0.118	0.123	0.060	-0.060
I	0.174	0.127	0.164	0.088	0.076	0.039	-0.025
Others	0.135	0.126	0.155	0.111	0.104	0.061	-0.008
Average	0.105	0.118	0.116	0.092	0.080	0.053	-0.020

*For SCS, a negative genetic progress is desirable.

Conclusion

Genetic evaluations for longevity are available in most countries, and three clusters of evaluation types were identified. Average genetic correlations among the two main clusters (survival analysis and survival at each lactation) were lower than within each cluster. The availability of those evaluations has allowed to include direct emphasis on longevity in national selection indices. Genetic progress for longevity has occurred also because of indirect selection for traits related to longevity, like conformation, female fertility, udder health and management traits. Use of indirect predictors of longevity is very important as it provides an early and accurate selection tool for newly proven bulls. As new indicators of longevity are investigated like body condition score, milk urea nitrogen, lactose, and particularly health traits, accuracy of longevity predictions is increasingly achieved.

References

Allaire, F.R. and J.P. Gibson, 1992. Genetic value of herd life adjusted for milk production. J. Dairy Sci., 75: 1349-1356.

Bascom, S.S. and A.J. Young, 1998. A Summary of the reasons why farmers cull cows J. Dairy Sci., 81: 2299-2305.

Beaudeau, F., V. Ducrocq, C. Fourchon and H. Seegers, 1994. Effect of disease on productive life of French Holstein dairy cows assessed by survival analysis. J. Dairy Sci., 78: 103-117.

Brotherstone, S., R.F. Veerkamp and W.G. Hill, 1998. Predicting breeding values for herd life of Holstein-Friesian dairy cattle from lifespan and type. Anim. Sci., 67: 405-411.

Carlén, E., E. Strandberg A. Roth, 2004. Genetic parameters for clinical mastitis, somatic cell score, and production in the first three lactations of Swedish Holstein cows. J. Dairy Sci., 87: 3062-3070.

Caraviello, D.Z., K.A. Weigel, G.E. Shook and P.L. Ruegg, 2005. Assessment of the impact of somatic cell count on functional longevity in Holstein and Jersey cattle using survival analysis methodology. J. Dairy Sci. 88: 804-811.

Chesnais, J., 2006. Forces of change. Proc. of Endless Performance Visioning Conference, Ottawa, ON, Canada, Nov. 2007. pp.15-19.

Cruickshank, J., K.A. Weigel, M.R. Dentine and B.W. Kirkpatrick, 2002. Indirect prediction of herd life in Guernsey dairy cattle. J. Dairy Sci., 85: 1307-1313.

De Vlieger, S., H.W. Barkema, G. Opsomer, A. de Kruif and L. Duchateau, 2005. Association between somatic cell count in early lactation and culling of dairy heifers using Cox frailty models. J. Dairy Sci., 88: 560-568.

Dekkers, J.C.M., 1994. Optimal breeding strategies for calving ease. J. Dairy Sci., 77: 3441–3453.

Ducrocq, V., 2002. A piecewise Weibull mixed model for the analysis of length of productive life of dairy cows. Proc. 7th World Congr. Genet. Appl. Livest. Prod., Montpellier, France. Communication No. 20–04.

InterBull, 2007. Interbull routine genetic evaluation for direct herd life, May 2007. http://www-inter-bull.slu.se/longevity/framesida-long.htm Accessed July, 2007.

Jairath, L., J.C.M. Dekkers, L.R. Schaeffer, Z. Liu, E.B. Burnside and B. Kolstad, 1998. Genetic evaluation for herd life in Canada. J. Dairy Sci., 81: 550-562.

Leitch, H.W., 1994. Comparison of international selection indices for dairy cattle breeding. Interbull Bull. No. 10

Miglior, F., B.L. Muir and B.J. van Doormaal, 2005. Selection Indices in Holstein Cattle of Various Countries. J. Dairy Sci., 88: 1255-1263.

Neerhof, H.J., P. Madsen, V.P. Ducrocq, A.R. Vollema, J. Jensen and I.R. Korsgaard, 2000. Relationship between mastitis and functional longevity in Danish black and white dairy cattle estimated using survival analysis. J. Dairy Sci., 83: 1064-1071.

Poso, J. and E.A. Mantysaari, 1996. Relationships between clinical mastitis, somatic cell score, and production for the first three lactations of Finnish Ayrshire. J. Dairy Sci., 79: 1284-1291.

Rogers, G.W. and B.T. McDaniel, 1989. The usefulness of selection for yield and functional type traits. J. Dairy Sci., 72: 187-193.

Roxstrom, A., V. Ducrocq and E. Strandberg, 2003. Survival analysis of longevity in dairy cattle on a lactation basis. Genet. Sel. Evol., 35: 305-318.

Samoré, A.B, J.A.M. van Arendonk and A.F. Groen, 2001. Impact of area and sire by herd interaction on heritability estimates for somatic cell count in Italian Holstein Friesian cows. J. Dairy Sci., 84: 2555-2559.

Sewalem, A., G.J. Kistemaker, F. Miglior and B.J. van Doormaal, 2004. Analysis of the relationship between type traits and functional survival in Canadian Holsteins using a Weibull proportional hazards model. J. Dairy Sci., 87: 3938-3946.

Sewalem, A., G.J. Kistemaker, V. Ducrocq and B.J. van Doormaal, 2005. Genetic analysis of herd life in Canadian dairy cattle on a lactation basis using a Weibull proportional hazards model. J. Dairy Sci., 88: 368-375.

Sewalem, A., F. Miglior, G.J. Kistemaker and B.J. van Doormaal, 2006. Analysis of the relationship between somatic cell score and functional longevity in Canadian dairy cattle. J. Dairy Sci., 89: 3609-3614.

Sewalem, A., F. Miglior, G.J. Kistemaker, P. Sullivan, G. Huapaya and B.J. van Doormaal, 2007. Modification of genetic evaluation of herd life from a three-trait to a five-trait model in Canadian dairy cattle. J. Dairy Sci., 90: 2025-2028.

Sewalem, A., F. Miglior, G.J. Kistemaker, P. Sullivan and B.J. van Doormaal, 2008. Relationship between reproduction traits and functional longevity in Canadian dairy cattle. J. Dairy Sci. 91: 1660-1668.

Short, T.H. and T.J. Lawlor, 1992. Genetic parameters for conformation traits, milk yield and herd life in Holsteins. J. Dairy Sci., 75: 1987-1998.

Van Arendonk, J.A.M., 1991. Use of profit equations to determine relative economic value of dairy cattle herd life and production from field data. J. Dairy Sci., 74: 1101-1107.

Van Doormaal, B.J., L.R. Schaeffer and B.W. Kennedy, 1985. Estimation of genetic parameters for stayability in Canadian Holsteins. J. Dairy Sci., 68: 1763-1769.

VanRaden, P.M. and E.J.H. Klaaskate, 1993. Genetic evaluation of length of productive life including predicted longevity of live cows. J. Dairy Sci. 76: 2758–2764.

VanRaden, P.M. and G.R. Wiggans, 1995. Productive life evaluations: calculation, accuracy and economic value. J. Dairy Sci., 78: 631-638.

VanRaden, P.M., 2004. Selection on Net Merit to improve lifetime profit. J. Dairy Sci., 87: 3125-3131.

Vollema, A.R. and A.F. Groen, 1998. A comparison of breeding value predictors for longevity using a linear model and survival analysis. J. Dairy Sci., 81: 3315-3320.

Weigel, K.A., T.J. Lawlor Jr., P.M. VanRaden and G.R. Wiggans, 1998. Use of linear type and production data to supplement early predicted transmitting abilities for productive life. J. Dairy Sci., 81: 2040-2044.

Heritability of lifetime milk yield and productive life and their relationship with production and functional traits in the Simmental, Swiss Fleckvieh and Red Holstein populations in Switzerland

M. Gugger[1], F. Ménétrey[1], S. Rieder[2] and M. Schneeberger[3]
[1]Swiss Simmental and Red&White Cattle Breeders' Association, P.O. Box 691, 3052 Zollikofen, Switzerland
[2]Swiss College of Agriculture, Länggasse 85, 3052 Zollikofen, Switzerland
[3]Institute of Animal Sciences, ETH, TAN D2, 8092 Zurich, Switzerland

Abstract

Costs of milk production can be reduced through lower replacement rate by increasing productive life of cows. Lifetime production (LP; milk production to 6th lactation) and productive life (PL; number of completed lactations) of 112,462 daughters of 766 test AI bulls were used to obtain daughter averages and to estimate heritabilities. Bulls belonged to three sections of the Swiss Simmental and Red and White cattle herd book, differing in percentage of Red Holstein genes. Correlations of daughter average LP and PL with sire EBVs for production and functional traits and with composite indices differed and, for the correlations with the composite index for meat production, changed sign among herd book sections which may be caused by different breeding objectives. The strongest correlations of LP were found with EBV milk (≥ 0.69 for the individual herd book sections), of PL with total merit index (0.47 to 0.59). Heritabilities were estimated using two sire models (without or with including a fixed effect of herd book section). The estimates were around 0.19 and 0.13 for LP (milk, fat and protein yield), and 0.11 and 0.09 for PL from the two models. Estimates obtained from the second model may be more appropriate because breeding objectives differ among herd book sections.

Keywords: longevity, lifetime production, productive life

Introduction

Reducing the average replacement rate by increasing productive life of cows is one of the options to reduce costs of milk production in dairy herds. Ducrocq and Soelkner (1998) reported estimated heritabilities of longevity between 0.05 and 0.1, obtained using a conventional linear model, and values between 0.15 and 0.2 when a Weibull proportional hazard model was employed, corresponding to the estimate of 0.18 by Caraviello *et al.* (2004). Several studies (Caraviello *et al.*, 2002; Cruickshank *et al.*, 2002; Hare *et al.*, 2006; Abdallah *et al.*, 2002; Sewalem *et al.*, 2004; Tsuruta *et al.*, 2005; Vukasinovic *et al.*, 2002) reported correlations between type traits and longevity. The strongest correlations were found with udder traits (suspension of fore and hind udder, udder depth, central ligament) and traits of feet, legs and claws.

The purpose of this study was to perform a preliminary investigation of factors affecting lifetime milk production and productive life, to estimate the heritability of these traits using a half sib structure, and, by means of simple correlations between daughter averages and sires' estimated breeding values, to obtain an idea of existing relationships between lifetime milk production and productive life and production and functional traits in the Simmental, Swiss Fleckvieh and Red Holstein populations.

Material and methods

The dataset included records of 112,462 daughters of 766 test AI bulls, born between 1985 and 1993. Records of daughters from the testing period of bulls with at least 50 daughters were used. Table 1 summarises the performance of these cows and Table 2 shows the distribution of the bulls on three sections of the Swiss Simmental and Red and White cattle herd book, i.e. Simmental (<14% Red Holstein genes, SI), Swiss Fleckvieh (14 to 74% Red Holstein, SF) and Red Holstein (≥75% Red Holstein, RH), together with their average number of daughters from the testing period, average daughters' life time milk production up to 6th lactation, average percentage of daughters with ≥6 lactations, and average culling rate of daughters in first lactation.

Lifetime production (LP) was defined as production up to 6th lactation (limited to 6th lactation because some of the daughters of the youngest bulls included in the analysis were still alive at the time of this investigation), productive life (PL) as number of completed lactations. In this investigation, data were not corrected for production level, thus the analysed traits do not represent effective lifetime production or functional longevity, and only limited comparison is possible with results of studies, were this correction was performed. Simple correlations were computed of daughters' means of LP and PL with sires' estimated breeding values for production and functional traits, as well as with composite estimated breeding values (indices). These correlations are neither pure genetic nor phenotypic measures, but may, nevertheless, give some indication on the relationships among the investigated traits.

Table 1. Summary of the performances of all cows.

Trait	Number of cows	Average	Standard deviation	Minimum	Maximum
Milk kg up to 6th lactation	112,318	15,768	13,295	7	80,687
Fat kg up to 6th lactation	112,305	649.9	549.1	1	3,475
Protein kg up to 6th lactation	112,307	509.3	429.4	1	2,504
Number of lactations	112,462	3.10	2.26	1	16

Table 2. Distribution of bulls on herdbook sections and daughter averages (standard deviations in parentheses).

Section of herdbook[1]	Number of bulls	Average number of daughters	Average milk production of daughters up to 6th lactation (kg)	Average percentage of daughters with ≥6 lactations (%)	Average culling rate from 1st to 2nd lactation (%)	Average number of lactations
SI	200	125 (41)	12,586 (2227)	14.5 (5.3)	37.9 (7.4)	2.91 (0.40)
SF	307	143 (41)	15,310 (2684)	15.4 (5.3)	33.0 (6.8)	3.07 (0.40)
RH	259	173 (50)	17,556 (2946)	15.9 (5.5)	29.5 (6.0)	3.18 (0.39)
Total	766	148 (48)	15,353 (3282)	15.3 (5.4)	33.1 (7.4)	3.07 (0.41)

[1]SI: Simmental (<14% Red Holstein genes); SF: Swiss Fleckvieh (14 to 74% Red Holstein); RH: Red Holstein (≥75% Red Holstein).

Heritabilities for LP (milk, fat and protein) and PL were estimated by a linear sire model. A fixed environmental effect was defined as the combination of two regional classifications (geographical region and altitude) and herd production level. The random sire effect was modelled in two alternative ways, nested or not nested within fixed herd book section. The MIXED procedure of SAS (SAS software version 8.1) was used to estimate sire and error variance components by REML. Standard errors of the estimated heritabilities were approximated by the formula given in Falconer and Mackay (1996):

$$\sigma_{h^2} = \sqrt{\frac{32h^2}{nN}}$$

where:
n is the number of sires and
N the average number of daughters per sire.

Results

The average culling rate from first to second lactation of daughters of all bulls was 33.1%, with the highest value (37.9%) for SI and the lowest (29.5%) for RH bulls, whereas the value found for SF bulls corresponded to the overall average (Table 2). Looking at the percentage of daughters producing during ≥6 lactations, the SF were again near the overall mean, the SI bulls had the lowest and the RH bulls the highest percentage. Average milk LP of daughters was highest for RH (17,556 kg) and almost 5,000 kg lower for SI bulls, reflecting the inferior milk production potential and the higher culling rate of daughters of SI bulls.

Table 3 gives correlations between daughter means of milk LP and PL with estimated breeding values for production and functional traits, as well as composite indices of bulls by herd book section.

Table 4 summarises the estimated heritabilities for LP and PL. The estimates were lower when the herd book section of the sires was included as a fixed effect in the linear model and the random sire effect was modelled nested within this fixed effect. The heritability estimates for LP were almost 0.2 using the model without nesting sires within section and around 0.13 for the model with the nested effects. The heritability estimates for PL of approximately 0.1 were in the range of other fitness traits.

Discussion

EBVs for milk of bulls were positively correlated (>0.6) with daughters' average LP in all herd book sections (Table 3). Cows producing high milk yield often reach high LP, but in fewer lactations than low producing cows. This is reflected in the lower correlations of EBV milk with average PL (0.35 to 0.53) compared to correlations between EBV milk of sire and LP of 0.69 to 0.76. The same tendency was observed for the correlations of LP and PL with the composite index for milk production (ILM). The largest difference between the correlations of EBV milk or ILM with LP and with PL was found for RH bulls. The reason could be higher culling due to insufficient milk production of daughters of SF and particularly SI sires than of daughters of RH sires. EBVs for fat and protein % were negatively correlated with both daughter averages, being in line with the generally negative relationship between milk yield and content.

The fitness index (IFI) was positively correlated with the daughter average for LP (0.21 and 0.24 for RH and SF bulls, but only 0.09 for SI bulls). A more pronounced correlation of IFI was found with PL (0.21 for SI, 0.38 for SF, and 0.48 for RH bulls). This stronger correlation is due to the high weight for the EBV for productive life in the IFI, the weight being 50% in the IFI for RH bulls. Only weak and non significant correlations (-0.04 to +0.08) were found between EBV for SCC and

Table 3. Correlations of daughter means of milk production to 6th lactation (LP) and number of lactations (PL) with estimated breeding values (EBV) for production and functional traits and composite indices of bulls belonging to three herd book sections, together with levels of significance (t test of hypothesis of zero correlation).

Herd book section[1]	RH		SF		SI		Total	
	LP	PL	LP	PL	LP	PL	LP	PL
EBV milk kg	0.69 ***	0.35 ***	0.76 ***	0.53 ***	0.69 ***	0.51 ***	0.45 ***	0.37 ***
EBV fat %	-0.35 ***	-0.26 ***	-0.15 **	-0.07 ns	-0.16 *	-0.09 ns	-0.12 ***	-0.11 **
EBV protein %	-0.33 ***	-0.22 ***	-0.31 ***	-0.18 **	-0.22 **	-0.12 ns	-0.21 ***	-0.16 **
EBV somatic cell count (SCC)	-0.02 ns	0.21 ***	0.08 ns	0.18 **	-0.04 ns	0.04 ns	0.03 ns	-0.16 ***
EBV persistency	-0.22 ***	-0.10 ns	-0.19 ***	-0.15 **	-0.26 ***	-0.24 ***	-0.17 ***	-0.15 ***
ILM (composite index milk)[2]	0.43 ***	0.17 **	0.50 ***	0.39 ***	0.48 ***	0.36 ***	0.29 ***	0.26 ***
IME (composite index type traits)[3]	0.35 ***	0.31 ***	0.18 *	0.25 ***	0.15 ns	0.20 ns	0.19 ***	0.26 ***
IVF (composite index meat production)[4]			-0.20 ns	-0.24 ns	0.36 **	0.37 **	0.07 ns	0.09 ns
IFI (composite index fitness)[5]	0.21 ***	0.48 ***	0.24 ***	0.38 **	0.09 ns	0.21 **	0.10 **	0.34 ***
Total Merit Index[6]	0.60 ***	0.48 ***	0.51 ***	0.48 ***	0.64 ***	0.59 ***	0.40 ***	0.47 ***

[1] SI: Simmental (with a maximum of 13% Red Holstein genes; SF: Swiss Fleckvieh (13 to 74% Red Holstein); RH: Red Holstein (≥75% Red Holstein).

[2] ILM: Index combining EBVs for milk kg, fat and protein kg and %.

[3] IME: Index combining EBVs for linear type traits.

[4] IVF: Index combining EBVs for growth and carcass conformation (not given for RH bulls).

[5] IFI: Index combining EBVs for SCC, productive life, persistency, reproduction (Non Return Rate and time to first service), production increase from first to later lactations, milkability and dystocia (fitness index).

[6] Index combining ILM, IME, IVF and IFI.

Table 4. Estimated heritabilities of milk, fat and protein production to 6th lactation (LP) and number of lactations (PL) with models taking or not taking into account herd book section of sires. The standard errors of the estimated heritabilities were all <0.01.

Trait	Including herd book section of sire	
	no	yes
LP milk kg	0.19	0.13
LP fat kg	0.19	0.12
LP protein kg	0.18	0.13
PL	0.11	0.09

daughter average LP, but were +0.18 and +0.21 with PL for SF and RH bulls. It has to be noted that EBVs for SCC are published as relative EBVs with high values indicating desirable EBVs, i.e. low SCC. The EBV for persistency, also a component of IFI, was negatively correlated with average LP and, to a lesser extent, with PL. The difference observed for the correlations between the index for meat production (IFV) and average LP and PL (negative and non significant for SF, positive and highly significant for SI) reflect the different breeding objectives for the two herd book sections (IFV is not estimated for RH bulls).

Positive relationships were obtained between the composite type index (IME) and LP and PL for all herd book sections with the strongest values for RH bulls. The total merit index, combining the above indices, and placing most weight on the milk production index, was again positively correlated with both, LP and PL. The different magnitudes of the correlations in the three herd book sections, again, reflect different breeding objectives and, thus, different index weights in the individual herd book sections.

The heritability of LP is moderate and somewhat lower than heritability estimates usually reported for lactation production (Table 4). This indicates that selection on this trait could be successful, but it would lead to an increased generation interval. The high correlations of average daughter LP with EBV milk indicate, however, that selection on lactation milk yield also tends to improve LP.

The heritability estimates for PL of around 0.1 correspond with results reported in the literature (Ducrocq and Soelkner, 1998). The positive correlations between average daughter PL and EBV for productive life or IFI lead to the conclusion that these measures, routinely available to breeders, provide a good opportunity for improving longevity. The heritability estimates obtained from the statistical model with nesting sire effects within fixed effects of herd book section may be more appropriate, given the fact, demonstrated in various cases in this analysis, that breeding objectives differ in the various herd book sections.

References

Abdallah, J.M., B.T. McDaniel and M.J. Tabbaa, 2002. Relationships of productive life evaluations with changes in evaluations for yields. J. Dairy Sci. 85: 677-681.

Caraviello, D.Z, K.A. Weigel and D. Gianola, 2002. Analysis of the relationship between linear type traits, inbreeding, and survival in US Jersey cows using a Weibull model. J. An. Sci. 80, Suppl. 1/J. Dairy Sci. 85, Suppl. 1: 88.

Caraviello, D.Z., K.A. Weigel and D. Gianola, 2004. Comparison between a Weibull Proportional Hazards Model and a Linear Model for predicting the genetic merit of US Jersey Sires for daughter longevity. J. Dairy Sci. 87: 1469-1476.

Cruickshank, J., K.A. Weigel, M.R. Dentine and B.W. Kirkpatrick, 2002. Indirect prediction of herd life in Guernsey dairy cattle. J. Dairy Sci. 85: 1307-1313.

Ducrocq, V. and J. Soelkner, 1998. Implementation of a routine breeding value evaluation for longevity of dairy cows using survival analysis techniques. In: Proc. 6th World Congr. Genet. Appl. Livest. Prod. 27: 447-448.

Falconer, D.S. and T.F.C. Mackay, 1996. Introduction to quantitative genetics. Fourth edition. Longman, Harlow, Essex.

Hare, E., H.D. Norman and J.R. Wright, 2006. Survival rates and productive herd life of dairy cattle in the United States. J. Dairy Sci. 89: 3713-3720.

Sewalem, A., G.J. Kistemaker, F. Miglior and B.J. van Doormaal, 2004. Analysis of the relationship between type traits and functional survival in Canadian Holsteins using a Weibull Proportional Hazards Model. J. Dairy Sci. 87: 3938-3946.

Tsuruta, S., I. Misztal and T.J. Lawlor, 2005. Changing definition of productive life in US Holsteins: Effect on genetic correlations. J. Dairy Sci. 88: 1156-1165.

Vukasinovic, N., Y. Schleppi and N. Künzi, 2002. Using conformation traits to improve reliability of genetic evaluation for herd life based on survival analysis. J. Dairy Sci. 85: 1556-1562.

Genetic correlation between persistency and calving interval of Holsteins in Japan

K. Hagiya[1], K. Togashi[2], H. Takeda[2], T. Yamasaki[2], T. Shirai[1], J. Sabri[1], Y. Masuda[3] and M. Suzuki[3]
[1]National Livestock Breeding Center, Nishigo-mura, Fukushima-ken 961-8511, Japan
[2]Hokkaido National Agricultural Station, Sapporo-shi 062-8555, Japan
[3]Obihiro University of Agriculture and Veterinary Medicine, Obihiro-shi, 080-8555, Japan

Abstract

Examined in this study were the effects of calving intervals (CI) on lactation curves. Genetic correlations among lactation yields, lactation persistency, increase in milk yield at an earlier stage of lactation and CI were estimated. Data included 31,227 cows with 293,982 test-day records obtained from the DHI program between 1997 and 2001. Persistency and increase in milk yield at an earlier stage of lactation were calculated for each cow defined as the difference between test day milk yields at 60 days in milk (DIM) and 150 DIM (calculated as milk yield at 150 DIM minus milk yield at 60 DIM). Increase in milk yields at an earlier stage of lactation (IEL) was defined as the difference between test day milk yields at 5 DIM and 35 DIM (calculated as milk yield at 35 DIM minus milk yield at 5 DIM). A random regression model (RRM) was applied to this analysis, and the GIBBS3F90 program was used to estimate the effects of CI classes on lactation curves. The RRM included the fixed effects of herd-test-day, age, calving season, and CI classes as well as random effects of animal, permanent environment and heterogeneous residuals with 10 intervals by DIM. The genetic correlations were estimated using the AIREMLF90 program with a model including fixed effects of herd, age, and calving season as well as random effects of animal and residual. Results indicated that lactation curves were affected by the length of CI when the DIM was longer than 150. A slightly positive genetic correlation was estimated between persistency and CI (0.20), but a slightly negative correlation was estimated between persistency and IEL (-0.19).

Keywords: Holstein cows, lactation persistency, calving interval

Introduction

Lactation persistency is generally defined as a rate of decline after a peak milk yield (Cole and VanRaden, 2006), or as a difference between peak yield and yield at a test day in a later stage of lactation (Ptak and Schaeffer, 1993). A cow with higher persistency is associated with more profit (Dekkers *et al.*, 1998), and allows greater use of cheap roughage (Sölkner and Fuchs, 1987). A cow with higher persistency has a lower peak yield than a cow with the same total milk yield but less persistency. Numerous studies have shown an antagonistic relationship between high milk yields, especially at peak and cow fertility (Kawashima *et al.*, 2007; Muir *et al.*, 2004). Bar-Anan and Wiggans (1985) reported a small positive genetic correlation between persistency and cow fertility. Muir *et al.* (2004) reported that slightly undesirable genetic correlations existed between persistency and calving intervals (CI) among first and second lactation. However, they also described the possibility of changing lactation curves by directing less energy toward the unborn calf with longer CI, also suggesting that the selection of persistency might improve total yields without increasing reproductive failures. Togashi and Lin (2006) demonstrated the construction of various indices for improving lactation milk yield and persistency.

Kawashima *et al.* (2007) examined the phenotypic relationships between daily milk yields and ovulation within 3 wk postpartum from 46 cows. They found cows with lower ratios of an increase in milk yield from the first to the peak week may have fewer fertility problems. The objectives of this paper were (1) to estimate effects of CI on lactation curves and (2) to estimate genetic correlations among lactation yields, persistency, increases in milk yields at an earlier stage of lactation (IEL), and CI.

Materials and methods

Data

Daily milk records of 293,982 test-days on 31,227 first lactation Holstein cows obtained from the DHI program between 1997 and 2001, and the pedigree records containing 74,361 animals were used. Calving records until 2006 were used for the calculation of CI.

Trait definition

Persistency and increase in milk yield at an earlier stage of lactation (IEL) were calculated for each cow. Persistency was defined as milk yield at 150 days in milk (DIM) minus milk yield at 60 DIM. The IEL was defined as milk yield at 35 DIM minus milk yield at 5 DIM. Lactation curves were estimated using the multiple trait prediction procedure (Schaffer and Jamrozik, 1996). The Wilmink's function (Wilmink, 1987) was used to estimate lactation yield, persistency, IEL, milk yield at 60 DIM, and DIM of peak yield (PeakD). In the analysis, cows were assigned to 7 subclasses for the length of CI (below 323 d, 324 to 344, 345 to 365, 366 to 428, 429 to 554, 555 to 743 and over 744 d from first calving date to second calving date), 15 subclasses for age at calving (18 to 25 mo, 26, 27, 28, 29, 30, 31, 32, 33, 34, 35, 36, 37, 38 to 39, and over 40 mo), and 12 subclasses for the month at calving.

Estimation of the effect of CI for lactation curves

The following random regression model was applied to estimate effects of CI on the lactation curves:

$$y_{ijklml} = HTD_i + \sum_{q=0}^{2} A_j z_q + \sum_{q=0}^{2} M_k z_q + \sum_{q=0}^{4} CI_l z_q + \sum_{q=0}^{2} u_m z_q + \sum_{q=0}^{2} pe_m z_q + e_{ijklml} \qquad (1)$$

where:
y_{ijklml} is test day milk yield,
HTD_i is a fixed effect of herd-test-day i,
A_j is a fixed effect of age at calving j,
M_k is a fixed effect of month at calving k,
CI is a fixed effect of CI classes l,
u_m is a random effect of animal m,
pe_m is a random effect of permanent environment (PE) m,
z_q is a set of Legendre polynomials of order q, and
e_{ijklml} is a effect of heterogeneous random residuals with 10 subclasses assigned by DIM (below 35 d, 36 to 65, 65 to 95, 96 to 125, 126 to 155, 156 to 185, 186 to 215, 216 to 245, 246 to 275, and 276 to 305 d at first calving).

Lactation curves were modeled using Legendre polynomials. The GIBBS3F90 program (Misztal *et al.*, 2002) was used to obtain the (co)variance components based on posterior means of 40,000 iterations after a burn-in of 10,000 iterations.

Relationships of the CI to persistency, milk yield at 60 DIM or PeakD

Genetic correlations among lactation yield, persistency, milk yield at 60 DIM, PeakD and CI were estimated using the AIREMLF90 program (Misztal *et al.*, 2002) with a model presented as:

$$y_{ijkl} = HY_i + A_j + M_k + u_l + e_{ijkl} \tag{2}$$

where:
y_{ijkl} is lactation yield, persistency, milk yield at 60 DIM, PeakD or CI,
HY_i is a fixed effect of herd-year i,
A_j is a fixed effect of age at calving j,
M_k is a fixed effect of month at calving k,
u_l is a random effect of animal l, and
e_{ijkl} is a random effect of residuals.

Results and discussion

Estimation of the effect of CI for lactation curves

Estimates from GIBBS3F90 were stabilised after 5,000 iterations. Therefore, 40,000 iterations after a burn-in of 10,000 rounds were enough to estimate for this analysis. Figure 1 shows estimates of variances on DIM. Curves of genetic variances and PE variances were higher at the beginning and end of lactations. The estimates of residual variance for milk yields decreased gradually with DIM.

Figure 2 shows estimated heritability and repeatability on DIM. Heritability estimates increased from 0.17 to 0.36 with DIM but decreased slightly at the end of lactations. These results agree with those reported in the literature (Olori *et al.*, 1999; Strabel and Jamrozik, 2006; Strabel and Misztal, 1999; Togashi *et al.*, 2007). Mean estimates of heritability and repeatability were 0.29 and 0.75, respectively.

The shapes of lactation curves differed due to the length of CI when DIM was more than 150 (Figure 3). However, when DIM was less than 150, the shapes of lactation curves were hardly changed within CI subclasses. Canavesi *et al.* (2006) reported that CI (or days open) may have an

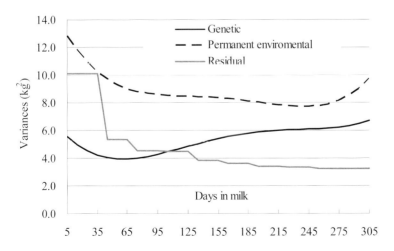

Figure 1. Variance estimates for milk yields on days in milk.

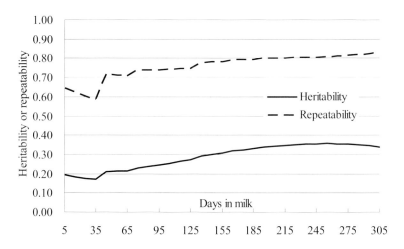

Figure 2. Heritability and repeatability estimates for milk yields on days in milk.

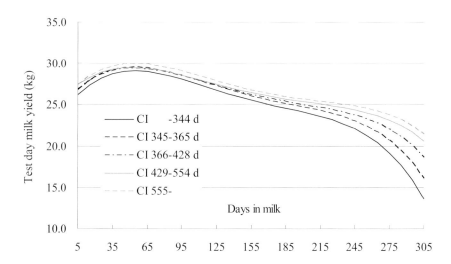

Figure 3. Effects of calving interval (CI) classes on days in milk.

effect on the last part of the lactation curve particularly. Trends of this study agree with their results obtained from Italian Hosteins. Although shorter CI decreased milk yields at the later part of the lactation curve, in this study, length of CI would hardly effect persistency because persistency was defined without use of the later part of lactation curve.

Relationships of the CI to persistency, milk yield at 60 DIM or PeakD

Table 1 shows mean, standard deviation, minimum and maximum of 305 d milk yields, persistency, IEL, and milk yields at 60 DIM and PeakD. For some cows, estimated PeakDs were equal to a test day at the beginning of the lactation (1 DIM). Trends at the beginning of the lactation curve may sometimes have poor accuracy because residuals of milk yields were higher at the beginning of lactation. However, milk yields at 60 DIM would be approximations of peak yields because the average of PeakD was very close to 60 DIM.

Breeding for robustness in cattle

Table 1. Mean, standard deviation (SD), minimum and maximum of 305 d milk yield, persistency, increase in milk yield at an earlier stage of lactation (IEL), milk yield at 60 days in milk (DIM) at peak yield (n=31,227).

	Mean	SD	Minimum	Maximum
305 d milk yield (kg)	7,774	1,422	2,197	15,742
Persistency[1] (kg)	-2.5	1.4	-11.1	4.3
IEL[2] (kg)	7.6	2.9	-11.8	23.9
Milk yield at 60 DIM (kg)	29.4	4.8	11.3	57.6
DIM at peak yield	59.1	18.8	1	304

[1] Defined as milk yield at 150 DIM minus milk yield at 60 DIM.

[2] Defined as milk yield at 35 DIM minus milk yield at 5 DIM.

The heritability estimate of persistency in this study (0.12) was in agreement with those of the first lactation reported in literature (Jakobsen *et al.*, 2002; Muir *et al.*, 2004; Togashi *et al.*, 2007; Weller *et al.*, 2006) (Table 2). The heritability estimate of IEL (0.07) was lower than that of persistency.

Phenotypic correlations of persistency to 305 d milk yield, milk yield at 60 DIM, and PeakD were 0.17, -0.06 and 0.39, respectively. Appuhamy *et al.* (2007) reported phenotypic correlations of persistency of 0.009 to 305 d milk yield, -0.04 to peak yield, and 0.70 to PeakD for Holstein cows in Virginia Tech and Pennsylvania State University. For PeakD, the estimated correlation in this study was relatively lower. However, results indicated that higher persistency was associated with the later PeakD rather than 305 d milk yield or milk yield at 60 DIM. Kawashima *et al.* (2007) used daily milk records and information on ovulations for 46 cows, and reported the ratio of increasing milk yields from the first week to the peak week in ovular cows was smaller compared with that of anovular cows. In this study, CI and IEL from monthly DHI records were uncorrelated (-0.02). Further study is required to provide correlations between fertility traits and trends of lactation curves at the beginning of lactation.

Table 2. Heritabilities (diagonal), genetic correlations (above diagonal) and phenotipic correlations (below diagonal).

	305 d milk yield	Persistency	IEL	Milk yield at 60 DIM	DIM at peak yield	Calving interval
305 d milk yield	0.35	0.56	0.37	0.98	0.40	0.37
Persistency[1]	0.17	0.12	0.28	0.37	0.78	0.20
IEL[2]	0.07	0.02	0.07	0.24	0.69	-0.19
Milk yield at 60 DIM[3]	0.93	-0.06	0.02	0.28	0.17	0.17
DIM at peak yield	0.05	0.39	0.52	0.02	0.08	-0.06
Calving interval	0.15	0.19	-0.02	0.04	0.12	0.07

[1] Defined as milk yield at 150 DIM minus milk yield at 60 DIM.

[2] Increasing in milk yield at a earlier stage of lactation: Defined as milk yield at 35 DIM minus milk yield at 5 DIM.

[3] Days in milk.

Estimates of genetic correlations between 305 d milk yield and other traits were positive, and the correlation with milk yield at 60 DIM was the highest (0.98). Genetic correlations of persistency with 305 d milk yield, milk yield at 60 DIM and PeakD were 0.56, 0.37 and 0.78, respectively. Several researchers have found genetic correlations between persistency and PeakD to be positive and high (Rekaya *et al.*, 2000; Muir *et al.*, 2004). Selection for persistency would increase lactation milk yield, and PeakD. A lower positive genetic correlation between persistency and CI was estimated (0.20). The result was similar to those presented by other researchers (Cole and VanRaden, 2006; Muir *et al.*, 2004). In general, longer CI is undesirable at an economic point of view. Therefore, a small antagonistic genetic correlation might exist between persistency and CI. Also small negative genetic correlation between CI and IEL (-0.19) was estimated. Togashi and Lin (2006) recommended simultaneous improvement of lactation milk and persistency. When the persistency is included in the selection index, fertility traits would be weighted to avoid the antagonistic selection response among persistency and CI.

Acknowledgements

The authors wish to thank M. Aihara of The Livestock Improvement Association Japan for supplying data.

References

Appuhamy, J.A.D.R.N., B.G. Cassell, C.D. Dechow and J.B. Cole, 2007. Phenotypic relationships of common health disorders in dairy cows to lactation persistency estimated from daily milk weights. J. Dairy Sci., 90: 4424-4434.

Bar-Ann, R.M. and G.R. Wiggans, 1985. Associations among milk yield, yield persistency, conception and culling of Israeli Holstein dairy cattle. J. Dairy Sci., 68: 382-386.

Canavesi, F., S. Biffani and F. Biscarini, 2006. Italian test day evaluation: Improving details that matter. Interbull Bulletin, 35: 17-21.

Cole, J.B. and P.M. van Raden, 2006. Genetic evaluation and best prediction of lactation persistency. J. Dairy Sci., 89: 2722-2728.

Dekkers, J.C.M., J.H. Ten Haag and A. Weersink, 1998. Economic aspects of persistency of lactation in dairy cattle. Livest. Prod. Sci., 53: 237-252.

Jakobsen, J.H., P. Madsen, J. Jensen, J. Pedersen, L.G. Christensen and D.A. Sorensen, 2002. Genetic parameters for milk production and persistency for Danish Holsteins estimated in random regression models using REML. J. Dairy Sci., 85: 1607-1616.

Kawashima, C., C. Amaya Montoya, Y. Masuda, E. Kaneko, M. Matsui, T. Shimizu, N. Matsunaga, K. Kida, Y.I. Miyake, M. Suzuki and A. Miyamoto, 2007. A positive relationship between the first ovulation postpartum and the increasing ratio of milk yield in the first part of lactation in dairy cows. J. Dairy Sci., 90: 2279-2282.

Misztal, I., Tsuruta, S., Strabel, T., Auvray, B., Druet, T. and D. Lee, 2002. BLUPF90 and related programs (BGF90). Proceedings of the 7th World congress on genetics applied to livestock production, Montpellier, France, CD-ROM Communication, 28: 07.

Muir, B.L., J. Fatehi and L.R. Schaeffer, 2004. Genetic relationships between persistency and reproductive performance in first-lactation Canadian Holsteins. J. Dairy Sci., 87: 3029-3037.

Olori, V.E., W.G. Hill, B.J. McGuirk and S. Brotherstone, 1999. Estimating variance components for test day milk records by restricted maximum likelihood with a random regression animal model. Livest. Prod. Sci., 61: 53-63.

Ptak, E. and L.R. Scaeffer, 1993. Use of test day yields for genetic evaluation of dairy sires and cows. Livest. Prod. Sci., 34: 232-24.

Rekaya, R., M.J. Carabano and M.A. Toro. 2000. Bayesian analysis of lactation curves of Holstein-Friesian cattle. J. Dairy Sci., 83: 2691-2701.

Strabel, T. and J. Jamrozik. 2006. Genetic analysis of milk production traits of Polish Black and White cattle using large-scale random regression test-day models. J. Dairy Sci., 89: 3152-3163.

Strabel, T. and I. Misztal, 1999. Genetic parameters for first and second lactation milk yields of Polish Black and White cattle with random regression test-day models. J. Dairy Sci., 82: 2805-2810.

Sölkner, J. and W. Fuchs, 1987. A comparison of different measures of persistency with special respect to variation of test-day milk yields. Livest. Prod. Sci., 16: 305-319.

Togashi, K. and C.Y. Lin, 2006. Selection for milk production and persistency using eigenvectors of the random regression coefficient matrix. J. Dairy Sci., 89: 4866-4873.

Togashi, K., C.Y. Lin, Y. Atagi, K.Hagiya, J. Sato and T. Nakanishi, 2007. Genetic characteristics of Japanese Holstein cows based on multiple-lactation random regression test-day animal models. Livest. Prod. Sci., in press.

Weller, J.I., E. Ezra and G. Leitner, 2006. Genetic analysis of persistency in the Israeli Holstein population by multitrait animal model. J. Dairy Sci., 89: 2738-2746.

Wilmink, J.B.M., 1987. Adjustment of test-day milk, fat and protein yield for age, season and stage of lactation. Livest. Prod. Sci., 16: 335-348.

Part 3
Secondary traits

Parameter estimation and genetic evaluation of female fertility traits in dairy cattle

Z. Liu, J. Jaitner, F. Reinhardt, E. Pasman, S. Rensing and R. Reents
VIT w.V., Heideweg 1, 27283 Verden, Germany

Abstract

A genetic evaluation system was developed for six fertility traits of dairy cattle: interval first to successful insemination (FSh) and non-return rate to 56 days (NRh) of heifers, and interval from calving to first insemination (CF), non-return rate to 56 days (NRc), and interval first to successful insemination (FSc) of cows. Using the two interval traits of cows, CF and FSc, as components, breeding values for days open (DO) were derived. A multiple trait animal model was applied to evaluate these fertility traits. Fertility traits of later lactations of cows were treated as repeated measurements. Genetic parameters were estimated by REML. Mixed model equations of the genetic evaluation model were solved with the Gauss-Seidel algorithm and iteration on data techniques. Reliabilities of estimated breeding values (EBV) were approximated with a multi-trait effective daughter contribution (EDC) method. Daughter yield-deviations (DYD) and associated EDC were calculated with a multiple trait approach. The genetic evaluation software was applied to the insemination data of dairy cattle breeds in Germany, Austria and Luxembourg. Small heritability estimates were obtained for all the fertility traits, ranging from 1% for NRh to 4% for CF. Genetic and environmental correlations were low to moderate among the traits. Notably, unfavourable genetic trends were obtained in all the fertility traits. Genetic correlations of the fertility with other traits were estimated by applying an approximate REML method to DYD of 5709 Holstein bulls. While NRh and NRc were almost uncorrelated with the production traits, the interval fertility traits, CF, FSh and FSc, were moderately correlated with the yield traits. These interval fertility traits were also moderately correlated with functional longevity and can thus be regarded as a reasonable predictor of longevity. Integrating fertility traits into total merit selection index can halt or reverse the decline of fertility in dairy cattle.

Keywords: REML, genetic correlations, daughter yield-deviations, effective daughter contribution

Introduction

In recent decades successful selection for milk production traits has led to a decline in female fertility in dairy cattle arising from an unfavourable, correlated selection response (Jorjani, 2006). In order to improve or at least slow the deterioration in fertility, more emphasis on fertility traits in selection is necessary (De Jong, 2005; Jamrozik *et al.,* 2005; Van Doormaal *et al.,* 2007; VanRaden *et al.,* 2004; Wall *et al.,* 2005). Germany has used a complete database to store all breeding data and has evaluated the first service non-return rate for more than ten years. A routine genetic evaluation of female and male fertility has been applied to non-return rate through 90 d since the early 1980's. Because the single trait genetic evaluation model doesn't consider some other useful fertility traits, such as CF or FS, a more up-to-date statistical model is required for accurately evaluating fertility of female and male animals.

Typically national genetic evaluations include the following fertility traits, two of them for heifers: non-return rate to 56 days NRh, and interval first to successful insemination FSh, and three traits for cow fertility: interval from calving to first insemination CF, non-return rate to 56 days NRc, and interval from first to successful insemination FSc. The five fertility traits fit well into the current Interbull grouping concept of fertility traits (Jorjani, 2006). As a combined trait, days open DO can be derived as the sum of CF and FSc (Jamrozik *et al.*, 2005). The change in the definition of the non-return rate from 90 to 56 days was made for international harmonisation of fertility traits. Therefore, the five fertility traits NRh and FSh of heifers and CF, NRc and FSc of cows were selected for genetic evaluation, the combined trait DO can be obtained by summing CF and FSc.

Total merit index (TMI), a function of EBV of economically important traits, are widely used in dairy cattle breeding programmes worldwide. A TMI typically contains information from several component traits: milk production, conformation including locomotion, health traits like somatic cell scores (SCS) und body condition score (BCS), longevity, calving traits, workability, and female fertility. Usually countries evaluate the trait groups separately using different statistical models, such as a random regression test-day model for production or SCS traits (Liu *et al.*, 2004), a non-linear survival model for longevity (Ducrocq, 2001, Tarrés *et al.*, 2006), or a multiple trait model for female fertility. A 2-step animal model using pre-corrected records or DYD (Ducrocq *et al.*, 2001) offers the following advantages over the selection index approach for setting up TMI. Indirect selection effect on correlated traits can be optimally considered. Higher genetic progress and proof reliabilities have been confirmed for the 2-step model (Lassen *et al.*, 2007). Double counting of some trait information is avoided, e.g. SCS proof included in combined longevity prediction as well as in TMI that contains again the same SCS information and combined longevity. All cows will also receive longevity proofs, whereas survival analysis model predicts longevity proofs only for bulls. Compared to average yield-deviations (AYD) of cows, DYD of bulls allow a more efficient estimation and more reliable estimates of genetic correlations among traits as a result of their much higher reliability. Therefore, it was decided to estimate genetic correlations among the component traits of the German TMI using DYD and associated EDC of bulls.

The objectives of this study were to develop a genetic evaluation system with a multiple trait animal model for the selected fertility traits, and to estimate genetic correlations between the fertility and other traits by applying an approximate REML method to DYD and EDC of bulls.

Materials and methods

Data for genetic evaluation

Insemination records of heifers and cows from first to six lactations were chosen for the fertility genetic evaluation. Early, less complete data were excluded by imposing the conditions that heifers must have been born from 1994 onwards and calving years of cows must have been 1995 or later. Female animals from all farms enrolled in milk recording programs were considered in the genetic evaluation, including those not in a herdbook system. Insemination data of both AI bulls and natural service bulls were analysed together. Table 1 has a summary of the fertility data used for April 2008 joint evaluation for dairy cattle breeds Holstein, Red Dairy Cattle and Jersey from Germany, Austria and Luxembourg. The genetic evaluation was conducted jointly for three breeds and across

Table 1. Descriptive statistics of the fertility data for April 2008 evaluation.

	No. of levels
Insemination records	25,526,858
Females with data	11,461,212
Cows with data	8,788,810
Animals in pedigree	16,525,323
Herd-year classes	522,030
Service sires	37,198
Total no. of equations	111,127,391

three countries. A total of 25.5 million insemination records of 11.5 million heifers and cows were evaluated. The total number of equations amounted to over 111 million.

Models and methods for genetic evaluation

Statistical models for genetic evaluation

The fertility traits were analysed with a multiple trait animal model, where all fixed effects are denoted in upper case and random effects in lower case:

$$y_{ijklmno} = H_i + L_j + B_o + F_m + t_{mn} + a_k + p_k + e_{ijklmno} \tag{1}$$

where:
$y_{ijklmno}$ denotes a fertility trait in the l-th lactation ($l = 0, 1, \ldots, 6$) of female animal k;
H_i is fixed effect of the i-th herd-year;
L_j is the j-th lactation group (heifer and 6 cow lactations) x age class (3 classes for heifers, 5 for the first three lactations, and 1 for the remaining lactations) x season (3 classes: January to March, April to August and September to December) and year of insemination x region (Eastern Germany and the rest);
B_o is the o-th effect of type of bull as service sire (proven sire or otherwise) x AI stud of service sire x AI stud servicing the female animal;
F_m represents fixed effect of service sire m;
t_{mn} represents random effect of insemination year class n within the service sire m;
a_k is additive genetic effect of female k;
p_k is random permanent environmental effect of female k; and
$e_{ijklmno}$ is the residual effect for this insemination record of the lactation.

Not all of the fertility traits were modelled with all the effects in the full model 1. For the interval trait of heifer FSh, only H_i, L_j and a_k were considered:

$$FSh_{ijk} = H_i + L_j + a_k + e_{ijk} \tag{2}$$

The sub-model for NRh of heifers contained B_o, F_m, t_{mn} plus to the effects for FSh:

$$NRh_{ijkmno} = H_i + L_j + B_o + F_m + t_{mn} + a_k + e_{ijkmno} \tag{3}$$

In comparison to NRh, non-return rate of the cow, NRc was evaluated with the additional random permanent environmental effect p_k, and its evaluation model is model 1. The two interval traits for cow fertility, CF and FSc, were analysed with a sub-model including H_i, L_j, a_k and p_k:

$$CF_{ijkl}, FSc_{ijkl} = H_i + L_j + a_k + p_k + e_{ijkl} \tag{4}$$

Solving the mixed model equations

The mixed model equation system of the multiple trait Model 1 was solved with a Gauss-Seidel algorithm together with the iteration on data technique. Phantom parents were grouped based on country of origin, four selection paths and birth year of animal; and the phantom parent groups were merged using pre-defined minimum number of animals per group. Trait values were scaled to a similar variance in order to make genetic evaluations numerically more stable; and (co)variance parameters were scaled accordingly in the iteration program. About 5.1 Gb RAM was required for solving the equations, and the CPU was about 55 seconds per round of iteration on a single central processing unit of a 64-bit Advanced Micro Devices Opteron Linux server.

Breeding for robustness in cattle

The multi-trait EDC method (Liu *et al.,* 2004) was applied to approximate reliability values of EBV of all the fertility traits, including the derived, combined trait DO and fertility indices. Using the method developed for test-day models (Liu *et al.,* 2004), AYD of female animals and DYD of bulls were computed together with their associated EDC. Two sets of EDC were made available for both single trait and multiple trait MACE evaluations (Jorjani, 2006).

Models and methods for estimating (co)variances of the fertility traits

Insemination records from 1999 to 2005 were selected for parameter estimation. Only Black-and-White and Red-and-White Holstein female animals were considered. Ancestors were traced back using the complete pedigree containing about 58 million animals. A multiple trait sire model was used for analysing a selected dataset. As a control, a single trait animal model, the model 1, was fitted to a considerably smaller sub-set of the data. We intended to determine if the sire model resulted in significantly lower heritability estimates than the animal model. Because fertility data are, by nature, highly selected (i.e. only cows that were fertile in the previous lactation can have insemination records in current lactation), sequential data selection steps were carefully performed. On the trait level, only fertility traits with valid trait values were selected. On the lactation level, only sequential missing trait patterns in the order of CF, NRc and FSc, were allowed. On the animal level, a female was included in the parameter estimation only if she had an insemination record as heifer.

In order to reduce computing requirements, herd-year classes were required to have at least 5 heifers or 10 cows. Each service sire was required to have a minimum number of 100 insemination records. A sire of females was included only if he had at least 50 daughters with fertility records. After these sequential selection procedures were imposed, 215,509 heifer and 282,183 cow insemination records remained for final parameter estimation with the sire model. In total, 2,437 bulls were evaluated. A total of 4,738 animals were included in the final pedigree file for parameter estimation. For the analyses with the animal model, a much smaller number of insemination records from a much higher number of females was used; and 30% herds were randomly chosen. Male ancestors were traced back to birth year of 1980, while female ancestors were traced to 1995. This was in contrast to the sire model analyses where all possible ancestors were considered. The animal model was applied only to single trait analyses because of computing restraints.

A REML method was used to estimate the (co)variance components using software package VCE 5 (Kovač *et al.,* 2002). A number of bi-variate analyses were conducted in order to use as much data and pedigree information as possible, because the five-trait analysis using the full dataset was computationally infeasible. The (co)variance estimates from the sub-analyses were averaged to get the final parameter estimates. The difference in heritability estimates between the sire and animal model was not significant. Therefore, parameter estimates from the sire model were chosen, when estimates of both models were available.

Model and methods for estimating genetic correlations with other traits

Data from April 2008 genetic evaluations for production, SCS, conformation, longevity and female fertility traits were used for estimating genetic correlations among the traits. Bulls' DYD of those traits were calculated following the multi-trait model procedure (Liu *et al.,* 2004), in addition to their associated EDC matrices or scalars, depending on genetic evaluation models of the analysed traits, e.g. a vector of DYD and a matrix of EDC for female fertility traits of bull. Because the three production and SCS traits were evaluated with a random regression model, DYD of bulls were expressed as random regression coefficients (RRC) of Legendre polynomials with three parameters. For longevity evaluated with a survival model, a pseudo-record of relative risk and its weight were calculated for each daughter of every bull following a procedure by Ducrocq (2001) and Tarrés *et al.* (2006), which were then used to compute DYD and EDC for the bull. The bulls were required to be present in Interbull's 010 file for production traits, 015 file for conformation, 016 file for SCS, 017 file for longevity, and 019 file for female fertility. However, due to much shorter history of data recording for locomotion and BCS, no restriction was imposed on availability of these two traits. As the multi-lactation random regression model (Liu *et al.*, 2004) provided DYD in first three lactations, adequate daughter information was needed for estimating genetic correlations between each of the lactations with the other traits. Therefore, bulls were required to have daughters' test-day records in all three lactations. Additionally, all bulls must not have fewer than 30 daughters with lactation passing 120 days in milk in each of the three lactations. No further selection was imposed on the remaining traits evaluated other than the test-day traits. A total of 5709 bulls with DYD remained after all the selection steps. Table 2 describes the final data set for the genetic correlation estimation. The number of the operational traits reached 49, with 9 RRC for each production or SCS trait.

Bull pedigree file from Interbull's April 2008 evaluations was reformatted from a sire, maternal grand-sire and maternal grand-dam format to a sire and dam format. Ancestors of the selected bulls with data were traced back from both sire and dam sides as far as possible. The final pedigree file contained 13,612 animals plus 18 phantom parent groups which were formed according to breed, country of origin, selection paths (son to sire, son to dam, daughter to sire and daughter to dam) and birth year of the animal. Small phantom groups were merged to ensure at least 200 animals assigned to each group. Among the 7,903 ancestors, 281 sires had also DYD data available.

The following statistical model was applied to estimate genetic correlations of the selected traits:

Table 2. Number of Black-and-White Holstein bulls and average number of daughters by birth year for estimating genetic correlations among the traits.

Year of birth	No. of bulls	Average no. of daughters in					No. of bulls	Average no. of daughters in	
		Milk, fat protein	SCS	Type traits	Longevity	Female fertility		Loco-motion	BCS
1986	63	6,219	6,214	701	4,429	2,606	2	21	22
1987	59	2,918	2,916	299	2,124	1,530			
1988	93	2,339	2,340	244	1,668	1,322	2	32	34
1989	110	3,918	3,918	523	2,931	2,471	9	455	545
1990	148	3,566	3,566	543	2,571	2,262	15	287	320
1991	186	2,715	2,716	408	2,026	1,822	32	66	75
1992	244	1,405	1,405	253	1,044	958	24	204	226
1993	639	570	570	107	426	386	41	169	199
1994	694	728	728	132	527	510	70	302	350
1995	716	211	211	66	154	144	47	147	181
1996	648	125	125	51	96	86	12	47	54
1997	661	115	115	52	88	79	6	20	21
1998	619	115	115	54	88	80	3	40	41
1999	575	123	123	59	95	87	260	30	36
2000	252	122	122	59	95	88	252	45	53
2001	2	99	99	67	84	67	2	52	66
All	5,709	690	690	126	505	555	777	91	107

$$\mathbf{q}_{ij} = \mathbf{f}_{jk} + \mathbf{a}_{ij} + \mathbf{e}_{ij} \qquad (5)$$

where:
$\mathbf{q}_{ij}$ is a vector of DYD of the i-th bull in trait j;
$\mathbf{f}_{jk}$ is a vector of fixed effects of birth year k in the j-th trait;
$\mathbf{a}_{ij}$ is a vector of additive genetic effects of bull i in trait j; and
$\mathbf{e}_{ij}$ is a vector of residual effects.

Adding a birth year effect in the model can provide more robust estimation of genetic trends (Lassen *et al.*, 2007). For traits evaluated with a single trait model, e.g. longevity, all the vectors become scalar. The (co)variance matrix of genetic effects of the m component trait blocks is denoted as:

$$\mathbf{G}_0 = \{\mathbf{G}_{0\,jl}\}_{j-1,\dots,m;l-1\dots,m} \qquad (6)$$

where:
$\mathbf{G}_{0\,jj}$ is genetic (co)variance matrix of trait j; and
$\mathbf{G}_{0\,jl}$ is the genetic covariance matrix between traits j and l.

The inverse of error (co)variance matrix of bull i in trait j is:

$$[\mathrm{Var}(\mathbf{e}_{ij})]^{-1} = \mathbf{\psi}_{ij} \qquad (7)$$

where:
$\mathbf{\psi}_{ij}$ is EDC matrix for bull i in trait j on animal basis, converted from reliability matrix contributed by his daughters' records in the j-th trait.

The multi-trait EDC method (Liu *et al.*, 2004) was applied to approximate matrix $\mathbf{\psi}$ for all the bulls. Similar to the parameter estimation for a multi-trait MACE model (Tarrés *et al.*, 2007), residual correlations between DYD of two traits were ignored, because it can be verified that the proportion of residual covariance in the covariance of DYD between two traits decreases with the number of daughters of the DYD. Even for two traits with high residual correlation, the residual covariance between the two DYD will become negligible when the number of daughters of the DYD is greater than 100. Therefore, the residual correlation of the DYD was not considered in the estimation of genetic (co)variances.

Mixed model equations of model 5 were solved using a pre-conditioned conjugate gradient algorithm and an iteration on data technique. An approximate expectation maximisation REML method was implemented to estimate the across-trait genetic correlations (Tarrés *et al.*, 2007). The iterative process of the parameter estimation was considered converged when the third decimal place of all the genetic correlation estimates no longer changed between two consecutive rounds of iteration.

Results and discussions

Selection of fertility traits for routine evaluation

In the previous German genetic evaluation system paternal genetic effect of service sire was fitted as a correlated trait to maternal genetic effect for the non-return rate to 90 days. Because a very low heritability estimate was obtained in the parameter estimation for the paternal genetic effect, 0.002, it was decided to remove this effect from the new fertility model for the NR traits. The decision of dropping this correlated paternal genetic effect also agreed with the international harmonisation of fertility traits (Jorjani, 2006). Paternal effects on NR are still included in the new model as the fixed effect of service sire. In the initial phase of this project DO and calving interval (CI) were chosen as directly evaluated traits within the multi-trait model. It was found that the EBV of DO or CI had much higher variation than a test run with a single trait model. The higher EBV variances were mainly caused by the fact that the data information of CF and/or FSc were double counted in the evaluation model, because CF and FSc are parts of DO and CI. Based on this finding, it was decided to include the component fertility traits CF and FSc in genetic evaluation and to subsequently derive EBV for DO as sum of the EBV of CF and FSc (Jamrozik *et al.*, 2005).

Estimates of genetic (co)variances of the fertility traits

Estimates of genetic parameters are in Tables 3 and 4. All of the fertility traits had low heritability estimates ranging from 1% to 4%. The heritability estimate of the interval trait CF was found to be almost three times as high as for the cow's conception traits NRc and FSc. Heritability estimates of the conception traits were similar for heifers and cows. Genetic and residual correlation estimates between both heifer traits were moderately negative. Genetic correlation estimate was moderately high between heifers and cows for the same trait NR (0.63) or FS (0.48). Genetic correlation between NRc and FSc is -0.39, indicating the limited accuracy of using NR56 for projecting conception. The NR56 traits and CF are found to have very low genetic correlations. Because NRc is significantly less correlated with DO than the interval trait FSc, the prediction of DO using NRc was less accurate than using CF. Among the cow fertility traits, both genetic and permanent environmental effects were correlated in similar magnitude and in the same direction. The two cow traits NRc and FSc have quite low percentage of permanent environmental variance, in contrast to the fertility trait CF with 11% of phenotypic variance attributed to the permanent environmental effect (Table 4). Despite the very low heritability estimates, the fertility traits had reasonably large genetic standard deviations, e.g. 9.8 days for DO, as a result of large phenotypic variation in the dairy cattle population. Approximate standard errors for the heritability estimates were from 0.11% to 0.24%. The standard errors of the genetic correlation estimates ranged from 0.015 to 0.035, similar range of standard errors were also found for the correlation estimates of permanent environment effects and the heifer residual correlation.

Table 3. Heritabilities on the diagonal, genetic correlations above diagonal, and residual correlation of heifer traits and permanent environmental correlations of cow traits below diagonal (units: % for NR and days for the interval fertility traits).

Trait	FSh	NRh	CF	NRc	FSc	DO	Genetic standard deviation
Interval first to successful insemination heifer (FSh)	0.014	-0.53	0.17	-0.25	0.48	0.37	7.44
Non-return rate to 56 days heifer (NRh)	-0.49	0.012	-0.02	0.63	-0.15	-0.09	4.77
Interval calving to first insemination cows (CF)			0.039	0.05	0.37	0.86	6.92
Non-return rate 56 days cow (NRc)			0.13	0.015	-0.39	-0.18	5.95
Interval first to successful insemination cows (FSc)			0.30	-0.37	0.010	0.78	4.88
Days open cow (DO)						0.026	9.83

Table 4. Estimated ratios of residual variance and cow permanent environmental (p.e.) in phenotypic variance of the fertility traits.

Trait	Ratio of residual variance	Ratio of cow p.e. variance
Interval first to successful insemination heifer (FSh)	0.986	
Non-return rate heifer (NRh)	0.988	
Interval calving to first insemination cows (CF)	0.851	0.110
Non-return rate 56 days cows (NRc)	0.949	0.036
Interval first to successful insemination cows (FSc)	0.961	0.029

The parameter estimates fell within the range of estimates found in literature from various populations (Berry *et al.*, 2007; Fogh *et al.*, 2003; Gredler *et al.*, 2007; Jamrozik *et al.*, 2005; VanRaden *et al.*, 2004; Wall *et al.*, 2005). But the heritability estimates were smaller than those found in Canada (Jamrozik *et al.*, 2005). In fact, the Canadian heritability estimates represented the upper bound of these parameter estimates. Similar heritability as well as genetic correlation estimates were also found in Nordic countries (Fogh *et al.*, 2003). For the NR traits, the auto-correlation parameter used in the genetic evaluation was 0.8 and variance of within-service-sire year effects was 1% of phenotypic variance. A comparison with the parameter estimates from other studies was not possible as no such estimates were available in literature.

Results of genetic evaluation of the fertility traits

Phenotypic trends of the five fertility traits were studied using 25.2 million insemination records of Holstein females. Cow insemination records with CF only were not used for calculating the trends, because such records tend to have shorter CF than insemination records of more complete lactations at the time of data extraction for the genetic evaluation. The interval first to successful insemination FS was set to missing, if no following calving was available. Figure 1 has phenotypic averages of NR traits in heifers and first six lactations of cows. The NR decreased as lactation number increased, with the largest difference of 0.05 between heifers and lactation six cows. Heifers have significantly

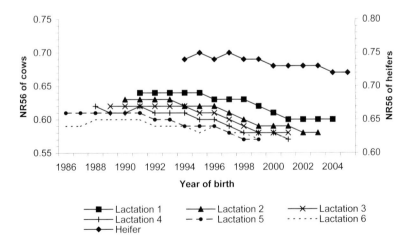

Figure 1. Phenotypic trend of non-return rate 56 days in German Holsteins.

Breeding for robustness in cattle

higher NR than cows. For either heifers or cows, negative, unfavourable trends could be observed in the last ten years, though the unfavourable trends were weaker for heifers than for cows. It can be seen in Figure 2 that FS increased significantly in the 1990's; but the phenotypic trends have become flatter or even slightly reversed ever since. As for NR traits, higher lactation number had longer FS interval, and heifers had much shorter FS than cows with an average difference of 12.7 d between heifers and lactation six cows. Similar phenotypic trends can be seen for CF in Figure 3 as for FS in Figure 2. From Figure 2, the interval first to successful insemination FS increased over time. In recent years there was a slight drop in phenotypic trend in FS caused by the fact that cows or heifers that had shorter, complete records of FS entered into genetic evaluations earlier at

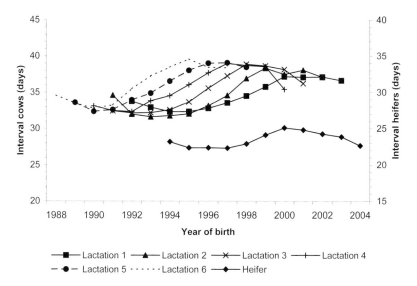

Figure 2. Phenotypic trend of interval from first to successful insemination in German Holsteins.

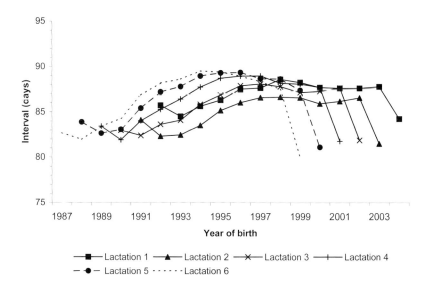

Figure 3. Phenotypic trend of interval from calving to first insemination in German Holstein cows.

the time of data extraction than females with longer FS. The trait CF had unfavourable phenotypic trends in 1990's, but the trends have become much flatter in recent years. Because of incomplete insemination records during the last year for respective lactations, CF average dropped dramatically in the last year, therefore the drop in the CF phenotypic trend should be interpreted as arising from data selection. In all three figures, there was a different pattern in phenotypic trends of first lactation of cow fertility traits than all later lactations. Overall, a clear deterioration of all the fertility traits in German Holstein population was noted, though the unfavourable phenotypic trends have become less severe for CF and FSc in recent years.

Figures 4 and 5 have estimated genetic trends of the fertility traits in Holstein AI bulls and Holstein female animals, respectively. The bulls were required to have at least 50 daughters in CF or the heifer trait NRh. All the female animals had heifer record as well as at least one cow insemination record. There were approximately 1000 bulls or 800,000 female animals in each of the years plotted

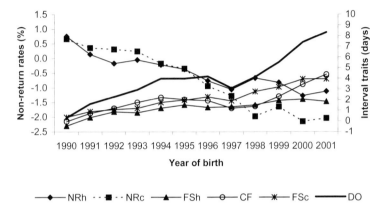

Figure 4. Genetic trends of heifer and cow fertility traits in German Holstein bulls with at least 50 daughters. [NRh/NRc is non-return rate of heifer/cow, FSh/FSc is interval first to successful insemination of heifer/cow, CF is interval calving to first insemination, and DO is days open].

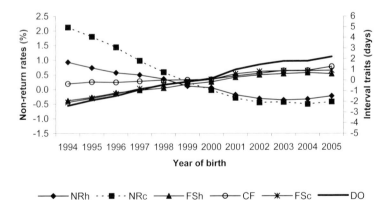

Figure 5. Genetic trends of heifer and cow fertility traits in German Holstein female animals with heifer and at least one cow insemination record. [NRh/NRc is non-return rate of heifer/cow, FSh/FSc is interval first to successful insemination of heifer/cow, CF is interval calving to first insemination, and DO is days open].

Breeding for robustness in cattle

in either figure. It can be clearly seen that unfavourable genetic trends existed in Holstein bulls and female animals for all the fertility traits. As a result of the larger number of animals in each year, the genetic trends in the Holstein females were much smoother than those of the Holstein bulls. The less smooth genetic trends in Holstein bulls can be explained by the fact that even a single popular bull sire with extreme fertility EBV could make the average EBV of bulls born in a particular year deviate much from the normal trends. Compared to the female animals, the Holstein AI bulls had much stronger, unfavourable trends in the fertility traits. For the same trait FS or NR, the cow fertility trait had stronger trend than the heifer trait, mainly caused by very little culling imposed on heifers. Between birth year 1990 and 2001 for the German Holstein bulls or between 1994 and 2005 for German Holstein female animals, there was an increase of about 8.0 d or 4.6 d of DO, respectively, being equivalent to 81% or 47% of genetic standard deviations. In the same time period, cow NRc EBV dropped 2.7% (45% genetic standard deviations) in the Holstein bulls and 2.5% in the Holstein females. The interval from first to successful insemination FSc increased by 3.6 d (74% genetic standard deviations) in the Holstein bulls and 3.0 d in the Holstein females in the 12-yr period. For the fertility trait CF, a similar and unfavourable genetic trend was found to be 4.4 d (64% genetic standard deviations) in the bulls and 1.6 d in the female animals during the same period. In summary, notably strong, unfavourable genetic trends were observed in both bulls and female animals for all the fertility traits. This fact may be explained by significant genetic progress achieved in production traits in the last decades.

Genetic correlation estimates of the fertility with other traits

The genetic correlation estimation was run on a 64-bit AMD Opteron Linux server. Our own custom-made REML program took about 700 Mb RAM and 32 CPU minutes per iteration round. A total of 349 rounds were needed to reach the pre-defined convergence criterion. The (co)variance estimates on RRC basis of production traits and SCS were converted to 305-day single lactations as well as combined lactation (Liu *et al.*, 2004). Genetic correlations of days open DO were derived from its two component traits CF and FS. Table 5 shows REML estimates of genetic correlations of the fertility traits with other traits: production, type traits, SCS, longevity, BCS and locomotion. Production traits were moderately correlated with interval fertility traits FS, CF and DO, whereas their genetic correlations with NR rates were close to 0 as shown in Table 5. Although SCS were almost uncorrelated with the fertility traits NR56 and the two heifer fertility traits, genetic correlation of about 0.20 was found between SCS and interval fertility traits CF or DO. Lower genetic correlation

Table 5. Genetic correlation estimates of the fertility with other traits.

	FSh	NRh	CF	NRc	FSc	DO
Milk yield	0.20	0.02	0.35	-0.04	0.37	0.45
Fat yield	0.16	-0.03	0.24	-0.08	0.30	0.33
Protein yield	0.19	-0.02	0.30	-0.08	0.36	0.41
Somatic cell scores (SCS)	0.06	-0.08	0.20	-0.01	0.11	0.19
Longevity (as relative risk)	0.21	-0.07	0.54	-0.01	0.40	0.58
Body depth	0.11	-0.13	0.09	-0.12	0.19	0.17
Rump angle	-0.03	-0.02	0.03	0.07	-0.05	-0.01
Udder depth	-0.09	0.03	-0.22	0.00	-0.16	-0.23
Overall udder score	-0.00	0.02	-0.06	-0.03	-0.00	-0.04
Overall feet & legs score	-0.01	0.02	-0.15	-0.08	-0.02	-0.10
Body condition score (BCS)	-0.09	-0.07	-0.25	-0.05	-0.18	-0.27
Locomotion	-0.03	0.09	-0.17	-0.05	-0.06	-0.14

was estimated for SCS with the cow interval fertility trait FS. Low genetic correlation estimates were obtained between the fertility traits and type traits, except that udder depth had a genetic correlation of about -0.22 with the fertility traits CF or DO. Among all the fertility traits, DO had the highest genetic correlation with BCS, -0.27. The three interval fertility traits had genetic correlations ranging from 0.40 to 0.58 with the relative risk. Based on the genetic correlation estimates with the relative culling risk, the interval fertility traits can be regarded as a good predictor of functional longevity.

A fertility selection index was composed of the five fertility traits, with double weights on the cow fertility traits compared to the two heifer traits. Within heifer or cows the two fertility traits of heifer or three fertility traits of cows, respectively, were given equal weights. A conception index was set up in the same way as the fertility index, except the cow fertility trait CF was excluded.

Among animal scientists and breeders there has been a debate about the optimal definition of the cow fertility trait FSc, interval from first to successful insemination. Some prefer the interval from first to last insemination, because using the interval first to successful insemination, FSc, excludes cows that never conceive or are culled. The reason for choosing the definition FSc is that using the interval from first to last insemination can lead to instable EBV over consecutive genetic evaluations for cows that have lactations in progress, because, for such cows, phenotypic value of the interval from first to last insemination would change between consecutive genetic evaluations. Although the incomplete insemination information is treated as a missing value until it is verified with the following calving, the multiple trait model can project EBV of this fertility trait reasonably well by optimally considering phenotypic information of this trait from previous lactations and the two correlated fertility traits CF and NRc from the same lactation of the same cow and her parental average of this fertility trait. However, our fertility model needs to be improved to consider insemination information of culled cows. In some countries, such as the Netherlands (De Jong, 2005), production traits are included in national fertility evaluations as correlated traits, the inclusion of the correlated production traits allows consideration of the selection in production traits on fertility EBV. Our fertility model does not take the correlated production traits into account, because the current international genetic evaluation is based on pure fertility information and cannot remove the influence of production traits on fertility proofs. Additionally, we will perform a two-step multiple trait genetic evaluation using yield deviations and EDC by including all relevant traits for setting up total merit index, e.g. production and fertility traits. The resulting fertility proofs from the multiple trait models would contain correlated information from production and other traits.

Conclusions

A genetic evaluation system based on a multi-trait animal model was developed for six fertility traits of dairy cattle, non-return rates to 56 d of heifer and cow, interval from calving to first insemination, and interval from first to successful insemination of heifer and cow as well as days open of cow. Genetic (co)variances of the fertility traits were estimated using the REML method. Heritability estimates were small for all the fertility traits, ranging from 1% to 4%. A very large mixed model equation system was efficiently solved using Gauss-Seidel algorithm in cooperation with iteration on data technique. Reliability values of fertility EBV were approximated using the multi-trait EDC approach. Average yield-deviations of female animals and DYD of bulls were calculated following a multi-trait method. About 26 million insemination records of female animals of Holstein, Red Dairy Cattle and Jersey from Germany, Austria and Luxembourg were jointly analysed using the newly developed fertility genetic evaluation software. Unfavourable genetic trends were found for all the fertility traits, with weaker trends for heifer traits than for cow traits. Considering the very low heritability values of the fertility traits, high correlations were obtained between DYD and EBV for bulls with a reasonable number of daughters, indicating a high consistence between DYD and EBV. Genetic correlations among a total of 49 traits were estimated by applying an approximate REML method to DYD of 5,709 bulls. This estimation procedure was proven to be efficient and led to reliable parameter estimates. The interval fertility traits can be used as a predictor for functional longevity as a result of moderate genetic correlations to the relative risk. It can be concluded that including fertility traits in selection index can halt the deterioration of fertility caused by correlated selection response of milk production and improve the longevity of dairy cattle.

References

Berry, D.P., S. Coughlan and R.D. Evans, 2007. Preliminary genetic evaluation of female fertility in Ireland. Interbull Bulletin, 37: 125-128.

Ducrocq, V., 2001. A two-step procedure to get animal model solutions in Weibull survival models used for genetic evaluations on length of productive life. Interbull Bulletin, 27: 147.

Ducrocq, V., D. Boichard, A. Barbat and H. Larroque, 2001. Implementation of an approximate multitrait BLUP to combine production traits and functional traits into a total merit index. 52nd Annual Meeting EAAP, Budapest, Hungary, 2001.

De Jong, G., 2005. Usage of predictors for fertility in the genetic evaluation, application in The Netherlands. Interbull Bulletin, 33: 69-73.

Fogh, A., A. Roth, O. Maagaard Pedersen, J.-Å. Eriksson, J. Juga, M. Toivonen, I.M.A. Ranberg, T. Steine, U. Sander Nielsen and G. Pedersen Aamand, 2003. A joint Nordic model for fertility traits. Interbull Bulletin 31: 52-55.

Gredler, B., C. Fuerst and J. Sölkner, 2007. Analysis of new fertility traits for the joint genetic evaluation in Austria and Germany. Interbull Bulletin, 37: 152-155.

Jamrozik, J., J. Fatehi, G.J. Kistemaker and L.R. Schaeffer, 2005. Estimates of genetic parameters for Canadian Holstein female reproduction traits. J. Dairy Sci., 88: 2199-2208.

Jorjani, H., 2006. International genetic evaluation for female fertility traits. Interbull Bulletin, 35: 42-46.

Kovač, M., E. Groeneveld, E. Carcía-Cortés and L.A. Carcía-Cortés, 2002. VCE-5, A package for the estimation of dispersion parameters. 7th World Congress on Genetics Applied to Livestock Production, August 19-23, 2002, Montpellier, France, Communication N° 28-06.

Lassen, J., M.K. Sorensen, P. Madsen and V. Ducrocq, 2007. An approximate multitrait model for genetic evaluation in dairy cattle with a robust estimation of genetic trends. Genet. Sel. Evol., 39: 353.

Liu, Z., F. Reinhardt, A. Bünger and R. Reents, 2004. Derivation and calculation of approximate reliabilities and daughter yield-deviations of a random regression test-day model for genetic evaluation of dairy cattle. J. Dairy Sci., 87: 1896-1907.

Tarrés, J., J. Piedrafita and V. Ducrocq, 2006. Validation of an approximate approach to compute genetic correlations between longevity and linear traits. Genet. Sel. Evol., 38: 65-85.

Tarrés, J., Z. Liu, V. Ducrocq, F. Reinhardt and R. Reents, 2007. Validation of an approximate REML algorithm for parameter estimation in a multitrait, multiple across-country evaluation model: a simulation study. J. Dairy Sci., 90: 4846-4855

Van Doormaal, B.J., G.J. Kistemaker and F. Miglior, 2007. Implementation of reproductive performance genetic evaluations in Canada. Interbull Bulletin, 37: 129-133.

VanRaden, P.M., A.H. Sanders, M.E. Tooker, R.H. Miller, H.D. Norman, M.T. Kuhn and G.W. Wiggans, 2004. Development of a national genetic evaluation for cow fertility. J. Dairy Sci., 87: 2285-2292.

Wall, E., I.M.S. White, M.P. Coffey and S. Brotherstone, 2005. The relationship between fertility, and selected type information in Holstein-Friesian cows. J. Dairy Sci., 88: 1521-1528.

Relationship between milk production traits and fertility in Austrian Simmental cattle

B. Gredler[1], C. Fuerst[2], B. Fuerst-Waltl[1] and J. Sölkner[1]
[1]University of Natural Resources and Applied Life Sciences Vienna, Department of Sustainable Agricultural Systems, Gregor Mendel Str. 33, 1180 Vienna, Austria
[2]ZuchtData EDV-Dienstleistungen GmbH, Dresdner Str. 89/19, 1200 Vienna, Austria

Abstract

The effects of milk urea nitrogen (MUN), fat-protein-ratio (F:P), milk lactose percentage (MLP) and test-day milk yield (Mkg) on fertility traits days to first service (DFS) and days open (DO) were analysed. In total, records of 12,828 dual purpose Simmental cows were examined. The test day record closest to the date of first insemination was used. For genetic parameter estimation trivariate analyses were run for milk production traits and fertility. Effects accounted for production traits were the fixed effects of herd, year and month interaction of test-day milk recording, AM/PM milking, a continuous effect of days in milk (linear and quadratic) and a random genetic animal effect. For days to first service and days open the fixed effects herd, year and season interaction of calving, calving age and lactation interaction and a random additive genetic animal effect were used. Heritabilities for MUN, F:P, MLP, Mkg, DFS and DO were 0.22±0.017, 0.10±0.014, 0.39±0.018, 0.19±0.017, 0.022±0.006 and 0.023±0.005, respectively. Genetic correlations were 0.65±0.13 and 0.75±0.09 between Mkg and DFS and DO, respectively. A slight negative genetic relationship was found between MUN and MLP and fertility, whereas small positive genetic correlations of 0.26±0.11 and 0.10±0.07 were estimated between MLP and DFS and DO.

Keywords: fertility, energy balance, milk production

Introduction

Poor reproductive performance is the main reason for involuntary culling in many countries all over the world. In Austria 23.2% of all culled cows were disposed due to reproductive disorders in 2007 (ZuchtData, 2007). There are many different reasons for impaired fertility as this represents a trait of a very complex nature. In early lactation dry matter intake may not meet the requirements of milk production resulting in negative energy balance as energy input is very low, while energy output is high (Wathes *et al.*, 2007). Monitoring energy balance may be conducted by analyses of blood metabolites (e.g. use of betahydroxybutyrate and non-esterified fatty acids for monitoring and testing ketosis and fatty liver, respectively), dietary evaluation, body condition scoring, and analyses of milk recording data (Mulligan *et al.*, 2006). Milk production traits such as milk fat, milk protein, milk lactose, and milk urea nitrogen are usually assessed during routine milk recording. The ratio between milk fat and milk protein (F:P) is suggested to be a useful indicator of negative energy balance, ketosis, and ovarian cysts (Heuer *et al.*, 1999; Buckley *et al.*, 2003) and therefore can be related to decreased fertility. Negative associations were found between milk urea nitrogen (MUN) and reproductive performance (e.g. Butler *et al.*, 1995; Guo *et al.*, 2004; Hojman *et al.*, 2004). Milk lactose content is also discussed in relation to fertility. Reksen *et al.* (2002) found that higher milk lactose content during the first weeks postpartum was positively related to an early resumption of luteal function after calving. Buckley *et al.* (2003) reported a positive relationship between milk lactose content and the pregnancy rates indicating that higher milk lactose content was associated with higher pregnancy rates. For these milk constituents considerable genetic variation was found in several studies (e.g. Wood *et al.*, 2003; Mitchell *et al.*, 2005; Miglior *et al.*, 2007). Friggens *et al.* (2007) found evidence that differences in energy balance among breeds and parities are based on different genetic backgrounds. Genetic correlations between energy balance measured in early and late lactation were close to zero indicating that energy balance in early and late lactations should be considered as genetically independent traits.

For improvement of the joint genetic evaluation of fertility in Austria and Germany (Fuerst and Egger-Danner, 2002) the project 'Development of genetic evaluations for fertility traits in cattle' was implemented. Many studies approve the relationship between MUN, F:P, MLP and fertility. Therefore, the objective of this study was to identify possible predictors of fertility for use in the genetic evaluation for fertility.

Material and methods

In total, records of 12,828 dual purpose Simmental cows calved between 2000 and 2006 in 1,505 herds in Lower Austria were used to investigate the relationship between milk production traits and fertility. Milk records, collected between 2000 and 2006, within the interval of ±30 days to the date of first insemination were used. Data were restricted to the milk record being closest to the date of first insemination. The first seven lactations were included. DFS and DO were calculated as number of days between calving and first insemination and number of days between calving and last insemination in a given lactation, respectively. MUN and MLP were routinely assessed during milk recording and F:P was computed from milk fat and milk protein percentages of each milk record. Data were excluded from the analysis if DFS and DO were outside the range of 20 to 200 and 20 to 365 days, respectively. MUN was restricted to 1 to 70 mg/dl, F:P to 0.5 to 2.5 and MLP 3 to 6%.

Heritabilities and genetic correlations of all traits were estimated by REML using VCE-5 (Kovač and Groeneveld, 2003). Pearson correlation coefficients for phenotypic values were calculated by means of the procedure CORR (SAS Institute, 2003). For genetic parameter estimation trivariate analyses were run based on an animal model. In each run Mkg, one of the auxiliary traits MUN, F:P or MLP and one fertility trait were included. The pedigree file consisted of 55,514 animals. The following statistical models were applied:

DFS and DO

$$y_{ijk} = \mu + HYSC_i + AGELACT_j + a_k + e_{ijk} \tag{1}$$

where:
y_{ijk} = the individual observation;
μ = the overall mean;
$HYSC_i$ = fixed effect of ith herd-year-season interaction of calving (i = 3,447);
$AGELACT_j$ = fixed effect of jth age at calving (in months)-lactation interaction (j = 33);
a_k = random additive genetic effect of animal;
e_{ijk} = random residual error term.

Mkg, MUN, F:P and MLP

$$y_{ijkl} = \mu + HYMT_i + LACT_j + AM/PM_k + b_1(DIM) + b_2(DIM)^2 + a_l + e_{ijkl} \tag{2}$$

where:
y_{ijkl} = the individual observation;
μ = the overall mean;
$HYMT_i$ = fixed effect of ith herd-year-month interaction of test-day of milk recording (i = 1,836);
$LACT_j$ = fixed effect of jth lactation (j = 7);

AM/PM$_k$ = fixed effect of kth AM/PM milking (k = 2, 1 = milking from 4 o'clock to 9 o'clock am, 2 = milking from 3 o'clock to 8 o'clock pm);
b$_1$ and b$_2$ = regression coefficients;
DIM = continuous effect of days in milk after calving (linear and quadratic);
a$_l$ = random additive genetic effect of animal;
e$_{ijkl}$ = random residual error term.

Genetic relationships among cows were included which gave the following variance-covariance structure for random effects:

$$Var\begin{bmatrix} a \\ e \end{bmatrix} = \begin{bmatrix} A\sigma_a^2 & 0 \\ 0 & I\sigma_e^2 \end{bmatrix} \tag{3}$$

Where:
A is the genetic relationship matrix among cows,
I is the identity matrix,
σ_a^2 is the additive genetic variance, and
σ_e^2 is the residual variance.

Heritabilities were calculated as follows:

$$h^2 = \frac{\sigma_a^2}{(\sigma_a^2 + \sigma_e^2)} \tag{4}$$

Results and discussion

In Table 1 arithmetic means and standard deviations are presented for all traits. Means of DFS and DO were 64.7 and 100.6, respectively. Average Mkg, MUN, F:P and MLP were 27.3 kg, 20.5 mg/dl, 1.24 and 4.88%, respectively. The distributions of Mkg, MUN, F:P and MLP are shown in Figure 1.

Estimated heritabilities with their standard errors for all traits are given in Table 2. As expected, rather low heritabilities of 0.022 and 0.023 were found for DFS and DO, respectively. The results for DFS agree with those found by Berry *et al.* (2003), whereas slightly higher heritabilities of 0.04 to 0.12 were estimated by Pryce *et al.* (2001). Gredler *et al.* (2007) also reported higher heritabilities of 0.06 for DFS and 0.04 for DO for the same population, but using a larger data set. For Mkg from a single test-day record a heritability of 0.187 was estimated. For MUN the estimated heritability was 0.216. Similar heritability results were found by Mitchell *et al.* (2005). Wood *et al.* (2003) calculated

Table 1. Arithmetic means (Mean) with standard deviations (SD) and Minimum (MIN) and Maximum (MAX) for all traits[1].

Trait	N	Mean	SD	Min	Max
DFS	12,828	64.7	22.6	20	199
DO	12,828	100.6	58.1	20	365
Mkg	12,828	27.3	7.1	4.8	69
MUN	12,828	20.5	9.4	1	68
F:P	12,828	1.24	0.22	0.69	2.43
MLP	12,828	4.88	0.17	3.5	5.5

[1] DFS = days to first service, DO = days open, Mkg = milk yield in kg, MUN = milk urea nitrogen in mg/dl, F:P = fat:protein ratio, MLP = milk lactose percentage.

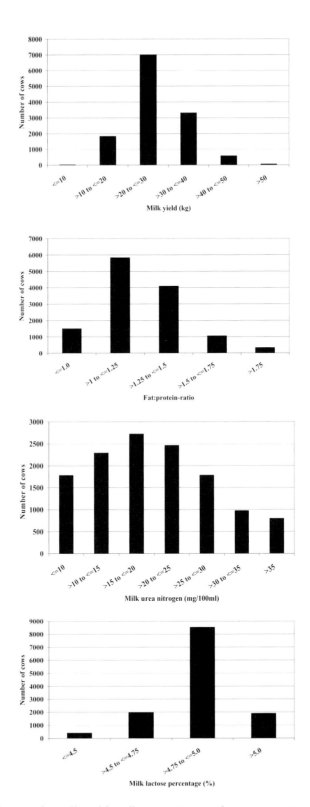

Figure 1. Distributions for milk yield, milk urea nitrogen, fat:protein-ratio and milk lactose percentage.

Table 2. Heritabilities (diagonal), genetic correlations (above diagonal) with their standard errors and phenotypic correlations (below diagonal) for all traits[1].

Trait	DFS	DO	Mkg	MUN	F:P	MLP
DFS	0.022	1.00	0.65	-0.21	0.26	-0.12
	±0.006	n.e.	±0.13	±0.10	±0.11	±0.07
DO	0.34	0.023	0.75	-0.14	0.10	-0.20
	***	±0.005	±0.09	±0.13	±0.07	±0.12
Mkg	-0.14	0.01	0.187	0.05	0.33	-0.26
	***	ns	±0.017	±0.06	±0.08	±0.05
MUN	0.04	0.00	0.11	0.216	0.06	0.12
	***	ns	***	±0.017	±0.05	±0.06
F:P	-0.07	0.00	0.00	0.11	0.095	0.00
	***	ns	ns	***	±0.014	±0.04
MLP	-0.12	-0.02	-0.04	0.03	-0.06	0.394
	***	ns	***	**	***	±0.018

[1] DFS = days to first service, DO = days open, Mkg = milk yield in kg, MUN = milk urea nitrogen in mg/dl, F:P = fat:protein ratio, MLP = milk lactose percentage.

higher heritabilities for MUN of 0.44, 0.59, and 0.48 in first, second and third lactation, respectively. Compared to Mkg and MUN a lower heritability of 0.095 was estimated for F:P. In a previous study Gredler *et al.* (2006) reported a slightly higher heritability of 0.14 for F:P. Comparable heritability estimates for Finnish Ayrshire cows for F:P in the range of 0.08 to 0.17 were calculated by Negussie *et al.* (2008) using a random regression model. Highest heritability of 0.394 was observed for MLP. Even higher heritabilities of 0.478, 0.506, and 0.508 were estimated by Miglior *et al.* (2007) in first, second, and third parity in Canadian Holstein cattle. A similar heritability for lactose percentage of 0.53 was reported by Welper and Freeman (1992).

Pearson correlation coefficients with significance level and genetic correlations between all traits are presented in Table 2. Between DFS and DO a correlation of 1, but without standard error, was estimated. Unfavourable genetic correlations of 0.65 and 0.75 were found between Mkg and DFS and DO, respectively. This confirms the antagonistic relationship between milk production and fertility as found in many other studies (e.g. Kadarmideen *et al.*, 2000; Roxström *et al.*, 2001). Low negative genetic correlations of -0.21 and -0.14 were obtained indicating that higher MUN concentrations are related to decreased DFS and DO. Mitchell *et al.* (2005) estimated a genetic correlation of -0.14 between MUN and DFS in first lactation Holstein cows, though they found positive correlations between MUN and DO in first and second lactation. F:P and the fertility traits DFS and DO were positively correlated (0.26 and 0.10) suggesting that higher F:P will increase DFS and DO. These results are in good agreement with results shown by Negussie *et al.* (2008). They found positive genetic correlations until 60 days in milk in the range of 0.14 to 0.28. Between MLP and fertility a favourable genetic relationship was observed. Genetic correlations were -0.12 between DFS and MLP and -0.20 between DO and MLP. All phenotypic correlations were zero or close to zero except the correlation of 0.34 between DFS and DO. Due to the low number of animals included in the analyses standard errors of estimated genetic correlations among all traits were high and some of the correlations were not significantly different form zero. Therefore results ought to be carefully interpreted.

Conclusions

Estimated heritabilities for fertility in dual purpose Simmental cattle were as expected rather low with values of 0.022 and 0.023 for DFS and DO, respectively. For Mkg, MUN, F:P and MLP heritabilities of 0.187, 0.216, 0.095 and 0.394 were estimated, respectively. Results observed in this study confirm that substantial genetic variation exists for the investigated auxiliary traits. Genetic relationships between fertility and traits from the routine milk recording indicate that MUN, F:P and MLP may be used as predictors in genetic evaluations for fertility. As heritabilities of fertility traits are very low, predictors will provide additional information on fertility and may increase accuracies of estimated breeding values. However, due to the low number of cows included in this study, standard errors for all genetic correlations are high. In order to obtain more reliable results a higher number of cows should be considered in the analysis.

Acknowledgements

This work was part of the project 'Development of genetic evaluations for fertility in cattle' (project no. 1426) which was financially supported by the Austrian Federal Ministry of Agriculture, Forestry, Environment and Water Management and the Federation of Austrian Cattle Breeders (ZAR).

References

Berry, D.P., F. Buckley, P. Dillon, R.D. Evans, M. Rath and R.F. Veerkamp, 2003. Genetic relationship among body condition score, body weight, milk yield, and fertility in dairy cows. J. Dairy Sci., 86: 2193-2204.

Buckley, F., K. O'Sullivan, J.F. Mee, R.D. Evans and P. Dillon, 2003. Relationships among milk yield, body condition, cow weight and reproduction in Spring-Calved Holstein-Friesians. J. Dairy Sci., 86: 2308-2319.

Butler, W.R., J.J.Calaman and S.W. Beam, 1995. Plasma and milk urea nitrogen in relation to pregnancy rate in lactating dairy cattle. J. Anim. Sci., 74: 858-865.

Friggens, N.C., P. Berg, P. Theilgaard, I.R. Korsgaard, K. L. Ingvartsen, P. Løvendahl and J. Jensen, 2007. Breed and parity effects on energy balance profiles through lactation: evidence of genetically driven body energy change. J. Dairy Sci., 90: 5291-5305.

Fuerst, C. and C. Egger-Danner, 2002. Joint genetic evaluation for fertility in Austria and Germany. Interbull Bulletin, 29: 73-76.

Gredler, B., C. Fuerst and J. Sölkner, 2006. Genetic relationship between body condition score, days to first service and production traits in Austrian Simmental cattle. 8th WCGALP Organizing Committee (ed.), CD-ROM: 11-05, 8th World Congress on Genetics Applied to Livestock Production, 13.-18.8.2006, Belo Horizonte, Brazil.

Gredler, B., C. Fuerst and J. Sölkner, 2007. Analysis of new fertility traits for the joint genetic evaluation in Austria and Germany. Interbull Bulletin, 37: 152-155.

Guo, K., E. Russek-Cohen, M.A. Varner and R.A. Kohn, 2004. Effects of milk urea nitrogen and other factors on probability of conception of dairy cows. J. Dairy Sci., 87: 1878-1885.

Heuer, C., Y.H. Schukken and P. Dobbelaar, 1999. Postpartum body condition score and results from the first test day milk as predictors of disease, fertility, yield, and culling in commercial herds. J. Dairy Sci., 82: 295-304.

Hojman, D., O. Kroll, G. Adin, M. Gips, B. Hanochi and E. Ezra, 2004. Relationships between milk urea and production, nutrition, and fertility traits in Israeli dairy herds. J. Dairy Sci., 87: 1001-1011.

Kadarmideen, H.N., R. Thompson and G. Simm, 2000. Linear and threshold model genetic parameters for disease, fertility and milk production in dairy cattle. Anim. Sci., 71: 411-419.

Kovač, M. and E. Groeneveld, 2003. VCE-5, User's guide and reference manual, Version 5.1.

Miglior, F., A. Sewalem, J. Jamrozik, J. Bohmanova, D.M. Lefebvre and R.K. Moore, 2007. Genetic analysis of milk urea nitrogen and lactose and their relationships with other production traits in Canadian Holstein cattle. J. Dairy Sci., 90: 2468-2479.

Mitchell, R.G., G.W. Rogers, C.D. Dechow, J.E. Vallimont, J.B. Cooper, U. Sander-Nielsen and J.S. Clay, 2005. Milk urea nitrogen concentration: Heritability and genetic correlations with reproductive performance and disease. J. Dairy Sci. 88: 4434-4440.

Mulligan, F.J., L. O'Grady, D.A. Rice and M.L. Doherty, 2006. A herd health approach to dairy cow nutrition and production diseases of the transition cow. Anim. Reprod. Sci., 96: 331-353.

Negussie, E., I. Strandén and E.A. Mäntysaari, 2008. Genetic association between test-day milk fat to protein ratio and fertility traits in dairy cows: A random regression model analysis. In: Proc. of Maataloustieteen Päivät 2008 10.-11.1.2008 Viikki, Helsinki. Esitelmä- ja posteritiivistelmät. Suomen maataloustieteellisen seuran tiedote, 24: 125.

Pryce, J.E., M.P. Coffey and G. Simm, 2001. The relationship between body condition score and reproductive performance. J. Dairy Sci., 84: 1508-1515.

Reksen, O., Ø. Havrevoll, Y.T. Gröhn, T. Bolstad, A. Waldmann and E. Ropstad, 2002. Relationship among body condition score, milk constituents, and postpartum luteal function in Norwegian dairy cows. J. Dairy Sci., 85: 1406-1415.

Roxström, A., E. Strandberg, B. Berglund, U. Emanuelson and J. Philipsson, 2001. Genetic and environmental correlations among female fertility traits and milk production in different parities of Swedish Red and White dairy cattle. Acta Agric. Scand. Sect. A, Animal Sci., 51: 7-14.

SAS Institute Inc., 2003. SAS/STAT® User's Guide, Version 9. Cary NC.

Wathes, D.C., N. Bourne, Z. Cheng, G.E. Mann, V.J. Taylor and M.P. Coffey, 2007. Multiple correlation analyses of metabolic and endocrine profiles with fertility in primiparous and mulitparous cows. J. Dairy Sci., 90: 1310-1325.

Welper, R.D. and A.E. Freeman, 1992. Genetic parameters for yield traits of Holsteins, including lactose and somatic cell score. J. Dairy Sci., 75: 1342-1348.

Wood, G.M., P.J. Boettcher, J. Jamrozik, G.B. Jansen and D.F. Kelton, 2003. Estimation of genetic parameters for concentrations of milk urea nitrogen. J. Dairy Sci., 86: 2462-2469.

ZuchtData, 2007. Jahresbericht 2007 – Zuchtprogramm und Leistungsprüfung. Polykopie ZuchtData, Wien.

Relationships between disorders of the bovine hoof and fertility in dairy cattle herds in Northern Germany

H.H. Swalve[1], H. Alkhoder[1] and R. Pijl[2]
[1]Institute of Agricultural and Nutritional Sciences, Group Animal Breeding, Martin-Luther-University Halle-Wittenberg, Adam-Kuckhoff-Str. 35, 06108 Halle, Germany
[2]Fischershäuser 1, 26441 Jever, Germany

Abstract

Claw disorders can be diagnosed at the time of hoof trimming. Findings were collected at hoof trimming using a personal digital assistant (PDA) with an interface to a data base on a PC and an interface to herd data stemming from the central milk recording computer. A total of around 50,000 records from 17,000 cows were collected over a period of seven years. Data comprised the pathological findings (sub-clinical and clinical), herd environment information, milk yields, pedigree information as well as records on all inseminations from which fertility parameters were derived. The most prominent disease of the claw was laminitis, found in around 33% of all cows. Other diseases were of lesser importance but summarising all findings for all 16 diseases recorded revealed that only around 39% of all records showed no claw disease at all. Relationships with fertility parameters were not as strong as anticipated. The number of inseminations per pregnancy was unaffected by claw diseases while days open showed tendencies for an increase for most diseases and significant differences (+9 days) for dermatitis interdigitalis. It may be concluded that only severe cases of claw diseases show a pronounced effect on fertility. The present study also is an example for general problems which arise when using field data on fertility, as strategies for data edits become more important than statistical modelling.

Keywords: hoof trimming, fertility, insemination records, claw disorders

Introduction

Disorders and diseases of the bovine hoof are a frequent problem in dairy herds and are a major cause for involuntary culling. Claw disorders can be diagnosed at the time of hoof trimming (e.g. Manske *et al.,* 2002; Pijl, 2004; Koenig *et al.,* 2005; Van der Waaij *et al.,* 2005, Emanuelson and Fall, 2007) and data stemming from a detailed collection of theses diagnoses can be subject to various kinds of analysis with respect to environmental and genetic factors.

The professional hoof trimmer René Pijl has been recording claw disorders at the time of hoof trimming in clients' herds electronically since 2000. The system used comprises of a personal digital assistant (PDA – pocket computer) equipped with custom-made software which is loaded from a data-base on a server at the office with all herd data necessary before the actual recording. This data contains information of all individual cows including their herd and stable identification. The server itself regularly connects to the central server of the milk recording agency to acquire the latest data stored (e.g. milk records, pedigree data, calving dates, etc.). On-farm, individual diagnoses are added and farm parameters such as housing system are updated when necessary.

It is important to note that hoof trimmings in the clients' herds is done on a regular basis and comprises the entire herds. Hence, the resulting data base contains all observations of all cows in the herds, being diseased or not, and each cow has contemporaries at a given herd-visit unit of observation. This unique data-base thus is an excellent basis for analyses of genetic and environmental effects as well as for relationships among different traits related to diseases of the bovine hoof. Disorders have proven to be frequent. Only around 39% of all trimmings remained without any recording of a disorder. Laminitis is the most prominent disease found with a prevalence of around 30%. Analyses so far have revealed strong environmental influences due to housing systems (type of cubicle and bedding), age and stage of lactation or milk yield at time of trimming. Heritabilities have been estimated for the six most frequent diseases and range from 6% to 12% using linear models and from 8% to 33% when applying threshold models (Swalve *et al.*, 2005; Pijl and Swalve, 2006; Pijl *et al.*, 2008).

In general, it has been widely accepted that the relationship between disorders or diseases of the bovine claw and fertility are negative. This statement appears to be quite plausible since a cow that has difficulties to walk will be reluctant to show pronounced signs of being in heat. Additionally, metabolic relationships between the causes of at least some disorders and fertility may exist. Chagas *et al.* (2007) give an extensive review on the metabolic relationships between fertility and diseases. However, the results from the literature of the relationship between fertility and claw problems are not quite clear. As Sogstad *et al.* (2006) point out, not many studies exist that differentiate between the causes of lameness when trying to investigate the relationship with fertility (e.g. Garbarino *et al.*, 2004; Melendez, 2003). However, some studies are available that specifically analyse the relationships with fertility with different diseases (Collick *et al.*, 1989; Hernandez *et al.*, 2001; Hultgren *et al.*, 2004; Sogstad *et al.*, 2006). But most studies were only able to include a limited amount of data. An example is the study by Melendez (2003) in which the fertility parameters of 65 lame cows out of a herd of 3,000 cows are analysed with the result that lame cows exhibit a lower rate of conception to first service. The results of the relatively large study by Sogstad *et al.* (2006), based on 2,583 cows in 112 herds does not find relationships with fertility for all types of claw diseases and mostly uses time derived parameters such as days to first service, days to last service, and calving interval. Hultgren *et al.* (2004) find negative relationships between sole ulcers and reproductive performance in their first study year but not in their second study year.

Aim of the present study was to evaluate the relationship between fertility parameters as derived from insemination records and the occurrence of a disorder observed at the time of hoof trimming using a large data set.

Material and methods

The data used for the present study originates solely from the collection of René Pijl while working as a professional hoof trimmer in around 100 herds located in North-Western and North-Eastern Germany. The data comprised only Holstein cows, mostly in family farms with herd sizes of 50 to 120 cows and housed in free-stall barns with cubicles and either slotted or solid flooring. The incidence rates for the 8 most important disorders found in the entire data base collected between 2002 and 2007 are described in Table 1.

The total number of observations was 49,875. Since the data originates from the routine work of a professional hoof trimmer, new clients are added as they newly enter the system and other clients may discontinue the service. Hence, not every cow can be followed throughout her life time. However, for most cows, repeated observations, mostly even twice per lactation, exist. Therefore, apart from the incidence rate of all observations, also incidence rates for cows which were trimmed by René Pijl for the first time (young cows, older cows for new clients) and for cows with a first trimming in 1st lactation are shown. Therefore, the number of observations in the second column is equal to the number of individual cows. The analysis presented here covers disorders and findings of the claws of the hind legs only. Disorders found at front legs are recorded but were found to be too rare to warrant further analysis. It should be noted that the incidence rates shown in Table 1 include sub-clinical as well as clinical cases. We therefore stick to the term 'disorders' rather than using the term 'diseases'.

Table 1. Incidence rates in three subsets of the data.

Name of disease	All observations (%)	All cows with a first examination done by R. Pijl (%)	First trimming in 1st lactation (%)
Laminitis	31.29	33.49	33.63
Dermatitis Digitalis	19.49	21.91	26.37
Dermatitis Interdigitalis	11.89	11.87	9.38
White line disease	13.78	14.38	13.85
Sole ulcer	6.63	5.50	4.73
Rotation	13.66	16.47	20.73
Interdigital growth (Tylom)	8.44	4.41	3.88
Thick hock	3.26	2.97	3.02
No. of observations	49,875	16,681	10,444

For most of disorders, their definition should be quite clear. Laminitis sometimes is also referred to as sole hemorrhage (e.g. Sogstad *et al.,* 2006), but is not to be confused with sole ulcer. A disorder rarely described in the literature is 'rotation'. We defined rotation as a twisting and dislocation of the medial claw easily seen when the leg is lifted for trimming. This again is not to be confused with the corkscrew condition. As can be seen from Table 1, laminitis is the most frequent disorder (31%), while other disorders vary between 3 and 20% of the total number of observations. In previous analyses (e.g. Pijl and Swalve, 2006), very pronounced effects of parity and stage of lactation were found. An example is laminitis which occurs mainly in the first half of the lactation and increases in frequency with age, whereas dermatitis digitalis clearly shows a decreasing trend with increasing parity. The rotation condition is mainly found in heifers in the first half of the lactation.

For the present study, data from the data base was augmented by further data from VIT, the central agricultural computing centre in charge for record keeping and processing of all Holstein cows in Germany. Individual insemination records were merged with the existing data base and fertility parameters, notably days to first service (DFS), days open (DO), and number of inseminations per pregnancy (NIPP) were calculated. All fertility parameters used were found to have a skewed distribution and furthermore also severe outliers were detected. This led to an editing strategy which left only records with values between 30 and 220 days for DFS, with values between 30 and 300 days for DO, and with values of 1 to 7 for NIPP. Lactation records showing values outside of these limits were deleted from the data set.

Selection of the final data set to be used for the subsequent analysis of the relationship between fertility and claw disorders required to work out a strategy: Claw trimmings outside 'the period of reproductive activities' are not informative, e.g. a cow with a sole ulcer at DIM (days in milk) = 300 which has been pregnant since DIM = 100 is clearly not informative. As it is well known that cows start cycling again around DIM = 30 and that most farmers, at least in the region studied, try to inseminate cows between DIM = 60 to 100, DIM <100 was defined as 'the period of reproductive activities' and only cows were retained in the data that had a trimming observation within this period. A further requirement was that DFS had to be > (trimming date DIM) – 20. This requirement ensures that the disorder, if found, clearly occurred around the time when 'reproductive activities' took place and allows for the likely condition that a disorder might have occurred a little earlier (up to 20 days) before the actual trimming date. Not many lactations (= service periods) were lost by this final requirement since this would have meant that the trimming would have taken place very early during lactation.

The final data set selected from the data base as described in Table 1 contained 6,283 service period observations from 4,827 cows in 106 herds. The number of herd-visits was 401. Herd-visits only were retained if they had more than three observations.

For 5,592 of the 6,283 total lactations, completed lactation milk yield records with >270 DIM were available. As an illustration for the level of production, average milk yield (kg) per lactation is given in Table 2. It has to be noted that the milk yields given are crude figures, and are not corrected for age in the form of mature equivalents (ME).

Statistical analysis was performed using the GLIMMIX and MIXED procedures of SAS (SAS, 2003). GLIMMIX was used for the analysis of the disorders as dependent variables which were coded as binary traits and hence a threshold model with a Logit-Link function was applied. In this model, fertility parameters, grouped into classes, were included as independent fixed factors. Other factors accounting for environmental factors described the housing system and accounted for year and season. However, it was not possible to additionally fit herd-visit effects directly so that this analysis was carried out across herds.

The MIXED procedure was used to model the continuous fertility parameters as dependent variables while including the binary disorder status along with environmental effects as independent fixed factors. For this model, herd-visits could be included to account for herd and time-dependent effects simultaneously.

Both models, applying MIXED or GLIMMIX, included a random cow effect to account for the repeated nature of the observations. Due to the strategy of data editing, i.e. keeping only observations with DIM <100, stage of lactation was not included in both models although in general this effect is of great importance when analysing claw trimming data. In detail, the following fixed effects were included for the two types of analysis:

Analysis I: Binary disorder status = dependent; GLIMMIX with logit-link function
- parity
- type of cubicle and bedding
- year-season
- grazing
- classes for DO
- classes for NIPP

Analysis II: Continuous fertility variables = dependent; MIXED
- herd-visit
- parity
- disorder status (disorders 1 to 6 as described in Table 1)

Table 2. Milk yield (kg) of the cows used in this study.

Parity	N	Mean	Std. dev.	Minimum	Maximum
1	2033	7,822.45	1,255.05	3,579	12,490
2	1538	8,949.04	1,577.16	4,451	13,736
3	1004	9,399.15	1,558.69	4,761	14,657
≥4	1017	9,433.36	1,513.45	4,860	14,767

Breeding for robustness in cattle

Results and discussion

The effect of levels (classes) of days open on the probability of a claw disorder (LSMEAN of incidence) from analysis I is given in Table 3. In the analysis of variance, significance was found for the disorders dermatitis digitalis ($P<0.05$), dermatitis interdigitalis ($P<0.01$) and rotation ($P<0.05$). This indicates that only these aforementioned disorders are likely to affect days open. However, with only very few exceptions, the lowest incidence rate throughout Table 3 is found for DO-class <80. For longer DO, especially for laminitis, white line defect, and sole ulcer, estimates do not show a clear trend at all. This may indicate that the nature of an effect of a claw disorder on fertility may be non-linear. For NIPP, results are presented in Table 4. The results are quite unexpected since for most disorders, their effect is not significant and in the case of dermatitis digitalis a high incidence of the disorder seems to be associated with a low number of inseminations per pregnancy.

Table 3. The effect of levels (classes) of days open on the probability of a claw disorder (LSMEAN of incidence) in analysis I (second lines = standard errors).

Classes DO	N (total = 6,243)	Laminitis	Dermatitis digitalis*	Dermatitis interdigitalis**	White line defect	Sole ulcer	Rotation*
≤80	1,541	0.3081	0.1169	0.0782	0.1175	0.0441	0.0799
		0.0218	0.0131	0.0111	0.0133	0.0079	0.0101
81-110	1,583	0.3047	0.1283	0.0930	0.1342	0.0414	0.0887
		0.0198	0.0125	0.0117	0.0133	0.0067	0.0098
111-140	1,058	0.3391	0.1417	0.1114	0.1334	0.0548	0.1164
		0.0216	0.0139	0.0140	0.0139	0.0085	0.0120
141-170	741	0.3082	0.1434	0.0995	0.1377	0.0557	0.0877
		0.0223	0.0152	0.0140	0.0152	0.0095	0.0112
171-200	554	0.3019	0.1604	0.0999	0.1422	0.0583	0.0780
		0.0246	0.0182	0.0157	0.01730	0.0011	0.0117
>200	766	0.3345	0.1813	0.1442	0.1359	0.0521	0.0986
		0.0237	0.0178	0.0187	0.0154	0.0095	0.0123

* ($P<0.05$); ** ($P<0.01$).

Table 4. The effect of number of inseminations per pregnancy (NIPP) on the probability of a claw disorder (LSMEAN of incidence) in analysis I (second lines = standard errors).

NIPP	N (total = 6,243)	Laminitis	Dermatitis digitalis**	Dermatitis interdigitalis	White line defect	Sole ulcer	Rotation
1	3,182	0.3266	0.1747	0.1224	0.1349	0.0606	0.0864
		0.0181	0.0128	0.0130	0.0113	0.0070	0.0081
2	1,615	0.3393	0.1498	0.1041	0.1155	0.0497	0.0849
		0.0189	0.0125	0.0118	0.0107	0.0066	0.0084
3	824	0.3045	0.1199	0.1021	0.1254	0.0418	0.1022
		0.0218	0.0134	0.0136	0.0139	0.0078	0.0118
≥4	622	0.2942	0.1342	0.0853	0.1608	0.0523	0.0909
		0.0253	0.0170	0.0144	0.0194	0.0108	0.0133

* ($P<0.05$); ** ($P<0.01$).

Table 5 displays the results from analysis II in which fertility parameters are the dependent variables in a linear model and the disorder status is fitted as a main environmental fixed effect. For the time-derived variables DO and DFS all comparisons show trends such that healthier cows exhibit shorter intervals. However, only for dermatitis interdigitalis and rotation a significant effect was found on DO, and laminitis, dermatitis digitalis, dermatitis interdigitalis and sole ulcer were significant for DFS. For the number of inseminations per pregnancy (NIPP) a significant effect is only found for rotation.

Both types of analysis agree quite well, i.e. antagonistic relationships between disorders of the bovine claw and fertility are found for time-derived parameters such as DFS and DO but not for the actual act of fertilisation once a cow was detected to be in heat and inseminated. As stated above, it should be noted that most studies that analysed the relationship between claw health and fertility used time-derived parameters to define fertility such as calving interval or days to first or last service (e.g. Sogstad *et al.*, 2006). Hence, the results in general seem to be in line with results from the literature. A delayed ovarian activity for lame cows was also observed by Garbarino *et al.* (2004) who used data of 238 cows closely monitored for ovarian activity by measuring plasma progesterone concentration.

However, in general, our results show a relatively weak relationship between claw disorders and fertility. Diseases of larger importance seem to be dermatitis digitalis, dermatitis interdigitalis and sole ulcer with around +5 days of DFS for the status 'yes'. Laminitis in this context is of a little bit less importance. A reason for these findings may be that we used data that included sub-clinical as well as clinical cases of lameness since our data comprises the entire herd at time of trimming.

On the other hand, our study appears to have a few distinct advantages over other studies found in the literature. These advantages are the size of the data set (6,283 first trimmings/lactations) and the detailed strategy of data editing. Our data is restricted to the 'period of reproductive activities' and does not include observations for which a biological relationship between disorder and fertility

Table 5. The effects of disorder status (0 = healthy, 1 = disorder) for six claw disorders on days open (DO), days to first service (DFS) and number of inseminations per pregnancy (NIPP) (second lines = standard errors; S = Significance).

	Days open (DO)			Days to first service (DFS)			No. of inseminations per pregnancy (NIPP)		
	0	1	S	0	1	S	0	1	S
Laminitis	131.14	133.53	-	100.70	103.34	**	1.76	1.73	-
	2.70	2.64		1.55	1.52		0.06	0.06	
Dermatitis digitalis	130.83	133.84	-	99.57	104.47	**	1.76	1.73	-
	2.48	2.98		1.43	1.71		0.05	0.06	
Dermatitis interdigitalis	127.75	136.92	**	99.51	104.53	**	1.72	1.77	-
	2.46	3.11		1.42	1.78		0.05	0.07	
White line defect	130.26	134.41	-	101.21	102.83	-	1.71	1.78	-
	2.41	3.07		1.39	1.77		0.05	0.07	
Sole ulcer	132.43	132.24	-	99.16	104.88	**	1.79	1.71	-
	2.08	3.73		1.20	2.14		0.04	0.08	
Rotation	129.08	135.59	**	100.90	103.14	-	1.70	1.79	*
	2.37	3.14		1.36	1.80		0.05	0.07	

* ($P<0.05$); ** ($P<0.01$).

is unlikely if not impossible. Furthermore, our data does not origin from selected, severe cases of diseased cows which then are compared to some sample of apparently 'healthy' cows. Rather, the main requirement of a performance recording procedure for animal breeding purposes is fulfilled: All cows are compared with contemporaries that have been subject to the same type of recording, i.e. in this case the recording at trimming.

Conclusions

Recording disorders of the bovine hoof at the time of hoof trimming is a method which is highly suitable for examinations of causes of disorders and their relationships with other traits. In this study, the relationships of claw disorders with fertility were analysed. The results show that even when clinical and sub-clinical cases are used without differentiation, adverse effects on claw disorders can be expected for fertility parameters which are time derived (e.g. days to first service, days open). Hardly any relationships were found between claw disorders and the number of inseminations per pregnancy. These findings could be interpreted in such a way that claw disorders delay the oestrus and the date of the successful insemination. However, a cow that does show heat will have the same chance of getting in calf, being diseased or not.

Acknowledgments

The authors are indebted to Vereinigte Informationssysteme Tierhaltung (VIT), Verden, Germany, for supplying parts of the data used in this study.

References

Chagas, L.M., J.J. Bass, D. Blache, C.R. Burke, J.K. Kay, D.R. Lindsay, M.C. Lucy, G.B. Martin, S. Meier, F.M. Rhodes, J.R. Roche, W.W. Thatcher and R. Webb, 2007. *Invited Review:* New perspectives on the roles of nutrition and metabolic priorities in the subfertility of high-producing dairy cows. J. Dairy Sci., 90: 4022-4032.

Collick, D.W., W.R. Ward and H. Dobson, 1989. Associations between types of lameness and fertility. The Veterinary Record, 125: 103-106.

Emanuelson, U. and N. Fall, 2007. Claw health in organic and conventional dairy herds. Paper presented at International Conference on Production Diseases in Farm Animals, Leipzig, Germany, July 30 - August 3, 2007; Published in Fürll, Prof. Manfred, Eds. Proceedings 13th International Conference on Production Diseases in Farm Animals, p. 611.

Garbarino, E.J., J. Hernandez, J.K. Shearer, C.A. Risco and W.W. Thatcher, 2004. Effect of lameness on ovarian activity in postpartum Holstein cows. J. Dairy Sci., 87: 4123-4131.

Hernandez, J., Shearer, J.K. and Webb, D.W., 2001. Effect of lameness on the calving-to-conception interval in dairy cows. JAVMA, 218: 1611-1614.

Hultgren, J., T. Manske and C. Bergsten, 2004. Associations of sole ulcer at claw trimming with reproductive performance, udder health, milk yield, and culling in Swedish dairy cattle. Prev. Vet. Med., 62: 233-251.

Koenig, S., A.R. Sharifi, H. Wentrot, D. Landmann, M. Eise and H. Simianer, 2005. Genetic parameters of claw and foot disorders estimated with logistic models. J. Dairy Sci., 88: 3316-3325.

Manske, T., J. Hultgren and C. Bergsten, 2002. Prevalence and interrelationships of hoof lesions and lameness in Swedish dairy cows. Prev. Vet. Med., 54: 247-263.

Melendez, P., 2003. The association between lameness, ovarian cysts and fertility in lactating dairy cows. Theriogenology, 59: 927-937.

Pijl, R., 2004. Results from claw trimming and electronic recording by one person. Proc. 5th Conf. Lameness in Ruminants, Maribor, Slovenia, pp. 14-15.

Pijl, R. and H.H. Swalve, 2006. An analysis of claw disorders diagnosed at claw trimming. Proc. 14th International Symposium & 6th Conference on Lameness in Ruminants, Colonia del Sacramento, Uruguay, Nov 8–11, pp. 34-36.

Pijl, R., H. Alkhoder and H.H. Swalve, 2008. The effect of management and genetics on claw disorders. 7th Conf. Lameness in Ruminants. Kuopio, Finland. 9-13 June, pp. 267-270.

SAS, 2003. SAS V 9.1 2002-2003 by SAS Institute Inc., Cary, NC, USA.

Sogstad, A.M., O. Osteras and T. Fjeldaas, 2006. Bovine claw and limb disorders related to reproductive performance and production diseases. J. Dairy Sci., 89: 2519-2528.

Swalve, H.H., R. Pijl, M. Bethge, F. Rosner and M. Wensch-Dorendorf, 2005. Analysis of genetic and environmental factors of claw disorders diagnosed at hoof trimming. Proc. 56th EAAP meeting, Uppsala, Sweden, Abstr. No. 263. In: Book of abstracts Bd. 56, Wageningen Academic Publishers, the Netherlands, p. 52.

Van der Waaij, E.H., M. Holzhauer, E. Ellen, C. Kamphuis and G. de Jong, 2005. Genetic parameters for claw disorders in Dutch dairy cattle and correlations with conformation traits. J. Dairy Sci., 88: 3672-3678.

Environmental and genetic effects on claw disorders in Finnish dairy cattle

A.-E. Liinamo, M. Laakso and M. Ojala
Helsinki University, Department of Animal Science, P.O. Box 28, 00014 Helsinki, Finland

Abstract

The aim was to study the environmental and genetic effects on the most common claw disorders in the Finnish dairy cattle population. The data were obtained through a nationwide claw health programme during 2003 and 2004. Altogether, 74,410 observations on 41,087 cows and heifers originating from 1,462 dairy farms were included in the data, representing 15% of the milk recorded dairy cows. Claw disorder information had been collected by claw trimmers during their routine visits to the dairy farms participating in the programme. Studied traits included sole hemorrhage, white line disease, heel erosion, screw claw, and all claw disorders combined. Environmental and genetic effects were preanalysed with LS-method using linear model, and genetic parameters were estimated with a linear approximation of the binary data using repeatability animal model and REML method. Breed, parity, lactation stage, 305 d milk production, claw trimming frequency, claw trimming season, feeding type, barn type, bedding type, manure removal method, herd size, herd, and claw trimmer all had a statistically significant effect on most of the studied claw disorders. Heritability estimates of the claw disorders varied between 0.01 and 0.07, while corresponding repeatability estimates varied between 0.05 and 0.33. Genetic and phenotypic correlations between different claw disorders were positive and low to moderate.

Keywords: dairy cattle, claw health, heritability, correlations, environmental effects

Introduction

Serious claw disorders are a frequent cause for having to remove a dairy cow from a herd prematurely, which has implications both for the economics of the herd as well as the welfare of the animals. With the increasing popularity of free-stall barns and milking robots the importance of good locomotion of dairy cows is growing, as a cow needs to be able to walk frequently to be fed and to be milked. In addition, claw and feet disorders cause economical loss for the farmers due to loss in milk production and increased veterinary and claw trimming costs and possibly extra labour. Claw disorders can cause fertility problems (e.g. Hernandez *et al.,* 2001) and may be connected with high somatic cell count in milk (Manske, 2002). A cow that has problems with its claws is also more vulnerable to teat injuries, which in turn are a serious risk factor for clinical mastitis (Elbers *et al.,* 1998).

Claw disorders are fairly frequent in dairy farms. In Sweden, Manske *et al.* (2002) observed 72% of the 4,899 dairy cows in the study to have at least one diagnosed claw disorder. However, studying the true claw disorder prevalence is difficult: even if a country would have a routine system for collecting veterinary treatment records from the dairy farms, a large proportion of the cases are treated either by the claw trimmers or the herd owners and not veterinarians. Routine collection of information via claw trimmers would enable getting a much more reliable picture on claw disorder prevalences in dairy herds. It would also enable utilising this information for studying claw disorders, and possibly also improving the claw health status of the national herd by genetic evaluation of dairy sires for claw disorders (Eriksson, 2006). However, thus far such data collection has not been implemented in most countries, including Finland.

The aim of this paper was to study a large number of environmental effects possibly associated with claw health of dairy cows diagnosed during claw trimming in Finnish dairy cattle farms, as well as to estimate the genetic parameters of the most frequent claw disorders in the Finnish dairy cattle population.

Materials and methods

Materials

The data for the study was obtained from the nationwide dairy cattle claw health programme 'Terveet Sorkat' (TS). This programme was established in 2002 by the companies Suomen Rehu and Vetman Oy, and the Union of Finnish claw trimmers. The programme was originally intended as an extra service to the dairy farms obtaining their feed from Suomen Rehu, to provide the participating farms both farm level and nationwide information on the claw health status of their dairy cows. Participation in the TS programme is voluntary for the dairy farms, and in the last years approximately 1,700 dairy farms (about 10% of all Finnish dairy farms) and 80% of the Finnish claw trimmers have been participating in the programme.

Claw health data is collected in the TS programme by specially trained claw trimmers during their normal routine visits to the participating dairy farms. The claw trimmers fill in claw health reports for all cows they have trimmed during their visit, and this data can later be combined with farm level information obtained from the client registry of Suomen Rehu, as well as information from the milk recording systems. The collected data is mainly used to provide feedback to the participating farms on the claw health status of their cows compared to other similar farms in the country, but the TS programme has also made the data available for research.

This study included data collected in the TS programme during the years 2003 and 2004. The data included 74,410 observations on 41,087 dairy cows, which represented about 10% of all dairy cows and 15% of all milk recorded dairy cows in Finland in the period. The cows originated from 1,462 different farms, and the data were collected by 43 claw trimmers.

The cows in the data represented all three dairy breeds in Finland: Ayrshire (71% of cows), Finncattle (1% of cows) and Holstein-Friesian (28% of cows). The animals had been born between 1987 and 2004, and originated from 2,381 dairy bulls (1,523 Ayrshire, 97 Finncattle and 761 Holstein-Friesian). The number of daughters per bull varied considerably in the data, and only 38% of Ayrshire bulls and 32% of Holstein-Friesian bulls had more than 10 daughters in the data.

Claw health status of each cow had been classified by a claw trimmer as yes/no at the moment of a routine claw trimming, and an animal had been considered as affected if it had a disorder diagnosed in any leg. Almost half of the studied cows had had at least some claw disorder diagnosed during the data collection period (Table 1). The most frequent diagnosis was sole hemorrhage, followed by white line disease, screw claw and heel erosion. Screw claws were defined in the reports as claws that were twisted more than 90° from the original plane. Sole ulcer, chronic laminitis and digital and interdigital dermatitis had noticeably low prevalences in the data. Consequently, only disorders with a prevalence above 5% of the observations in the original data were studied individually, including sole hemorrhage, white line disease, screw claw and heel erosion. All claw disorders were also studied jointly as one trait (in addition to the previous disorders, also digital and interdigital dermatitis, chronic laminits, sole ulcer and other unspecified claw disorders).

In addition to the claw health status, the data also included information on several environmental effects related to the cow itself and the farm where it had been producing (Table 2). The farm data included information on barn type, bedding type, manure removal system type and type of energy

Table 1. The prevalence of the diagnosed claw disorders in the studied data.

Claw disorder	% of observations (n = 74,410)	% of animals (n = 41,087)
Sole hemorrhage	28.2	30.0
White line disease	10.6	9.0
Screw claw	9.2	10.0
Heel erosion	8.1	7.8
Sole ulcer	3.5	3.7
Chronic laminitis	1.7	1.4
Digital dermatitis	0.9	0.6
Interdigital dermatitis	0.2	0.1
Other claw disorders	0.8	0.7
All claw disorders	45.2	45.8

Table 2. Classification of the studied environmental effects and numbers of observations in their subclasses in the claw health data[1].

Effect	Class	Observations
Breed	Ayrshire	52,991
	Finncattle	490
	Holstein-Friesian	20,929
Feed	compound feed	25,624
	protein supplement	30,497
	high CP prot. suppl.	10,226
	total mixed ration	5,676
	no information	2,386
Barn	tie-stall	44,322
	cold free-stall	1,332
	warm free-stall	21,221
	no information	7,535
Bedding	hard	9,547
	rubber mat	55,380
	other cover	919
	no information	8,564
305 d milk (kg)	<6,000 kg	3,503
	6,001-7,000	7,902
	7,001-8,000	12,969
	8,001-9,000	14,280
	9,001-10,000	11,388
	10,001-11,000	7,036
	>11,000 kg	5,792
	no information	11,540
Claw trimming season	spring	25,569
	summer	14,867
	autumn	21,718
	winter	12,256

Table 2. Continued.

Effect	Class	Observations
No. of claw trimmings in data	1	20,778
	2	22,606
	3	16,857
	4	11,624
	>4	2,545
Parity	heifer	323
	1	25,888
	2	19,454
	3	12,043
	4	6,458
	5	3,203
	>5	2,754
	no information	4,287
Lactation stage during claw	0-60	12,771
trimming (d after calving)	61-120	12,239
	121-180	11,212
	181-240	10,679
	241-300	10,079
	301-360	7,570
	no information	9,860
Manure removal	dry	23,977
	liquid	42,477
	no information	7,956
Herd size	5-20	14,467
	21-30	23,672
	31-40	16,340
	41-60	13,675
	61-180	6,256

[1] In addition, the effect of the claw trimmer was studied as a fixed effect including 43 subclasses and 25 to 7 077 observations per each claw trimmer.

and protein supplements fed at the farm (protein supplement and high crude protein supplement were fed together with grain and forage in different proportions, and compound feed was fed in addition to forage; total mixed ration included all feedstuffs mixed together). The claw trimmer identity, the routine claw trimming frequency at the farm and the season of claw trimming had also been recorded in the data. The pedigree, breed, age, and calving and milk production information were obtained for the cows in the claw health data from the milk recording registry kept by the ProAgria Agricultural Data Processing Centre Ltd.

Methods

The environmental effects on the claw health status of the dairy cows presented on Table 2 were studied with Least Squares-analyses and F-test using WSYS-L -programme (Vilva, 2007). All effects were treated as fixed in this model:

$$y_{ijklmnopqrstu} = \mu + breed_{i=1..3} + parity_{j=1..8} + lactation\ stage_{k=1..7} + milk\ production_{l=1..8} +$$
$$number\ of\ claw\ trimmings_{m=1..5} + claw\ trimming\ season_{n=1..4} + feed_{o=1..6} + barn_{p=1..4} +$$
$$bedding_{q=1..4} + manure\ removal_{r=1..3} + herd\ size_{s=1..5} + claw\ trimmer_{t=1..43} + \varepsilon_{ijklmnopqrstu} \qquad (1)$$

where:

$y_{ijklmnopqrstu}$ = the studied claw disorder status of a cow;
μ = the general mean;
$\varepsilon_{ijklmnopqrstu}$ = the residual effect;

and the other effects with subclasses as indicated in Table 2.

Estimates of heritability, repeatability and phenotypic and genetic correlations for the claw disorders were estimated with Restricted Maximum Likelihood method using VCE4-programme (Groeneveld, 1997) and repeated observations. The mixed animal model included breed, parity, lactation stage, claw trimming season, and claw trimming year as fixed effects, and herd, claw trimmer, permanent environment related to a cow, additive genetic effect of a cow and residual effect as random effects:

$$y_{ijklmnop} = \mu + breed_{i=1..3} + parity_{j=1..8} + lactation\ stage_{k=1..7} + claw\ trimming\ season_{l=1..4}$$
$$+ claw\ trimming\ year_{m=1..2} + herd_n + claw\ trimmer_o + a_p + pe_p + \varepsilon_{ijklmnop} \qquad (2)$$

where:

$y_{ijklmnop}$ = the studied claw disorder status of a cow;
μ = the general mean;
a_p = the additive genetic effect associated with animal p;
pe_p = the permanent environmental effect associated with animal p; and
$\varepsilon_{ijklmnop}$ = the residual effect.

When estimating genetic parameters, the data consisted of 41,087 dairy cows with own observation on claw health status, as well as their pedigree five generations backwards, including information on 94,450 additional animals. The distributions of a, pe and ε were assumed to be multivariate normal with zero means and with $Var(a) = A\sigma_a^2$, $Var(pe) = I\sigma_{pe}^2$, and $Var(\varepsilon) = I\sigma_\varepsilon^2$. Covariances among a, pe and ε were assumed to be zero.

Results and discussion

Prevalence of the claw disorders

In the data, 54.8% of the cows had been considered having fully healthy claws, i.e. 45.2% of the cows had had some claw disorder during trimming (Table 1). Although high in general, the prevalence of claw disorders in the Finnish data is still considerably lower than reported e.g. in Sweden (72% of cows diagnosed with some claw disorder; Manske et al., 2002) or the Netherlands (70% of cows diagnosed with some claw disorder; van der Waaij et al., 2005). The difference in claw disorder prevalences between these three studies is mostly explained by the very low frequency of infectious claw disorders (heel erosion, digital dermatitis and interdigital dermatitis) in the Finnish data compared to the Swedish and the Dutch studies. The joint prevalence of the infectious claw disorders in the Finnish data is below 10%, whereas Manske et al. (2002) reported that 41% of cows were diagnosed with heel erosion, 27% of cows were diagnosed with interdigital dermatitis and 2% of cows were diagnosed with digital dermatitis in their data. In the study by van der Waaij et al. (2005), the prevalence of interdigital dermatitis/heel erosion was 38% and digital dermatitis 21%. Thus, it would seem that the infectious claw disorders either are still considerably more rare in Finnish dairy herds than in other countries, or that the prevalence of these disorders is underestimated by claw trimmers for some reason.

Environmental effects on claw disorders

Breed had a statistically highly significant effect on all other studied claw disorders except for heel erosion (Table 3). Holstein-Friesians had the highest prevalence of sole hemorrhage, white line disease and all claw disorders combined, while Ayrshires had the highest prevalence of screw claw. The lowest claw disorder prevalences were observed with the Finncattle animals, but there were very few animals in the breed compared to the other two breeds. The breed effect may have been caused e.g. by the differences in the average mature weight and the claw quality between the studied breeds. Finncattle, like Jerseys, are smaller in mature size than Holstein-Friesians and Ayrshires, and their claws tend to be dark. Pigmented claws have been shown to be harder and more resistant than light claws (Vermunt, 2004). The claw health of Holstein-Friesians has also been observed to be worse than in Ayrshires or Jerseys in other studies before (e.g. Alban, 1995; Huang *et al.*, 1995).

Parity had a statistically highly significant effect on all studied claw disorders (Table 3). In general, the first parity cows had a high prevalence of most claw disorders, and the prevalences increased again with the number of parities from the third parity onwards. The exception was the screw claw, where the prevalence was highest in 1st and second parity cows and actually decreased in older animals. Screw claw is an easily observable disorder, and it is likely that the worst affected animals are culled early on from the dairy herd. As to the other claw disorders, the observed increase in disorder prevalences with the number of parities probably reflects the cumulative effect of physiological stress caused by repeated lactations and previous claw injuries and disorders. Similar pattern of increased risk for claw disorders in first and again third parity onwards was observed by Alban (1994) in Danish dairy cattle population. Koenig *et al.* (2005), Manske (2002), Somers *et al.* (2005) and Van der Waaij (2005) also reported higher risk for claw disorders when animals were getting older.

Lactation stage had a statistically highly significant effect on sole hemorrhages as well as all claw disorders combined (Table 3). The highest prevalence of claw disorders was observed from 60 to 180 days after calving, while the lowest prevalence was observed in the very end of the lactation. This effect may be due to the cows having had a higher risk to obtain subclinical laminitis during their early lactation, when their body weight and diet change rapidly and the lactation stress is beginning to manifest. Claw material grows slowly, so subclinical laminitis can be observed as sole hemorrhages only after approximately two months from the onset of the laminitis. In the study of Manske (2002), the highest risk for claw disorders was reported at 61 to 150 days after calving, and van der Waaij *et al.* (2005) observed that sole hemorrhage mainly occurred from the second until the fifth lactation month.

The cows that had a 305 d milk production over 11,000 kg had the highest prevalence of claw disorders, especially sole hemorrhage, and the lowest claw disorder prevalences were observed in cows that had a 305 d milk production under 6,000 kg (Table 3). High milk production level stresses the cow physiologically and may lower its immunological resistence to infectious diseases. There are also more easily problems with maintaining the correct energy balance in high productive cows. Fleischer *et al.* (2001) and Manske (2002) also observed that high productive cows had a higher risk to get sole hemorrhage or other claw disorders.

The cows that had had their claws trimmed five times or more during the two years of data collection (corresponding to claw trimming more often than twice per year) had the highest prevalence of white line disease, screw claw, heel erosion and all claw disorders combined (Table 3). Similar effect was also observed by Huang *et al.* (1995), who reported the worst scores for all claw disorders for cows that had had the most frequent claw trimming. However, in this case the cause and effect are not particularly clear: it is likely that the most frequently trimmed cows had been actually under treatment for claw disorders, and thus had been trimmed more often than the norm.

Table 3. Effects of breed, parity, lactation stage, milk production level, number of claw trimming sessions in the data and claw trimming season on the studied claw disorders*.

Effect Class	Sole hemorrhage	White line disease	Screw claw	Heel erosion	All claw disorders
Breed					
Ay	0	0	0	0	0
Fc	-10.2	0.2	-5.7	-0.7	-13.2
HF	6.8	4.2	-1.2	0	7.8
Parity					
heifer	0	0	0	0	0
1	12.3	4.2	4.0	0.8	16.6
2	2.3	5.4	3.4	1.7	10.5
3	2.7	8.9	2.7	2.4	13.5
4	4.5	12.3	2.5	2.3	17.0
5	6.5	16.0	1.1	2.9	21.6
>5	7.7	19.2	-0.9	3.3	25.2
no information	3.1	3.5	2.2	13.1	5.8
Lactation stage, d					
0-60	0	0	0	0	0
61-120	16.8	-0.2	0.2	0.2	14.0
121-180	15.0	0.4	1.8	0.1	13.0
181-240	4.3	0.5	1.2	-0.2	5.2
241-300	-2.5	0.9	0.8	-0.2	-0.1
301-360	-6.8	1.0	0.4	-0.5	-3.4
no information	-7.5	0.8	-0.4	-0.4	-4.1
305 d milk production, kg					
<6,000	0	0	0	0	0
6,001-7,000	-0.2	0.1	0.5	1.1	0.9
7,001-8,000	0.5	0.0	0.4	1.0	1.8
8,001-9,000	-0.4	0.1	0.4	1.3	1.3
9,001-10,000	0.5	0.3	-0.7	1.6	1.7
10,001-11,000	1.8	0.1	-0.9	1.8	3.0
>11,000	4.3	0.3	-1.2	2.1	5.1
no information	1.7	0.6	-1.3	0.1	1.8
No. claw trimmings					
1	0	0	0	0	0
2	-1.2	0.5	0.7	-0.4	-1.4
3	-0.5	0.1	-0.4	0	-1.5
4	-1.2	1.3	-0.6	-1.3	2.4
>4	-0.2	2.6	2.5	2.6	4.9
Claw trimming season					
spring	0	0	0	0	0
summer	0.8	2.8	1.0	0.2	2.8
autumn	1.3	3.8	-0.8	-1.1	1.7
winter	1.7	1.4	0.3	3.7	2.9

* Effects expressed as deviation in %-units from the first class.

The highest prevalence of sole hemorrhage and heel erosion were observed when claws were trimmed in winter, which may be due to the cows having had to stay indoors and also probably many of them having been in their peak lactation then (Table 3). On the other hand, the highest prevalence of white line disease was observed in autumn, which may be related to the mechanical stress on claws during and after the grazing season. Also Huang et al. (1995) have reported seasonal differences in claw disorder prevalence, and Somers et al. (2005) observed more heel erosion in cows during housing period.

Of the effects related directly to the buildings, the barn type had the most significant effect on the prevalence of claw disorders (Table 4). All claw disorder prevalences were observed to be much higher in free-stall barns compared to tie-stall barns, which is in accordance with Alban (1995) and Manske (2002). Manske (2002) and Somers et al. (2005) suggested that hard floors were a risk for claw disorders, and a similar effect could also be observed in this study with regard of the bedding type (Table 4). Barns with liquid manure removal system had a lower prevalence of heel erosions, screw claws and all claw disorders combined compared to dry manure system barns, which might indicate better hygienic conditions.

Table 4. Effects of barn, feed, bedding, manure removal system and herd size on the studied claw disorders[*].

Effect Class	Sole hemorrhage	White line disease	Screw claw	Heel erosion	All claw disorders
Barn					
tie-stall	0	0	0	0	0
cold free-stall	5.9	4.0	11.0	11.3	16.2
warm free-stall	6.9	11.6	8.9	16.0	24.6
no information	2.0	4.6	2.5	2.7	5.4
Feed					
compound feed	0	0	0	0	0
protein suppl.	0.3	-1.4	-0.5	0.8	-1.0
high CP prot. suppl.	-0.6	-1.8	-0.5	0.9	-1.0
total mixed ration	-0.3	-1.8	2.2	4.2	3.8
no information	-1.6	-2.2	0.7	3.8	-0.6
Bedding					
hard	0	0	0	0	0
rubber mat	-2.0	-1.0	-1.4	-2.6	-4.2
other cover	-2.8	-0.8	-5.4	-1.8	-4.7
no information	2.1	1.0	-0.6	0.3	0.0
Manure removal					
dry	0	0	0	0	0
liquid	0.0	0.5	-1.0	-1.9	-1.5
no information	-2.0	-1.8	0.8	0.3	-0.9
Herd size					
5-20	0	0	0	0	0
21-30	-0.7	-0.5	0.2	-0.8	-1.1
31-40	-0.4	-0.1	0.2	-0.1	-1.6
41-60	-2.1	0.1	3.1	1.1	-0.3
61-180	-2.9	-1.1	0.3	3.8	-0.7

[*] Effects expressed as deviation in %-units from the first class.

Breeding for robustness in cattle

The effect of feed on claw disorders was somewhat puzzling in that the herds feeding total mixed ration feed were observed to have the highest prevalence of screw claw, heel erosion and all claw disorders combined (Table 4). This result might reflect more the effect of the herd size rather than the effect of the actual feed given to cows, as the mixed ration system is only used in the largest herds in Finland. Total mixed ration feeding is also a relatively new technique for feeding dairy herds in Finland, so the farms using mixed ration feeding might not yet have fully optimised their feeding schemes at the time of the data collection.

The effect of herd size in itself was statistically significant for heel erosion prevalence, with higher prevalences in the largest herds (Table 4). Since heel erosion is an infectious disease this effect is logical, and similar effect was also observed by Wells *et al.* (1999) with respect of digital dermatitis, another infectious claw disorder. Finally, the claw trimmer who had made the initial diagnosis of the claw health of a cow had a statistically highly significant effect on all studied claw disorders, and large differences were observed in scoring between individual claw trimmers in spite of most of them having been specially trained to collect data for the TS programme (data not shown).

Genetic parameters

When treated as a random effect, the herd effect explained 4 to 22% of the total phenotypic variation observed in various claw disorders in the data, with an especially strong effect on heel erosion (Table 5). This result reflects the importance of herd management in controlling the prevalence of heel erosion and similar contagious claw disorders, as also suggested by e.g. Wells *et al.* (1999). The random claw trimmer effect explained from 5 to 10% of the total phenotypic variation observed in the studied claw disorders, which result was very similar to that obtained when treating claw trimmer as a fixed effect in Model 1.

The heritability estimates for the various claw disorders were very low, varying from 0.01 (heel erosion) to 0.06 (all claw disorders combined) (Table 5). This result is in line with e.g. van der Waaij *et al.* (2005), although in some studies also higher estimates of heritability have been obtained for individual claw disorders. E.g. Huang and Shanks (1995) obtained heritability estimate (0.13) for heel erosion, and Koenig *et al.* (2005) obtained heritability estimate 0.10 for white line disease. However, the data sets in these two studies were much smaller than in this study, and in case of Huang and Shanks (1995) came from a single herd. Lyons *et al.* (1991) obtained heritability estimate of 0.11 for foot problems in general.

The repeatability estimates for the claw disorders were higher than the corresponding heritability estimates, especially for screw claw (Table 5). Screw claw is known to be a chronic claw growth problem that requires frequent trimming to be maintained under control. The repeatability estimates are mostly in line with those obtained by Huang and Shanks (1995) in their study, except for heel

Table 5. Proportion of phenotypic variance observed in claw disorders explained by herd (c^2_{herd}) and claw trimmer effects ($c^2_{trimmer}$), and estimates of heritability (h^2) and repeatability (r) of the studied traits.

Trait	c^2_{herd}	$c^2_{trimmer}$	$h^2 \pm$ s.e.	r
Sole hemorrhage	0.04	0.10	0.05±0.004	0.11
White line disease	0.06	0.05	0.04±0.004	0.19
Screw claw	0.09	0.07	0.05±0.004	0.30
Heel erosion	0.22	0.07	0.01±0.002	0.05
All claw disorders	0.09	0.09	0.06±0.005	0.17

erosion for which they obtained repeatability of 0.20. However, the prevalence of heel erosion in the data of this study was much lower than in Huang and Shanks (1995).

The phenotypic correlations between the individual claw disorders were positive and moderate to strong, while the corresponding genetic correlations were mostly low (Table 6). This indicates that these disorders have a largely different genetic background, although phenotypically a fairly strong association can be observed between e.g. the incidences of sole hemorrhage and white line disease and white line disease and screw claw. The phenotypic correlations observed in this study are considerably higher than obtained in most other studies, e.g. Manske *et al.* (2002), Holzhauer *et al.* (2004) and van der Waaij *et al.* (2005), while the corresponding genetic correlations are much more in line with other studies.

Table 6. Phenotypic (below diagonal) and genetic correlations with their standard errors (above diagonal) between the individual claw disorders.

	Sole hemorrhage	White line disease	Screw claw	Heel erosion
Sole hemorrhage		0.46±.07	0.29±0.07	0.07±0.07
White line disease	0.73		0.24±0.08	-0.04±0.10
Screw claw	0.48	0.75		0.18±0.10
Heel erosion	0.46	0.29	0.48	

Conclusions

The most significant environmental effects on claw disorders in this study were the breed, the parity and the lactation stage of the cow, and the barn type that the cow had been living in. Additionally, the effect of the claw trimmer, who had made the initial diagnosis of the claw health status, was found to be statistically highly significant on all studied disorders. When treated as random factor, claw trimmer explained from 5 to 10% of the total phenotypic variance observed in different claw disorders. In addition, the herd had a very large effect on claw disorders, especially on infectious heel erosion where the herd effect explained alone 22% of the total phenotypic variance observed in the trait. In contrast, the heritability estimates for the individual claw disorders were low, indicating that it will be difficult to obtain genetic progress when breeding for these traits. The genetic correlations between different claw disorders were mostly low to moderate, indicating that they have largely different genetic backgrounds.

References

Alban, L., 1995. Lameness in Danish dairy cows: frequency and possible risk factors. Prev. Vet. Med., 22: 213-219.
Elbers, A.R.W., J.D. Miltenburg, D. de Lange, A.P.P. Crauwels, H.W. Barkema and Y.H. Schukken, 1998. Risk factors for clinical mastitis in a random sample of dairy herds from the southern part of The Netherlands. J. Dairy Sci., 81: 420-426.
Eriksson, J.Å., 2006. Swedish sire evaluation of hoof diseases based on hoof trimming records. Proceedings of the 2006 Interbull Meeting Kuopio, June 4-6, 2006, Finland. Interbull Bull., 35: 49-52.
Fleischer, P., M. Metzner, M. Beierbach, M. Hoedemaker and W. Klee, 2001. The relationship between milk yield and the incidence of some diseases in dairy cows. J. Dairy Sci., 84: 2025-2035.
Groeneveld, E., 1997. VCE4 User's guide and reference manual. Institute of Animal Husbandry and Animal Behaviour, Federal Agricultural Research Centre, Germany.

Hernandez, J., J.K. Shearer and D.W. Webb, 2001. Effect of lameness on the calving-to-conception interval in dairy cows. J. Am. Vet. Med. Ass., 218: 1611-1614.

Holzhauer, M., B.H.T. Borne, E.A.M. van den Graat and C.J.M. Bartels, 2004. Preliminary results of prevalence and correlations between major rear claw disorders in 348 Dutch dairy herds. Proceedings of 13th International Symposium and 5th Conference on Lameness in Ruminants. Available at: http://www.ruminantlameness.com/end/proceedings2004.pdf., pp. 40-43.

Huang, Y.C. and R.D. Shanks, 1995. Within herd estimates of heritabilities for six hoof characteristics and impact of dispersion of discrete severity scores on estimates. Livest. Prod. Sci., 44: 107-114.

Huang, Y.C., R.D. Shanks and G.C. McCoy, 1995. Evaluation of fixed factors affecting hoof health. Livest. Prod. Sci., 44: 115-124.

Koenig, S., A.R. Sharifi, H. Wentrot, D. Landmann, M. Eise and H. Simianer, 2005. Genetic parameters of claw and foot disorders estimated with logistic models. J. Dairy Sci., 88: 3316-3325.

Lyons, D.T., A.E. Freeman and A.L. Kuck, 1991. Genetics of health traits in Holstein cattle. J. Dairy Sci., 74: 1092-1100.

Manske, T., 2002. Hoof lesions and lameness in Swedish dairy cattle. Prevalence, risk factors, effects of claw trimming, and consequences for productivity. Doctoral Thesis. Swedish University of Agricultural Sciences.

Available at: http://diss-epsilon.slu.se/archive/00000081/01/Ram_Manske.pdf

Manske, T., Hultgren, J., Bergsten, C. 2002. Prevalence and interrelationships of hoof lesions and lameness in Swedish dairy cows. Prev. Vet. Med., 54: 247-263.

Somers, J.G.C.J., K. Frankena, E.N. Noordhuizen-Stassen and J.H.M. Metz, 2005. Risk factors for interdigital dermatitis and heel erosion in dairy cows kept in cubicle houses in The Netherlands. Prev. Vet. Med., 71: 23-34.

Van der Waaij, E.H., M. Holzhauer, E. Ellen, C. Kamphuis and G. de Jong, 2005. Genetic parameters for claw disorders in Dutch dairy cattle and correlations with conformation traits. J. Dairy Sci., 88: 3672-3678.

Vermunt, J., 2004. Herd lameness - a review, major causal factors, and guidelines for prevention and control. Proceedings of 13th International Symposium and 5th Conference on Lameness in Ruminants. Available at: www.ruminantlameness.com/end/proceedings2004.pdf. pp. 3-18.

Vilva, V., 2007. WSYS-L, a data editing system for statistical data sets. Department of Animal Science, Helsinki University, Finland.

Wells, S.J., L.P. Garber and B.A. Wagner, 1999. Papillomatous digital dermatitis and associated risk factors in US dairy herds. Prev. Vet. Med., 38: 11-24.

Genetic analysis of claw disorders and relationships with production and type traits

S. König[1], X.-L. Wu[2], D. Gianola[2] and H. Simianer[1]
[1]Institute of Animal Breeding and Genetics, University of Göttingen, 37075 Göttingen, Germany
[2]Department of Animal Sciences, University of Wisconsin, Madison 53076, USA

Abstract

Generalised linear mixed models (GLMMs) with a logit link function as well as linear and threshold models in a Bayesian framework were applied for the genetic analysis of four different claw disorders, and to infer relationships among disorders, milk yield, and type traits. The data set comprised test day production records and claw disorders collected in 2005 from 5,360 Holstein cows. Estimates of heritabilities were in the range from 7.3% to 18.3%, and all claw disorders were highly correlated among each other. Higher heritabilities for disorders were obtained from sire models compared to animal models. Relationships between claw disorders and test day milk yield were analysed in a Bayesian framework by fitting recursive threshold models. A two-way causal path was postulated describing first the influence of test day milk yield on claw disorders and, secondly, the effect of the disorder on milk yield at the following test day. For all disorders, the coefficient was positive in the range from $\lambda_{21} = 0.016$ to $\lambda_{21} = 0.042$ indicating an increase of incidences with increasing milk yield. Structural coefficients λ_{32} ranged from -0.12 to -0.46 predicting that one unit increase in the incidence of any disorder reduces milk yield at the following test day by up to 0.67 kg. For 79 sires with at least 30 daughters in the claw database, correlations between estimated breeding values of linear type traits and breeding values of claw disorders were calculated. Claw disorders were genetically positively correlated with conformation traits related to feet and legs. However, the correlations among type traits and claw disorders are not large enough to achieve a substantial reduction of laminitis or of other claw disorders within the dairy cattle population via indirect selection on conformation traits as discussed by various selection strategies.

Keywords: claw disorders, milk yield, type traits

Introduction

Functional traits are defined as those characters of an animal that increase efficiency not by higher outputs of products, but by reduced costs (Groen et al., 1997). As pointed out by Bishop et al. (2002), the economic burden only due to infection diseases embraces nearly 20% of the total output value in animal production. Beyond theses economic reasons, animal welfare, directions of the law (§1 German law for animal breeding; §11b German law for animal protection) as well as demands of consumers enforce the implementation of functional traits in breeding goals (Simianer and König, 2002).

As shown by the annual statistics published by the German Cattle Breeders Federation, the average production level in lactation milk yield increased by in average 100 kg per year and reached a current level of 8,524 kg (ADR, 2005). However, milk production in dairy cows has a high metabolic priority and is clearly maintained at the costs of other reproductive and metabolic processes. Fleischer et al. (2001) examined the relationship between milk yield and incidence of certain disorders in German Holstein cows. An increase of milk yield was generally associated with an increased risk for the occurrence of any disease. Results were verified by recently published trends for production traits and diseases in Norwegian dairy cattle obtained from a long-term selection experiment (Heringstad et al., 2007). Increased income based on higher milk yield per cow and year can be eroded due to increased costs. For example in the case of claw disorders, Esslemont and Kossaibati (1996) calculated an economic loss due to lameness of 615 € per cow and year: 65 € additional veterinary costs, 30 € for increased labour time, 240 € replacement costs, 205 € for the longer calving interval, 50 € for the decrease in milk yield, and 25 € for discarded milk. However, several efforts are in progress for the inclusion of additional functional traits in a combined breeding goal for dairy cattle. When comparing breeding goals over the last two decades, the emphasis of dairy cattle breeding objectives has gradually shifted from production traits towards functional traits such as fertility, longevity and calving traits (Mark, 2004). The main problem in the past for the inclusion of functional traits in the breeding goal was the lack of appropriate data, and in most cases, indirect measurements were used. In the case of udder health, somatic cell score is used as an indicator for mastitis, but Heringstad *et al.* (2000) found an average estimated genetic correlation between somatic cell count and clinical mastitis of 0.6 based on several values from the literature. The main problem is the lack of appropriate data for the genetic evaluation of functional traits which are mainly based on indirect measurements. A substantial improvement for functionality in acceptable time can only be achieved via direct selection on various diseases as recently shown by König and Swalve (2006) when evaluating different selection strategies. The direct accurate recording of health traits is a prerequisite for genetic evaluation, and such a system is practiced in the Nordic countries for more than 20 years. The Nordic countries have a unique position in registration and collection of information about functional traits. Health data from veterinary systems are computerised and integrated in the national databases. A detailed overview about the recording system for health traits and the procedure from data collection up to genetic evaluation is given by Heringstad *et al.* (2000). Modern technology gives new recording opportunities for functional traits, also in the case for claw disorders in Germany (Landmann *et al.*, 2006).

Data from the claw data base was used for the following investigations:
1. Comparison of genetic parameters when applying different methodologies for the analysis of categorical traits.
2. Estimation of relationships between test day milk yield and claw disorders applying recursive models.
3. Estimation of correlations among claw disorders and type traits, and to asses the impact of claw data on genetic gain in a combined breeding goal.

Materials and methods

Recording system and data

A prerequisite for genetic evaluation of claw disorders is an efficient recording system as well as the ability of claw trimmers for a uniform identification of disorders. To cope with this problem, the Association for Claw Trimming and Hygiene (German Agriculture Society; DLG), the Teaching and Research Facility for Animal Husbandry Echem (LVA Echem), and the Cooperative of Claw trimmers Saxony and VIT PC-Software Paretz Co. developed a computer supported documentation and analysis system as described by Landmann *et al.* (2006). This system was shortly extended to an automatic creation of suitable data formats and data transfers for statistical analysis applying logistic or threshold models at the institute of Animal Breeding and Genetics at Göttingen University. A special interface of the program enables the direct contact to several statistical tools. A second possibility is the digital pen. Handwritten, digitalised hoof management offers the professional hoof trimmer, farmer or veterinarian a new way to document hoof care. Any type of written data can be quickly and easily compiled using a digital pen and the paper that goes with it. An infrared camera in the pen records the motion of the pen on the paper and converts it to digital information. The compiled data can be transmitted to a database or other applications (e.g. hoof care software) by Internet or mobile phone from anywhere in the world in just a few seconds. The compiled data can then be viewed or downloaded from a secure website by the hoof trimmer or farmer. The data can also be exported to extraneous software.

The data set used here comprised test day production records, and claw and foot disorders collected in 2005 from 5,360 Holstein cows. Additionally, national official estimated breeding values (EBV) of sires for all conformation traits were available. The average number of daughters for the 511 different sires was 10.46, and 79 sires had more than 30 daughters scored for claw disorders. Cows of all parities were included. Claw and foot disorders were divided into four different conditions: digital dermatitis (DD), sole ulcer (SU), wall disorder (WD) and interdigital hyperplasia (IH), and scored separately as 'all or none' traits. IH and and to a large degree, DD, are foot disorders that do not directly affect the medial or distal claw on each foot, whereas SU and WD belong to classical claw disorders. Wall disorder mainly describes the different types of white-line-disease and further lesions along the wall of the claw. A few specific cases of heel erosion were considered together with digital dermatitis, because both disorders are caused by bacteria. A detailed description of the individual disorders is given by König *et al.* (2005). The period of observation spanned 200 days, starting at calving. If a cow had the foot problem within this period in one or both rear legs for the respective disorder, she was given a score of 1; otherwise she was scored 0. Mean incidences for DD, SU, WD, and IH were 13.67%, 16.51%, 9.78%, and 6.72%, respectively.

Statistical models

For the estimation of genetic parameters, three different models were used. Model M1 was a standard linear model assuming a Gaussian distribution for the disorders. Model M2 was a threshold model treating the claw disorder as a binary trait. In the threshold model (Gianola, 1982; Gianola and Foulley, 1983), it is assumed that an underlying continuous variable, liability (l_{i2}), exists such that the observed binary variable y_{i2} takes a value of 1 if l_{i2} is larger than a fixed threshold $\kappa = 0$. Both, models M1 and M2, were sire models, and analysed in a Bayesian framework. Furthermore, recent developments in statistical analyses allowing application of a generalised linear mixed model (GLMM) technique (Schall, 1991) can be used to analyse data with appropriate distributions, such as a Binomial. Generalised linear models (GLMM) were originally developed by Nelder and Wedderburn (1972). The GLMM is an extension of the linear model. The main feature is a link function that allows the mean of a population to depend on a linear predictor and therefore the application to a wider range of data analysis problems, e.g. data which are not normally distributed or data that show an observed increase of the variances with the mean.

The linear component $\eta_i = x_i'\beta$ of a GLMM (x_i = column vector of covariates or explanatory variables for observation i, β = unknown coefficients to be estimated) is defined as known for traditional linear models. A link function g describes the relationship between the expected value of the response variable y_i and the linear predictor η_i:

$$g(\eta_i) = x_i'\beta \tag{1}$$

The variance of the response y_i is linked to the mean μ by a variance function. For binary data, linear logistic models were suggested. In this case, the link function is:

$$\eta = \log\left(\frac{\mu}{1 - \mu}\right) \tag{2}$$

and the associated variance function is:

$$V(\mu) = \mu(1 - \mu) \tag{3}$$

The GLMM-Logit model was defined as model M3 and genetic relationships among all animals were used. Parity and herd were considered as fixed effects in models M1, M2, and M3, and the random effect was the sire (in models M1 and M2) or the animal itself (model M3). Estimated breeding values for sires from model M2 for claw disorders were correlated with official breeding values for conformation traits.

Recursive models (Gianola and Sorensen, 2004) in a Bayesian framework were used to depict the impact of claw disorders and test day milk yield in detail. A lagged progressive path involving three different traits was postulated. The first path described the influence test day milk yield has on claw disorders, and the second path pertained to the effect of the disorder on milk production level at the following test date. This definition involved the following traits: test day milk yield before occurrence of the disorder (MY1 = trait 1); the disorder itself (trait 2), and test day milk yield after occurrence of the disorder (MY2 = trait 3). Cows without disorders were assigned a value of 0 for trait 2 at a general dummy date of day 100 within their lactation. The nearest test day observation for healthy cows before day 100 was defined as MY1 and the nearest test day observation after day 100 was MY2. Bayesian modelling and MCMC sampling procedures for recursive models are described in details in the users' manual of the SIR-BAYES software package (Wu, 2007).

Results

Heritabilities and genetic correlations among claw disorders

Table 1 gives an overview of the heritabilities found for claw disorders when applying GLMMs (logit link function) and for the threshold as well as for standard linear models in the Bayesian framework. Apart from sole ulcer, heritabilities were generally highest when applying the Bayesian threshold model, especially for disorders showing low incidences lower than 10% such as wall disorders or interdigital hyperplasia. Higher heritabilities on the liability scale obtained from threshold models compared to results from standard linear models is what theory for analysis of categorical traits leads one to expect (Dempster and Lerner, 1950). This was also found in other studies analysing categorical data with different models (e.g. Weller and Ron, 1992; Andersen-Ranberg et al., 2005). Theoretically, nonlinear models are more appropriate for statistical analysis of categorical traits than linear methods (Thompson, 1979; Gianola, 1982).

Differences in estimated heritabilities form the logit model M3 and the probit (= Bayesian threshold) model M2 could be due to other differences in the statistical model. Model M3 was an animal model, model M1 was a sire model. However, the difference in heritabilities seemed mainly to be associated with the link function. In extended studies, König and Swalve (2006) merged the claw databases from their projects and they estimated heritabilities for laminitis applying threshold-animal and threshold-sire models. Results were nearly identical. From the theoretical point of view, sire models are better for the analysis of categorical data compared to animal models. Some simulation studies with threshold animal models have shown convergence problems with the Gibbs sampler

Table 1. Heritabilities for four different claw disorders applying GLMM-logit, and threshold and linear models in a Bayesian framework.

Model	Claw disorder			
	Sole ulcer	D. digitalis	Wall disorder	Interdigital hyperplasia
Bayesian-linear (M1)	0.077	0.100	0.101	0.112
Bayesian-threshold (M2)	0.088	0.134	0.136	0.186
GLMM-logit (M3)	0.086	0.073	0.104	0.115

(e.g. Hoeschele and Tier, 1995; Luo *et al.,* 2001). This problem is related to the extreme category problem, which occurs when all observations within a level of an effect are in the same category. It turned out that sire threshold models are more reliable than animal threshold models.

In our study, most genetic correlations between disorders were large and positive (Table 2). Genetically, health problems appear to occur in clusters. The genetic correlations suggest that cows genetically susceptible to some type of health problems are likely to be susceptible to other health problems as well. Correlations between EBVs for different diagnoses of claw disorders using substantial hoof trimming data in Sweden were also positive in a study by Eriksson (2006), especially when focussing on dermatitis and heel horn erosion.

According to the theoretical expectation, the genetic correlations between disorders were very similar, regardless the applied methodology. Vinson and Kluwer (1976) have shown that the genetic correlation estimated from multi- or binomial phenotypes of related animals is equal to normally distributed variables, and vice versa. There are several other applications in animal breeding focusing on the comparison of genetic correlations estimated via threshold methodology or via linear models (e.g. König *et al.,* 2005).

Effect of test day milk yield on claw disorders and vice versa

The structural equation coefficient λ_{21} is the gradient of the liability of the respective claw disorder with respect to test day milk yield (milk yield 1) after calving. For all disorders, the coefficient was positive in the range from $\lambda_{21}= 0.016$ to $\lambda_{21}= 0.042$ indicating an increase of incidences with increasing milk yield. The rate of change in test day milk yield (milk yield 2) with respect to the previous claw disorder is given by λ_{32}. Structural coefficients λ_{32} ranged from -0.12 to -0.46 predicting that one unit increase in the incidence of any disorder reduces milk yield at the following test day by up to 0.67 kg. Figure 1 provides the relationships among test day milk yield and claw disorders for SU, DD, WD, and IH.

Ongoing discussions for the improvement of genetic evaluation of production traits, i.e. if to correct production test day records or account statistical models for pregnancy of cows or not, should also evaluate such recursive or simultaneous modelling. There are different possibilities to select the most appropriate model. In the present study, comparison of models was done using the Bayesian Information Criterion (BIC) (Schwarz, 1978). Wu *et al.* (2007) suggested a method to calculate BIC by contrasting a SIR model with a standard mixed model. If the standard threshold-linear model M2 is taken as the baseline model, for example, the BIC for the recursive threshold-linear model M4 is calculated as:

Table 2. Genetic correlations among claw disorders (Results from the GLMM-logit model above the diagonal, results from the linear model below the diagonal).

	Digital dermatitis	Sole ulcer	Wall disorder	Interdigital hyperplasia
Digital dermatitis		0.56	0.34	0.39
Sole ulcer	0.53		0.44	0.49
Wall disorder	0.29	0.40		0.67
Interdigital hyperplasia	0.44	0.48	0.60	

Milk yield 1	$\xrightarrow{\lambda_{21} = 0.016}$	Sole ulcer	$\xrightarrow{\lambda_{32} = -0.44 \text{ kg}}$	Milk yield 2
Milk yield 1	$\xrightarrow{\lambda_{21} = 0.042}$	D. Digitalis	$\xrightarrow{\lambda_{32} = -0.40 \text{ kg}}$	Milk yield 2
Milk yield 1	$\xrightarrow{\lambda_{21} = 0.039}$	Wall disorder	$\xrightarrow{\lambda_{32} = -0.12 \text{ kg}}$	Milk yield 2
Milk yield 1	$\xrightarrow{\lambda_{21} = 0.034}$	I. hyperplasia	$\xrightarrow{\lambda_{32} = -0.46 \text{ kg}}$	Milk yield 2

Figure 1. Effect of test day milk yield after calving (milk yield 1) on incidences of disorders indicated by the structural equation coefficient λ21 and the recursive effect of incidences of disorders on milk yield at the following test day (milk yield 2) indicated by the structural equation coefficient λ32.

$$BIC_{M4} = 2(\bar{l}_{M2} - \bar{l}_{M4}) - (d_{M2} - d_{M4}) \log n \qquad (4)$$

where:

$\bar{l} = \frac{1}{c} \sum_{t=1}^{c} \log p\,(\mathbf{y} \mid \theta^{(t)}, M)$ is the average of MCMC sampled log-likelihoods;

c is the number of saved MCMC samples;

d is the dimension of the corresponding parameter vector $\theta = \{\lambda, \beta, u, G_0, R_0\}$ with the differences $(d_{M2} - d_{M4}) = -2$; and

n is the sample size.

G_0 and R_0 are the 'system covariance matrices' for the and sire and residual effects, and G^*_0, R^*_0, and P^*_0 are the sire, residual and phenotypic variance-covariance matrices, respectively.

β is a vector of 'fixed' effects in the Bayesian sense, and

vector u represents sire effects.

Table 3 gives BIC values for model M4, which was contrasted to model M2. Negative BIC values indicate that the recursive threshold-linear model M4 was better supported by the data than the standard threshold-linear model M2. Likewise, the recursive threshold model M4 received more support than the standard threshold model M2. A recursive threshold model seems to provide an appealing statistical specification for genetic evaluation of traits that are affected in a manner similar to that shown in Figure 1. Based on results from other recently conducted studies (e.g. Lopez de Maturana *et al.*, 2007; Wu *et al.*, 2007), the real nature among phenotypes in dairy cattle breeding can be depicted much more accurately when considering recursive models.

Genetic correlations among claw disorders and conformation traits

For 79 sires with at least 30 daughters in the claw database, correlations between EBVs of linear type traits (official national German EBVs from 02/2005) and breeding values of claw disorders were calculated. The restriction of the data was done to ensure reliable estimates for this analysis. Correlations between estimated breeding values for claw disorders and official breeding values for

Table 3. Bayesian information criterion (BIC) calculated for the recursive model M4 contrasted against the reduced model M2 without recursive effects.

	Dermatitis digitalis	Sole ulcer	Wall disorder	Interdigital hyperplasia
BIC M4	-278.01	-263.90	-271.72	-269.30

type traits of bulls are presented in Table 4. However, correlations between breeding values are not identical with genetic correlations unless accuracies of estimated breeding values are close to one. Therefore, results should only be interpreted as general trends, keeping in mind that correlations between breeding values are always shrinked to zero. According to the official procedures, all EBVs were standardised to a mean of 100 points and a standard deviation of 12 points. Genetically favourable bulls are indicated by EBVs above 100; characterised by less disorders or the absence of disorders and high scores for type traits.

Not surprisingly, genetic correlations were mostly positive between all traits belonging to the feet and leg composite and individual claw disorders. This is in fair agreement when comparing these results to genetic correlations among claw disorders and type traits estimated from logistic models (König *et al.,* 2005). Values less than 100 indicate steep legs which seemed to be favourable. The only negative correlation in Table 4 was found for the combinations of rear leg side view and claw disorders. This indicates that bulls transmitting straighter legs had fewer daughters with claw disorders. The correlations between rear leg rear view and disorders were in the range from 0.31 for sole ulcer to 0.50 for wall disorders indicating the advantage of a parallel position of rear legs.

Correlations among EBVs for claw disorders and stature, dairy character, body depth, and strength, were close to zero, but without exception negative; indicating an advantage for smaller cows and lower scores for body depth, strength, and dairy character as well. For practical applications, it is important to know that improved quality of feet and legs, e.g. better conformation scores for traits such as hocks, foot angle, rear leg rear view, and rear leg side view, is genetically associated with fewer incidences of claw disorders. Also the slight negative correlation between angularity, stature, and body strength with claw disorders supports some findings by Buenger *et al.* (2001) when analysing relationships between conformation traits and longevity.

Conventional German dairy cattle breeding programs only include a selection for improved claw health via indirect selection on EBVs of four different conformation traits. These conformation traits are foot angle, rear legs rear view, rear legs side view, and the quality of the hocks. Due to the moderate correlations among EBVs for feet and leg scores and EBVs for claw disorders, direct selection on disorders is strongly recommended. As shown in this study, genetic parameters of various claw disorders enable the possibility for direct selection strategies on these traits. Different scenarios for the calculation of selection response for different selection strategies were elaborated by König

Table 4. Correlations among breeding values for type traits and breeding values for claw disorders (Results for claw disorders are posterior means for sire effects from the threshold-linear model M2).

Linear scored type trait	Claw disorders			
	Dermatitis digitalis	Sole ulcer	Wall disorder	Interdigital hyperplasia
Feet and leg score	0.31	0.34	0.54	0.30
Rear leg side view	-0.38	-0.47	-0.51	-0.26
Foot angle	0.36	0.29	0.28	0.18
Hocks	0.03	0.22	0.57	0.12
Rear leg rear view	0.37	0.31	0.50	0.49
Stature	-0.16	-0.22	-0.08	-0.11
Dairy character	-0.05	-0.10	-0.15	-0.09
Body depth	-0.06	-0.23	-0.17	-0.27
Strength	-0.08	-0.32	-0.20	-0.21

and Swalve (2006). The aim of their study was to quantify the relative importance of different index traits with respect to selection response for the trait laminitis resistance. Hence, applying selection index theory, the trait in the breeding goal was laminitis and index traits for EBVs of bulls in laminitis were laminitis observations and linear scores for hock quality of daughters and one claw measure of the bull (hardness of the dorsal wall). As one result, selection response in laminitis resistance per generation and accuracy of EBVs of bulls in laminits could be more then doubled when laminitis observations of 50 daughters were included as index traits compared with the 'baseline' scenario. The baseline scenario is the one currently used in Germany: about 50 daughters per bull are routinely scored for conformation traits. Finally, the correlations among conformation traits and claw disorders are not large enough to achieve a substantial reduction of laminitis or of other claw disorders within the dairy cattle population via indirect selection on conformation traits.

However, when including health traits such as claw disorders in a combined breeding goal, economic weights for all these traits have to be known. The objective of a current dairy cow profitability project by König *et al.* (2007) is to simulate the individual variability of cows in 212 different traits. Changes during life in production, growth, feed intake and some type traits were modelled through random regression coefficients. The influence of management practices and environmental effects on all these traits and diseases and their interactions in combination with prices and costs over the whole cow's lifespan are used to determine net returns per cow and day. Net returns per cow and day can be regressed on their true breeding values to determine relative economic values for each trait. This method also allows the derivation of economic weights for 'new traits' like claw disorders, assuming that genetic parameters are available.

Conclusions

The current analyses revealed the possibility to include claw disorders in a combined breeding goal for dairy cattle. Heritabilities were in moderate range, and correlations between claw disorders and type traits were genetically favourable. However, as discussed, direct selection on claw disorders will ensure additional genetic gain for health traits. The most critical point is the routine implementation of the claw data recording system. An essential step for the successful implementation of new recording systems is the establishment of co-operator herds for progeny testing (PT). The strong need for highly accurate phenotyping of additional functional or health traits such as claw disorders is impossible to conduct across an entire population. Based on the PT information for important health traits, EBVs for young bulls can be computed. For the multitude of functional or health traits included, recursive models should be applied for genetic evaluations. The current analysis utilising recursive threshold models provided reliable estimates and was in favour compared to the standard threshold model when using BIC. In conclusion, there are several possibilities from data recording up to highly sophisticated statistical models towards a general improvement of functionality in dairy cattle.

References

ADR, 2005. Annual statistics published by the German Cattle Breeders Federation.

Andersen-Ranberg, I.M., B. Heringstad, D. Gianola, Y.M. Chang and G. Klemetsdal, 2005. Comparison between bivariate models for 56-day nonreturn and interval from calving to first insemination in Norwegian Red. J. Dairy Sci., 88: 2190-2198.

Bishop, S.C., J. Chesnai and J. Stear, 2002. Breeding for disease resistance: issues and opportunities. In: 7th World Congress on Genetics Applied to Livestock Production. CD-ROM communication, 13-01.

Buenger, A., V. Ducrocq and H.H. Swalve, 2001. Analysis of survival in dairy cows with supplementary data on type scores and housing systems from a region of Northwest Germany. J. Dairy Sci., 84: 1531-1541.

Dempster, E.R. and M. Lerner, 1950. Heritability of threshold characters. Genetics, 35: 212-286.

Esslemont, R.J. and M.A. Kossaibati, 1996. Incidence of production diseases and other health problems in a group of dairy herds in England. Veterinary Record, 139: 486-490.

Eriksson, J.A., 2006. Swedish sire evaluation of hoof diseases based on hoof trimming records. Interbull Bulletin, 35: 49-52.

Fleischer, P., M. Metzner, M. Beyerbach, M. Hoedemaker and W. Klee, 2001. The relationship between milk yield and the incidence of some diseases in dairy cows. J. Dairy Sci., 84: 2025-2035.

Gianola, D.,1982. Theory and analysis of threshold characters. J. Anim. Sci., 54: 1079-1096.

Gianola, D. and J.L. Foulley, 1983. Sire evaluation for ordered categorical data with a threshold model. Genet. Sel. Evol., 15: 201-223.

Gianola, D. and D. Sorensen, 2004. Quantitative genetic models for describing simultaneous and recursive relationships between phenotypes. Genetics, 167: 1407-1424.

Groen, A.F., T. Steine, J.J. Colleau, J. Pedersen, J. Prybil and N. Reinsch, 1997. Economic values in dairy cattle breeding, with special reference to functional traits. Report of an EAAP-working group. Livest. Prod. Sci., 49: 1-21.

Heringstad. B., G. Klemetsdal and J. Ruane, 2000. Selection for mastitis resistance in dairy cattle–a review with focus on the situation in Nordic countries. Livest. Prod. Sci., 64: 95-106.

Heringstad, B., G. Klemetsdal and T. Steine, 2007. Selection responses for disease resistance in two selection experiments with Norwegian Red cows. J. Dairy Sci., 90: 2419-2426.

Hoeschele, I. and B. Tier, 1995. Estimation of variance components of threshold characters by marginal posterior modes and means via Gibbs sampling. Genet. Sel. Evol., 27: 519-540.

König, S. and H.H. Swalve, 2006. A model calculation on the prospects of an improvement of claw health in dairy cattle via genetic selection. Züchtungskunde, 78: 345-356.

König, S., A.R. Sharifi, H. Wentrot, D. Landmann, M. Eise and H. Simianer, 2005. Genetic parameters of claw and foot disorders estimated with logistic models. J. Dairy Sci., 88: 3316-3325.

König, S., J. Fatehi, H. Simianer and L.R. Schaeffer, 2007. Assessment of dairy cow profitability. Book of abstracts 58th annual eeting of the EAAP, Dublin, Ireland. Wageningen Academic publishers, the Netherlands, p. 58.

Landmann, D., J. Burmester, S. König and H. Simianer, 2006. Utilizing data from PC – supported documentation to reveal the impact of housing systems on claw diseases. 14th International Symposium on Lameness in Ruminants, Colonia, Uruguay.

Lopez de Maturana, E., A. Legarra, L. Varona and E. Ugarte, 2007. Analysis of fertility and dystocia in Holsteins using recursive models, handling censored and categorical data. J. Dairy Sci., 90: 2012-2024.

Luo, M.F., P.J. Boettcher, L.R. Schaeffer and J.C.M. Dekkers, 2001. Bayesian inference for categorical traits with an application to variance component estimation. J. Dairy Sci., 84: 694-704.

Mark, T., 2004. Applied genetic evaluations for production and functional traits in dairy cattle. J. Dairy Sci., 87: 2641-2652.

Nelder, J. A. and R.W.M. Wedderburn, 1972. Generalized linear models. J. Roy. Statist. Soc. Ser. A, 135: 370-384.

Schall, R., 1991. Estimation in generalized linear models with random effects. Biometrika, 78: 719-727.

Schwarz, G., 1978. Estimating the dimension of a model. Ann. Stat., 6: 461-464.

Simianer H. and S. König, 2002. Ist Zucht auf Krankheitsresistenz erfolgreich? Züchtungskunde, 74: 413-425.

Thompson, R., 1979. Sire evaluation. Biometrics, 35: 339-346.

Vinson, W.E. and R.W. Kluwer, 1976. Overall classification as a selection criterion for improving categorically scored components of type in Holstein. J. Dairy Sci., 59: 2104-2114.

Weller, J.I. and M. Ron, 1992. Genetic analysis of fertility traits in Israeli Holsteins by linear and threshold models. J. Dairy Sci., 75: 2541-2548.

Wu, X-L., 2007. SIR-BAYES: Computing quantitative trait models with simultaneous and recursive relationships between phenotypes in a Bayesian framework. Available at: https://mywebspace.wisc.edu/xwu8/programs/sir-bayes. Accessed Aug. 10, 2007

Wu, X-L., B. Heringstad, Y-M, Chang, G. de los Campos and D. Gianola, 2007. Inferring relationships between somatic cell score and milk yield using simultaneous and recursive models. J. Dairy Sci., 90: 3508-3521.

Milk production, udder health, body condition score and fertility performance of Holstein-Friesian, Norwegian Red and Norwegian Red×Holstein-Friesian cows on Irish dairy farms

N. Begley[1,2], K. Pierce[2] and F. Buckley[1]
[1]*Moorepark Dairy Production Research Centre, Teagasc, Moorepark, Fermoy, Co. Cork, Ireland*
[2]*School of Agriculture, Food and Veterinary Medicine, University College Dublin, Belfield, Dublin 4, Ireland*

Abstract

The objective of this study was to compare the milk production performance, udder health, body condition score and reproductive efficiency of Holstein-Friesian (HF), Norwegian Red (NR) and Norwegian Red×Holstein-Friesian (F1) cows. The study animals were distributed across 46 herds. In year 1 (all cows in first lactation) milk production data was available for 1,407 cows: 771 HF, 325 NR and 311 F1. Predicted 305 d yield data and somatic cell count data were obtained from the Irish Cattle Breeding Federation. The 305 d milk yields of the HF and F1 were similar at 5,358 kg and 5,331 kg, respectively. The NR produced slightly less milk at 5,151 kg ($P<0.001$). Fat content was higher ($P<0.05$) for the HF at 4.00%, compared to the NR (3.94%). That of the F1 was intermediate at 3.96%. Protein content was similar for all breeds at 3.46%, 3.45% and 3.45% for HF, F1 and NR, respectively. The NR and F1 showed superior udder health based on somatic cell score. Compared to the HF (2.04), values for the F1 and NR cows were at 1.97 ($P<0.01$) and 1.92 ($P<0.001$), respectively. Based on information provided by participating herd owners the NR and F1 cows also had slightly less cows recorded with mastitis at least once during lactation. BCS at breeding was 2.85, 2.98 and 3.03 for the HF, F1 and NR, respectively, significantly higher for the NR and F1 ($P<0.001$) compared to the HF. Reproductive efficiency was excellent for all breed groups during year 1 of the study, with small differences tending to favour the NR and F1 cows. In year 2 of the study differences in milk production performance were consistent with year 1. However, potential differences in reproductive efficiency became more evident.

Keywords: Norwegian Red, crossbreeding, milk, fertility, udder health, condition score

Introduction

Until recently, in the world of dairy cattle breeding, the term 'high genetic merit' was synonymous with high milk production potential. Now it is acknowledged that the term 'high genetic merit' should reflect as many characteristics as are required to reflect total economic profitability. In this regard, two of the greatest challenges facing dairy cattle breeding currently are (1) to halt/overcome the decline in reproductive efficiency that has been observed in the Holstein-Friesian (Veerkamp *et al.*, 2001) as a result of past selection programs geared towards maximising production potential and (2) to improve udder health. Although many countries have diversified their breeding goals to include measures of survivability or functionality (Miglior *et al.*, 2005), it is arguable that few have weighted fertility sufficiently to truly counteract the decline. Achieving high reproductive efficiency is fundamental to profitability across all dairy production systems. In seasonal grass based dairy systems such as in Ireland, the relative importance of fertility is greater than in confinement and year-round calving systems, because breeding and calving are restricted to a limited time period of the year (Veerkamp *et al.*, 2002). The rationale of this strategy is to obtain a concentrated calving pattern in spring (February to April) that enables grass growth to match food demand. This is achieved by attaining high pregnancy rates within a short interval after the start of the breeding season. Calving intervals of 365-370 days and culling rate for infertility of less than 10% are required for optimal financial performance within a seasonal dairy system (Esslemont *et al.*, 2001). Inferior udder health can result in reduced milk yield and milk quality (Bartlett *et al.*, 1991), changes in milk composition (Auldist *et al.*, 1995) as well as increased involuntary culling (Berry *et al.*, 2005), and veterinary and treatment costs (Berry and Amer, 2005). Furthermore, in Ireland as in many other countries, milk pricing is influenced by monthly arithmetic mean bulk SCC with penalties imposed on milk with high SCC. Heritability estimates for SCC (Mrode and Swanson, 1996) indicate that udder health is under modest genetic control. In 1994, the US introduced genetic evaluations for lactation SCS (Schutz, 1994) that were subsequently incorporated into a net merit index (Van Raden, 2004), while in Ireland, SCC was incorporated into an economic breeding index in 2006 (Berry and Amer, 2005).

Crossbreeding offers a potentially attractive avenue for farmers to improve economic efficiency through:
1. The introduction of favourable genes from another breed selected more strongly for traits of interest (otherwise referred to as breed complimentarity).
2. Removal of the negative effects associated with inbreeding depression.
3. Capitalising on what is known as heterosis or hybrid vigour (HV) – enhanced performance of crossbred animals relative to the average of their parental breeds.

This tends to be especially true for traits related to health and fertility, traits with low heritability. While many studies have documented the potential role of crossbreeding in the dairy industry the majority are dated (Fohrman, 1946; Bereskin and Touchberry, 1966; Brandt *et al.*, 1974; Rincon *et al.*, 1982), and because of the divergence historically between the 'secondary' breeds and the Holstein-Friesian for production, the crossbred cows in general were deemed to offer little advantage (Willham and Pollak, 1985). The notion of crossbreeding within the context of dairy breeding therefore until very recently has been essentially shunned. Interest now due to genetic improvement for milk production in other dairy breeds and a decline in fertility/survival within the Holstein-Friesian. One exception, however, has been New Zealand where crossbreeding has been used to a large extent to capitalise on the benefits of HV (Harris, 2005; Lopez-Villalobos *et al.*, 2000). In New Zealand HV values in the region of 5-6% have been documented for production traits and values of up to 18% for reproduction and health traits. Consequently, crossbred cows now make up a significant proportion of the national dairy herd (Lifestock Improvement, 2006). Concern for decreased additive genetic merit for fertility and health traits of Holsteins, as well as concern for increased potential for inbreeding, has resulted in renewed global interest in crossbreeding (Hansen, 2006; Walsh *et al.*, 2007).

Genetic differences exist between breeds because of their aetiology. In their country of origin, dairy cow breeds have evolved based on traits considered to be of economic importance such as milk production, health and fertility. The breeding program of the Norwegian Red (NR) dairy breed, for example, is well known for its broad selection goals. In Norway the Total Merit Index (TMI) philosophy has been implemented for almost five decades (Heringstad *et al.*, 2000; Lindhe and Philipsson, 2001). Accurate recording and selection against clinical mastitis since 1978, has contributed to a lower incidence of clinical mastitis in the national population (Heringstad *et al.*, 2003). Research at Teagasc Moorepark Ireland, has demonstrated that the NR breed have superior udder health (Walsh *et al.*, 2007) and reproductive efficiency (Walsh *et al.*, 2008) compared to Holstein-Friesians. The studie of Walsh *et al.* (2007, 2008) was based on a limited data set. Consequently, a larger on-farm study involving 46 dairy farms was established in order to conclusively evaluate the merits of the NR breed and to determine the potential suitability of crossbreeding with the NR as a breeding strategy for Irish dairy farmers. An important aim of the study was to generate data that could be used by the Irish Cattle Breeding Federation (ICBF) to provide breeding value estimates for the NR breed in Ireland through an across breed evaluation. The objective of this paper is to provide some initial results from this study.

Materials and methods

Data were available from 46 Irish dairy herds, with a total of 1,407 spring calving first lactation cows HF (n=771), NR (n=325) and NR×HF (F1) (n=311). With the exception of three herds, which had HF and NR cows only, all herds had cows of each breed group. All farms on the study were participants of the Dairy Management Information System (DairyMIS) at Teagasc Moorepark. DairyMIS is a recorder-based computerised database, where detailed information regarding stock, farm inputs and production events are recorded on a regular basis. All cows were born during spring 2004 and calved as two-year-olds during the spring of 2006. The NR and F1 cows were sired by a total of 10 proven NR sires, with an average total merit index (TMI) of 2 (www.geno.no; 2008). The HF cows were sired by a total of 103 bulls of North-American HF, New Zealand Friesian and British Friesian genetics. The HF and F1 cows were born on the study farms, while the NR cows were imported from Norway in 2004 as calves and reared on the study farms along side their HF and F1 contemporaries.

Milk production, udder health and body condition score

Cows were milk recorded on average 5.45 (±2.02) times during lactation. Individual test day milk production and SCC records were obtained from the ICBF, with 305 d predicted yields derived using the SLAC method (Olori and Galesloot, 1999) which adjusts for calving month, age at calving, parity and season of calving. The SCC data were transformed to SCS using log10 transformation for normalisation of the residuals. The occurrence of clinical mastitis (available for 42 of the 46 farms) was also recorded by the individual farmers. Two or more mastitis incidences were recorded as one occurrence for the purpose of the analyses. Herds were visited in May/June to measure body condition score (BCS). Body condition score (BCS) was recorded by a single operator on a scale of 1 (emaciated) to 5 (extremely fat) with increments of 0.25 (Lowman et al., 1976).

Fertility

Seven key indicators of reproductive efficiency were defined: calving to first service (interval in days from calving to first service), pregnancy rate to first service (pregnancy rate to first service confirmed by pregnancy diagnosis in late autumn), six week in-calf rate (pregnancy rate within 6 weeks of the start of breeding), thirteen week in-calf rate (pregnancy rate within 13 weeks of the start of breeding i.e. standardised breeding season), final pregnancy rate (pregnancy rate at the end of the breeding season), calving to conception interval (number of days from calving to conception, otherwise known as days open) and number of services per cow (total number of services recorded). The breeding season lasted on average 16 weeks across the 46 farms. Pregnancy diagnosis was carried out in late autumn to determine pregnancy by transrectal ultrasound examinations (Aloka 210 * II, 7.5 MHz).

Statistical analysis

Statistical analysis was carried out using SAS (SAS, 2006). Analysis of variance was conducted using proc GLM. Data for 2006 and 2007 were analysed separately. For milk production variables the following model was used:

$$Y = \mu + Herd + Breed + e$$

Continuous fertility variables as well as the body condition score records were analysed using a similar model but with calving date also included. Binary fertility and mastitis incidence were analysed using the LOGISTIC procedure of SAS assuming a logit link function.

Results

Milk production and udder health details for first and second lactation are presented in Tables 1 and 2, respectively. The 305 day predicted milk yield of the HF and F1 was similar at 5,358 kg and 5,331 kg, respectively. That of the pure NR was slightly lower at 5,151 kg. Fat content was higher ($P<0.05$) for the HF at 4.00%, compared to the NR (3.94%). That of the F1 was intermediate at 3.96%. Protein content was similar for all breeds at 3.46%, 3.45% and 3.45% for HF, F1 and NR, respectively. Milk protein content was not different across groups with values averaging 3.46%, 3.45% and 3.45% for the HF, F1 and NR cows, respectively. Udder health, as indicated by SCS was superior for the pure NR at 1.93 (equating to an SCC value of 131,000 cells/ml) ($P<0.001$) and F1 1.97 (equating to an SCC value of 133,000 cells/ml) ($P<0.01$) cows compared to the HF at 2.04 (equating to an SCC value of 188,000 cells/ml). Based on information provided by participating herds the NR and F1 also had slightly better udder health as indicated by a lower proportion of cows recorded with mastitis at least once during lactation. Fourteen percent of the HF cows had mastitis at least once during lactation, compared to 9.5% for both the F1 ($P=0.063$) and the NR ($P<0.05$).

Relative 305 d predicted milk production results for 2007 (second lactation) were consistent with that observed in 2006. While the NR cows maintained their udder health superiority compared to the HF cows, the advantage of the F1 was not obvious in year 2.

Table 1. Effect of breed group on 305 d milk production and udder health during lactation 1.

	HF	F1	NR	s.e.	*P*-value
Milk yield (kg)	5,358[a]	5,331[a]	5,151[b]	33.5	<0.001
Fat (%)	4.00[a]	3.96[ab]	3.93[b]	0.018	<0.01
Protein (%)	3.46	3.45	3.45	0.008	NS
Lactose (%)	4.71[a]	4.68[b]	4.65[c]	0.006	<0.001
Fat + protein yield (kg)	398[a]	394[a]	379[b]	2.2	<0.001
Lactation average SCS	2.04[a]	1.97[a]	1.93[b]	0.018	<0.001
(SCC in parentheses)	(188,000)	(133,000)	(131,000)		
Incidence of mastits (%)	14[a]	9.5[b]	9.5[b]		<0.05

[a,b,c] Within-row means not sharing a common superscript differ significantly (*P*<0.05).

Table 2. Effect of breed group on 305 d milk production and udder health during lactation 2.

	HF	F1	NR	s.e.	*P*-value
Milk yield (kg)	6,194[a]	6,081[a]	5,867[b]	42.4	<0.001
Fat (%)	3.95[a]	3.89[b]	3.90[b]	0.021	<0.05
Protein (%)	3.48	3.49	3.49	0.010	NS
Lactose (%)	4.59[a]	4.56[b]	4.51[c]	0.007	<0.001
Fat + protein yield (kg)	458[a]	447[b]	432[c]	2.7	<0.001
Lactation average SCS	2.08[a]	1.99[b]	2.07[a]	0.019	<0.001
(SCC in parentheses)	(186,000)	(153,000)	(179,000)		
Incidence of mastits (%)	13	11	10		NS

[a,b,c] Within-row means not sharing a common superscript differ significantly (*P*<0.05).

Breeding for robustness in cattle

Body condition score (BCS) recorded once during May/early June 2006 to represent body condition at breeding was higher ($P<0.001$) with the NR at 3.03 compared to the HF at 2.85. That of the F1 (2.98) cows was similar to the NR but higher ($P<0.001$) than the HF. Findings for 2007 were similar.

Reproductive efficiency was excellent for all breed groups during year 1 of the study, with small differences tending to favour the NR and F1 cows, compared to the HF (Table 3).

In year 2, large differences in pregnancy rates were observed (Table 4). Pregnancy rate to first service was 45% for the HF, and 55% and 56% for the NR and F1, respectively. The proportion of cows pregnant after 6 weeks was 59% for the HF compared to 69% for the NR and 72% for the F1. In-calf rate at the end of breeding was 85%, 88% and 87% for the HF, F1 and NR cows, respectively. However, had the breeding season on each herd been restricted to a maximum of 13 weeks (standardised breeding season), the in-calf rates of the HF, F1 and NR cows would have decreased by 4%, 2% and 2%, respectively.

Discussion

Over the two lactations the predicted 305 d milk yield of the NR cows was 95% of that produced by the HF cows. Thus, while significantly lower than the HF, the production potential of both breeds from a practical perspective under grazing was not greatly divergent. Consequently, despite an

Table 3. Effect of breed group on reproductive efficiency during first lactation.

	HF	F1	NR	s.e.	P-value
Mean calving date 2007	Feb-20	Feb-20	Feb-21		
Calving to 1st service interval (days)	80	79	79	0.8	NS
Pregnancy rate to 1st service (%)	57	60	58		NS
6 week in-calf rate (%)	67^a	75^b	73^b		<0.05
13 week in-calf rate (%)	88	91	92		NS
Final in-calf rate (%)	91	93	95		NS
Calving to conception interval (days)	93^a	90^a	92^{ab}	1.3	<0.05
Number of services per cow	1.59	1.53	1.58	0.045	NS

[a,b,c] Within-row means not sharing a common superscript differ significantly ($P<0.05$).

Table 4. Effect of breed group on reproductive efficiency during second lactation.

	HF	F1	NR	s.e.	P-value
Mean calving date 2007	Feb-28	Feb-25	Feb-25		NS
Calving to 1st service interval (days)	74^a	72^b	71^b	0.7	<0.05
Pregnancy rate to 1st service (%)	45^a	56^b	55^b		<0.01
6 week in-calf rate (%)	59^a	72^b	69^b		<0.001
13 week in-calf rate (%)	81^a	86^{ab}	85^a		0.07
Final in-calf rate (%)	85	88	87		NS
Calving to conception interval (days)	92^a	83^b	85^b	1.4	<0.001
Number of services per cow	1.82^a	1.64^b	1.60^b	0.052	<0.01

[a,b,c] Within-row means not sharing a common superscript differ significantly ($P<0.05$).

apparent low level of HV for milk yield between the two breeds (in the region of 1% based on the current study) Norwegian Red crossbred (F1) cows appear to exhibit a capability to produce to a similar production level to the HF. In the US, Heins *et al.* (2006a) demonstrated that Scandinavian Red×Holstein cows were capable of similar fat plus protein yields to HF cows under a high input regime. Similar to the study of Heins *et al.* (2006b), no adjustment was made to production values for differences in pregnancy between the breed groups. If appropriate adjustments were made differences in predicted 305 d production may be closer. The similarity in production between the HF and F1 cows, in particular, together with the observed improvements in reproductive efficiency with the F1 may offer positive implications for production potential at farm level. Differences in production expressed as 305 d predicted values are likely to be different to actual yields delivered, particularly under seasonal production circumstances, as differences in actual production will arise as a consequence of changes in (1) calving pattern and (2) the proportion of cows surviving to develop into maturity. Therefore, a crossbred herd may in fact be more productive compared to a HF herd where production potential is compromised by poor reproductive efficiency. Similarly, lifetime performance is likely to be increased with crossbred cows (McAllister *et al.,* 1994; McDowell *et al.,* 1974).

In 2007 (second lactation) the F1 cows were only marginally more favourable (though not significantly so) compared to the HF, while the data from year 1 indicated a distinct advantage with the crossbred compared to the HF. According to the results, SCC with the HF did not increase with age. However, a subsequent examination of the data revealed that cows culled at the end of first lactation for reasons other than reproductive failure had mean SCS (SCC in parentheses) values of 2.09 (349,000 cells/ml), 1.89 (181,000 cells/ml) and 1.94 (138,000 cells/ml) for the HF (n=80), NR (n=18) and the F1 (n=24) cows respectively. Of these cows, 15%, 6% and 6% for the HF, NR and the F1, respectively, had mean SCC values during first lactation in excess of 400,000 cells/ml. This data suggests that greater culling for poor udder health took place within the HF breed group. More over, recent preliminary breeding values obtained from the ICBF (February 2008) also provide evidence of additive genetic improvement from crossbreeding with the NR with health sub index (weighted 75% towards udder health and 25% towards locomotion) values of €-1.67, €4.6, €1.9 for the HF, NR and F1 cows, respectively, across these study herds.

With the exception of Heins *et al.* (2006), the authors are unaware of other studies that have compared Norwegian or 'Scandinavian' Red crossbred cows with Holstein-Friesians. However, Walsh *et al.* (2007, 2008) have demonstrated the favourable udder health and reproductive efficiency characteristics of pure NR compared to HF cows. The current study is unique having all three (both pure bred and crossbred cows) contemporaneous on each farm. The crossbred cows in the current study displayed superior reproductive efficiency compared to the HF. This is in line with many previous studies investigating reproductive efficiency/survival of crossbred cows (Dickinson and Touchberry, 1961; Harris, 2005; Heins *et al.,* 2006b; Hocking *et al.,* 1988; Lopez-Villalobos *et al.,* 2000). The superior fertility performance observed in the current study is likely due in part to the additive genetic superiority of the NR breed as well as the contribution of hybrid vigour. Differences in calving to conception intervals between the breed groups indicate a slippage of approximately 7 days in calving interval for the HF in second lactation. However, in total, a difference of almost two weeks has now developed between the HF and the crossbred cows in terms of calving date in 2008. Both the pure NR and crossbred cows are expected to maintain a 365-day calving interval.

Imperative to the utilisation of crossbreeding as an alternative breeding strategy at farm level is the necessity for an across breed evaluation, thus enabling comparisons to be made with HF sires in a given environment. More detailed analyses involving the current data set will aid the Irish Cattle Breeding federation in Ireland in their bid to estimate accurate breeding values for Norwegian Red and other alternative breed sires.

Breeding for robustness in cattle

Conclusion

Initial data has been presented from a 3 year research study carried out by Moorepark, the objective of which is to evaluate the potential of dairy crossbreeding for Irish dairy farmers. Such studies are uncommon. Given the localised nature of dairy cattle breeding, studies of this type are useful to compare differences that may arise in traits such as milk yield, fertility, health and survival of different breeds/crossbreeds. The decision to crossbreed for many will likely be borne out of the frustration of poor herd health/fertility, though improved production potential of crossbred cows relative to pure parent breeds is possible. While an economic comparison has not been presented, considerable evidence indicates that crossbreeding with the Norwegian Red represents a viable option for dairy farmers in terms of improving herd profitability, in particular arising from improved reproductive efficiency. This is likely being obtained from a combination of additive genetic improvement and hybrid vigour. Differences in production expressed as 305 d predicted values are likely to be different to actual yields delivered as differences arise between the breed groups in terms of (1) calving pattern and (2) the proportion of cows surviving to develop into maturity. Thus compensating for differences in milk yield that may exist compared to pure HF herds.

Acknowledgements

The technical assistance of Ann Geoghegan, Billy Curtin and Tom Condon at Moorepark, as well as Sean Coughlan and Rachel Wood at ICBF is gratefully acknowledged. The commitment and efforts of the farmers involved in the Norwegian Red crossbreeding study is to be commended. Milk recording was provided free of charge for the experimental cows on the study. This support, provided by Progressive Genetics, Munster Cattle Breeding Group and ICBF is very much appreciated.

References

Auldist, M.J., S. Coats, G.L. Rogers and G.H. McDowell, 1995. Changes ion the composition of milk from healthy and mastitic dairy cows during lactation cycle. Aust. J. Exp. Agric., 35: 427-436.

Bartlett, P.C., J. van Wijk, D.J. Wilson, C.D. Green, G.Y. Miller, G.A. Majewski and L.E. Heider, 1991. Temporal patterns of lost milk production following clinical mastitis in a large Michigan Holstein herd. J. Dairy Sci., 74: 1561-1572.

Bereskin, B. and R.W. Touchberry, 1966. Crossbreeding dairy cattle. III. First lactation production. J. Dairy Sci., 49: 659-667.

Berry, D.P., B.L. Harris, A.M. Winkleman and W. Montgomerie, 2005. Phenotypic associations between traits other than production and longevity in New Zealand dairy cattle. J. Dairy Sci., 88: 2962-2974.

Berry, D.P. and P.R. Amer, 2005. Derivation of a helath sub-index for the economic Breeding Index in Ireland. Technical report to the Irish Cattle Breeding Federation (August). Available at: www.icbf.com/publications/files/genetics_of_udder_health_in_ireland_doc. Accessed May 5 2007.

Brandt, G.W., C.C. Brannon and W.E. Johnston, 1974. Production of milk and milk constituentsby Brown Swiss, Holsteins, and their crossbreds. J. Dairy Sci., 57: 1388-1393.

Dickinson, F.N. and R.W. Touchberry, 1961. Livability of purebred versus crossbred dairy cattle. J. Dairy Sci. 44: 879-887.

Esslemont, R.J., M.A. Kossaibati and J. Allock, 2001. Economics of fertility in dairy cows. In: M.G. Diskin (ed.) Fertility in the high producing dairy cow. British Society of Animal Science, Penicuik, Midlothian, Scotland, UK. Occasional publication No. 26, pp. 19-29.

Fohrman, M.H., 1946. A crossbreeding experiment with dairy cattle. BDIM-INF-30. USDA, Bureau of Dairy Industry, Washington DC, USA.

Hansen, 2006. Monitoring the worldwide genetic supply for dairy cattle with emphasis on managing crossbreeding and inbreeding. 8th World Congress on Genetics Applied to Livestock Production, August 13-18, 2006, Belo Horizonte, MG, Brazil.

Harris, B.L., 2005. Breeding dairy cows for the future in New Zealand. New Zealand Veterinary Journal, 53: 384-389

Heins, B.J., L.B. Hansen and A.J. Seykora, 2006a. Production of pure Holsteins versus crossbreds of Holstein with Normande, Montbeliarde, and Scandinavian Red. J. Dairy Sci., 89: 2799-2804.

Heins, B.J., L.B. Hansen and A.J. Seykora, 2006b. Fertility and survival of pure Holsteins versus crossbreds of Holstein with Normande, Montbeliarde, and Scandinavian Red. J. Dairy Sci., 89: 4944-4951.

Heringstad, B., G. Klemetsdal and J. Ruane, 2000. Selection for mastitis resistance: a review with focus on the situation in the Nordic countries. Livest. Prod. Sci., 64: 95-106.

Heringstad, B., Y.M. Chang, D. Gianola and G. Klemetsdal, 2003. Genetic analysis of longitudinal trajectory of clinical mastitis in first- lactation Norwegian Red cattle. J. Dairy Sci., 86: 2676-2683.

Hocking, P.M., A.J. McAllister, M.S. Wolynetz, T.R. Batra, A.J. Lee, C.Y. Lin, G.L. Roy, J.A. Vesely, J.M. Wauthy and K.A. Winter, 1988. Factors affecting length of herdlife in purebred and crossbred dairy cattle. J. Dairy Sci. 71: 1011-1024.

Lindhe, B. and J. Philpsson, 2001. The Scandinavian experience of including reproductive traits in breeding programmes. In: M.G. Diskin (ed.) Fertility in the high producing dairy cow. British Society of Animal Science, Penicuik, Midlothian, Scotland, UK. Occasional publication No. 26, pp. 251-261.

Livestock Improvement, 2006. Dairy Statistics 2005 – 2006. Livestock Improvement Corporation, Hamilton, New Zealand.

Lowman, B.G., N. Scott and S. Somerville, 1976. Condition scoring of cattle. Rev. ed. East of Scotland College of Agriculture. Bulletin no. 6. Edinburgh, UK.

Lopez-Villalobos, N.; D.J. Garrick, C.W. Holmes, H.T. Blair and R.J. Spelman, 2000. Profitability of some mating systems for dairy herds in New Zealand. J. Dairy Sci., 83: 144-153.

McAllister, A.J., A.J. Lee, T.R. Batra, C.Y. Lin, G.L. Roy, J.A. Vesely, J.M. Wauthy and K.A. Winter, 1994. The influence of additive and nonadditive gene action on lifetime yields and profitability of dairy cattle. J. Dairy Sci. 77: 2400-2414.

McDowell, R.E., J.A. Velasco, L.D. van Vleck, J.C. Johnson, G.W. Brandt, B.F. Hollon and B.T. McDaniel, 1974. Reproductive efficiency of purebred and crossbred dairy cattle. J. Dairy Sci., 57: 220-234.

Miglior, F., B.L. Muir and B.J. van Doormaal, 2005. Selection indices in Holstein cattle of various countries. J. Dairy Sci., 88: 1255-1263.

Mrode, R.A. and G.J.T. Swanson, 1996. Genetic and statistical properties of somatic cell count and its suitability as an indirect means of reducing the incidence of mastitis in dairy cattle. Anim. Breed. Abstr., 66: 847-857.

Olori, V.E. and J.B. Galesloot, 1999. Projection of partial lactation records and calculation of 305-day yields in the Republic of Ireland. Interbull Bulletin, 22: 149-154

Rincon, E.J., E.C. Schermerhorn, R.E. McDowell and B.T. McDaniel, 1982. Estimation of genetic effects on milk yield and constituent traits in ccrossbred dairy cattle. J. Dairy Sci., 65: 848-856.

SAS, 2006. User's Guide Version 9.1: Statistics. SAS Institute, Cary, NC.

Schutz, M.M., 1994. Genetic evaluation of somatic cell scores for United States dairy cattle. J. Dairy Sci., 77: 2113-2129.

VanRaden, P.M., 2004. Invited review: Selection on net merit to improve lifetime profit. J. Dairy Sci., 87: 3125-3131.

Veerkamp, R.F., E.P.C. Koenen and G. de Jong, 2001. Genetic correlations among body condition score, yield, and fertility in first parity cows estimated by random regression models. J. Dairy Sci., 84: 2327-2335.

Veerkamp, R.F., P. Dillon, E. Kelly, A.R. Cromie and A.F. Groen, 2002. Dairy cattle breeding objectives combining yield, survival and calving interval for pasture-based systems in Ireland under different milk quota scenarios. Livest. Prod. Sci. 76: 137-151.

Walsh, S., F. Buckley, D.P. Berry, M. Rath, K. Pierce, N. Byrne and P. Dillon, 2007. Effect of breed, feeding system and parity on udder health and milking characteristics. J. Dairy Sci., 90: 5767-5779.

Walsh, S., F. Buckley, K. Pierce, N. Byrne, J. Patton and P. Dillon, 2008. Effects of breed and feeding system on milk production, bodyweight, body condition score, reproductive performance and postpartum ovarian function. J. Dairy Sci., 91: 4401-4413.

Willham, R.L. and E. Pollak, 1985. Theory of heterosis. J. Dairy Sci., 68: 2411-2417.

Part 4
Energy balance

Derivation of direct economic values for body tissue mobilisation in dairy cows

E. Wall[1], M.P. Coffey[1] and P.R. Amer[2]
[1]Sustainable Livestock Systems Group, Scottish Agricultural College, Bush Estate, Penicuik, Midlothian, EH26 0PH, United Kingdom
[2]Abacus Biotech Limited, P.O. Box 5585, Dunedin, New Zealand

Abstract

This study presents a simplified schema for body energy mobilisation defining three traits to describe how body energy lost/gained in dairy cows. A theoretical framework was developed to derive economic weights for these traits accounting for changing feed costs during lactation. Results show that the economic values for body tissue mobilisation is dependent on the calving system employed. For example, the economic value for early lactation body mobilisation is positive (+11p) in an autumn calving system and negative (-14p) in a spring calving system. Any loss in early lactation in a spring calving system will need to be repaid towards the end of lactation when feed is more expensive. The opposite is true in an autumn calving system when it is economically sensible for a cow to lose body energy when feed is expensive in the winter and regain it when turned out to grass. This suggests that the economic cost of body tissue mobilisation is different dependent on the system of production. If it is, it may be necessary to consider customised indices allowing farmers to choose bulls on an index that is suitable for their system.

Keywords: energy balance, economic value, body tissue mobilisation

Introduction

Many studies have examined mobilisation of body tissue and partitioning of nutrients in dairy cows (Friggens and Newbold, 2007). It has been suggested that the increased capacity for milk production has resulted in the partial 'shift' of nutrient intake towards production output away from maintaining functional fitness (Veerkamp *et al.,* 2001). Cows that have higher milk production either have to eat more food to produce this milk (and therefore increase feed costs) or utilise their own body energy or both, potentially at the expense of their metabolic function. This information may be harnessed for selection purposes in terms of developing a breeding goal trait with an economic cost. The effective energy given up by the dairy cow during her lactation can be combined with the economic costs of replenishing that body energy on each day of lactation. Simply put, the selection goal trait is the maintenance of energy balance across first lactation with the outcome being a financial breeding value describing the cost of body mobilisation, which maybe related to profitability.

If there is a genetic component to the partitioning of energy to production and other functions versus maintenance of body fat reserves there is the potential for including body tissue mobilisation in national breeding programmes, although it's inclusion may not be simple. Firstly, breeding values for a trait(s) that underlie body tissue mobilisation have to be calculated and secondly, the economic value of body tissue mobilisation needs to be estimated in order to include it in a profit index.

The aim of this study was to derive descriptors of first lactation daughter body energy curves for sires. A method for the derivation of economic weights for these traits, which describe the feed costs associated with animals losing/gaining body energy at different times during the lactation when feed costs vary with time, was also developed.

Materials and methods

Figure 1 describes key reference points of body tissue mobilisation during the calving interval (lactation and dry period) of a dairy cow: (1) the body energy content (BEC_base) at the start of the lactation, (2) the minimum level of body energy content (BEC_min) which occurs at time LDAY_min, (3) and the body energy content at the end of a standard 305 day lactation BEC_305. Changes between these key reference points are assumed to be linear with respect to time. Thus there are linear representations of the average rates of body energy mobilisation (b_early) and body energy recovery (b_late) during the lactation. While this representation is simplistic, when computing economic values, it is the modelling of the change in the profiles due to a genetic change that is most relevant, and this is addressed below.

From this model three genetic traits were defined, which characterise the body tissue mobilisation profile. These are:
- *BEC_min - BEC_base*: the point of minimum body energy balance for the cow (MJ/EE/day). This can be described as the maximum absolute change in body energy content.
- *b_early*: the average rate of loss of body energy from the start of lactation to the point of minimum body energy balance (MJ/EE/day). This can be described as the rate of body energy content loss.
- *b_late*: the average rate of recovery of body energy from the point of minimum body energy balance until the start of the dry period (MJ/EE/day). This can be described as the rate of body energy content recovery.

From Figure 1 and the energy cost pattern through the lactation, the calving interval can be divided into three segments that deal with body tissue mobilisation. These three segments are:
- The period where BEC is reducing until it reaches a minimum. During this period, feed costs are essentially saved because the energy obtained from mobilisation would have to have been obtained from feed (assuming feed intake is not limiting).
- The period where BEC of the national herd is increasing back towards base BEC until the end of the lactation. During this period, there are additional feed costs associated with the recovery of body energy content.

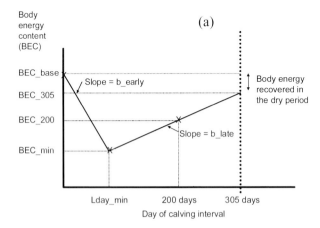

Figure 1. Stylised national average change in cow body energy content (BEC) through the lactation with key reference points including BEC at the start (base) and DIM 200 and 305 of first lactation, minimum BEC (BEC_min), rate of decline to BEC_min (b_early) and rate of recovery from BEC_min (b_late).

- The dry period where the remaining shortfall of BEC_305 relative to BEC_base has to be recovered using feed supplied during the dry period. As for segment 2, during this period, there are additional feed costs associated with the recovery of body energy content.

It is assumed that feed costs change in a stepwise manner per unit of metabolisable energy supplied and that the feed costs are assumed to be constant for days within the three periods day 0 to LDAY_min (EC_early), LDAY_min until the dry period (EC_late) and the dry period (EC_dry). A theoretical framework was developed to derive economic values for the three traits of body tissue mobilisation (for further details see Wall *et al.*, 2008).

Information on feed costs and practices in the UK to parameterise the feed cost model were based on the assumptions of Stott *et al.* (2005). The national profile of body energy mobilisation were derived from the study of Wall *et al.* (2007). Other population parameters required for the model were derived using production and fertility data from nationally milk recorded cows from January 2003 – December 2005. These population parameters were then combined with the model framework to derive economic weights for body tissue mobilisation during lactation. The model was firstly run using the calving pattern of UK first lactation cows from 2003-2005 in which cows calve all year round (Wall *et al.*, 2008). For comparison the model was then run assuming two different calving systems - a system that cows are managed so calves are born in the Spring months and therefore cows have a forage based diet opposed to a system that calves are born in the Autumn months.

Results and discussion

From the national profile of body energy mobilisation in cows it was estimated that the day that cows reached the lowest BEC was 57.9 days (s.d. = 26.16), which is close to the peak of first lactation. The mean drop in BEC from calving (min_BEC) was 330 MJ (s.d. = 229.78). The average rate of body energy loss until minimum was reached (b_early) was 6.09 MJ/day (s.d. = 2.89) and the average rate of recovery (b_late) was 3.09 MJ/day (s.d. = 1.32). A previous study should that the national profile of body energy mobilisation estimates that cows are predicted to end first lactation with a higher body energy content than at the start (Wall *et al.*, 2007). Previous work in experimental herd data (Coffey *et al.*, 2004) showed similar curves in terms of body tissue mobilisation from start to end of first lactation with the average cow predicted to increase body tissue energy across first lactation.

Figure 2 shows the economic values for the three traits that define body tissue mobilisation at 1st calving date starting from the 1st of September. If the economic weights are averaged following the annual calving pattern for heifers in the UK the economic value for the maximum drop in body energy content (min_BEC) is small and negative at -0.14 pence (p). The economic value for the rate of loss of body energy in early lactation (b_early) is also negative (-3.1p) and for the rate of body energy gain in later lactation (b_late) is large and positive (19.7p).

Figure 2 shows how the economic values for the three traits of body tissue mobilisation change dramatically depending on the date of calving. The economic value for min_BEC for the majority of the calving year was negative with the lowest value (-0.39p) occurring for cows calving in early September. The economic value for min_BEC was only positive for cows calving from mid March to mid April reaching a maximum of to 0.09p. The economic value for b_early also varies conditional on date of first calving ranging from –15.13p for cows calving between April and June inclusive to 10.92p for all other cows. The largest economic value was for b_late and ranged from -66.34p for calvings from September to the end of January to 96.08p for calvings from April to June.

The results suggest that the economic cost, and therefore economic value, for body tissue mobilisation is very dependent on the calving system employed (Figure 2). Generally, when the economic value for b_early is positive the economic value for b_late is negative. The calving pattern by month in

Breeding for robustness in cattle

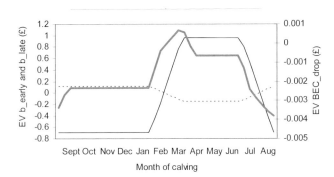

Figure 2. Economic values for components of body energy mobilisation during first lactation (min_BEC, b_early and b_late) dependent on day of calving from the initial model calving date of 1st of September.

the UK has flattened out over time as calving intervals increase and farmers move away from strict seasonal calving systems. Therefore, the average economic value for body energy mobilisation traits based on the flatter national calving pattern is appropriate. However, some farmers may still operate predominantly spring or autumn seasonal calving systems and the economic values differ depending on the type of system farmers employ. The economic value, for example, for b_early is positive in an autumn calving system and negative in a spring calving system. It would not be cost effective for a cow to lose body energy in a spring calving system when the cost of energy from feed is low as any loss in early lactation will need to be repaid towards the end of lactation when feeding indoors and energy is more expensive. The opposite is true for autumn calving when it is economically sensible for a cow to lose body energy when feed is expensive in the winter and regain it when turned out to grass.

The feed costs model used assumes that cows are grazing for part of the year (see Stott *et al.,* 2005 for more detail) and therefore feed costs rise when cows are indoors and the source of forage is silage. Concentrates are added to the diet to meet energy requirements based on lactation yield state. However, there will be large variation in the practices of farmers with some potentially not grazing cows at all, with many alternative feeding systems being practiced throughout the country. The different feeding systems will have differing costs of a MJ of metabolisable energy, which may differ at different times of the year of vary from year to year depending on markets. However, the framework can accommodate the costs of differing energy sources throughout the year for different systems. Developing a model framework that allows for variations in milk yield, feed energy costs and calving pattern will allow farm system specific economic values for body tissue mobilisation. It will also be easily adapted to account for any fluctuations in the feedstuffs market.

Many studies have shown a link between cow condition score with a decline in fitness traits (e.g. Veerkamp *et al.,* 2001). A proposed biological mechanism for this is that, with selection for production traits, that the modern dairy cow has 'shifted' nutrient partitioning towards production and away from functional fitness. Many breeding indices around the world now incorporate fitness traits in their national dairy indices (Miglior *et al.,* 2005). However, the direct incorporation of body energy mobilisation in such indices is not routine. This methodology provides a mechanism for including body energy mobilisation, accounting for differences in management practices, in an economic index to help breeders select for an increased range of robustness traits.

Breeding for robustness in cattle

Conclusions

This study simplifies the description of body energy mobilisation across first lactation into three parameters for which breeding values can be estimated. This study also develops a model that allows for the derivation of economic values for the traits which were shown to vary according to a number of key parameters (e.g. seasonality of calving system, feed costs). The framework provides the first step in selecting sires based on an index of body energy mobilisation of their daughters.

References

Coffey, M.P., G. Simm, J.D. Oldham, W.G. Hill and S. Brotherstone, 2004. Genotype and diet effects on energy balance in the first three lactations of dairy cows. J. Dairy Sci., 87: 4318-4326.

Friggens, N.C. and J.R. Newbold, 2007. Towards a biological basis for predicting nutrient partioning: the dairy cow as an example. Anim., 1: 87-97.

Miglior, F., B.L. Muir and B.J. van Doormaal, 2005. Selection Indices in Holstein Cattle of Various Countries. J. Dairy Sci., 88: 1255-1263.

Stott, A.W., M.P. Coffey and S. Brotherstone, 2005. Including lameness and mastitis in a profit index for dairy cattle. Anim. Sci., 80: 41-52.

Veerkamp, R.F., E.P.C. Koenen and G. de Jong, 2001. Genetic correlations among body condition score, yield, and fertility in first-parity cows estimated by random regression models. J. Dairy. Sci., 84: 2327-2335.

Wall, E., M.P. Coffey and S. Brotherstone, 2007. The relationship between body energy traits and production and fitness traits in UK dairy cattle. J. Dairy Sci., 90: 1527-1537.

Wall, E., M.P. Coffey and P.R. Amer, 2008. A theoretical framework for deriving direct economic values for body tissue mobilization traits in dairy cattle. J. Dairy Sci., 91: 343-353.

The link between energy balance and fertility in dairy cows

G.E. Pollott[1] and M.P. Coffey[2]
[1]Royal Veterinary College, Royal College Street, London, NW1 0TU, United Kingdom
[2]Sustainable Livestock Systems Research Group, SAC, Sir Stephen Watson Building, Bush Estate, Penicuik, Midlothian, EH26 0PH, United Kingdom

Abstract

The declining energy balance characteristics of modern dairy cows have been implicated in a reduction of fertility throughout a number of countries. The link between various aspects of energy balance in early lactation and fertility has been difficult to define using traditional statistical methods. One reason for this may be that both traits are continuously varying over time in different patterns and it could be the sequence of events that is more important than the quantitative link between them. Cows from a research herd were progesterone profiled for the first 140 days of lactation to identify luteal cycling activity. Behavioural oestrus was also recorded. Cows were weighed daily and condition scored weekly and these data used to monitor energy balance. Characteristics of both energy balance and luteal cycles were used to identify a key sequence of events which may precipitate the day of first observed heat. In 58% of lactations first heat was observed once cows had gone into positive energy balance before the start of the preceding luteal cycle, had a high level of progesterone in that cycle and were in positive energy balance at the time of oestrus. A further 12% of cows with a high energy balance nadir or a fast rate of recovery to positive energy balance showed first heat just prior to the return to positive energy balance. Certain health conditions delayed the onset of luteal cycling and day of first oestrus and so analyses were carried out for healthy cows separately from those with a health condition. This aligning of fertility and energy balance cycles approach holds out more hope of elucidating the link between energy balance and fertility than traditional statistical methods.

Keywords: energy balance, fertility, health effects, silent heats

Introduction

The declining fertility in dairy herds, both at the phenotypic and genetic level, appears to be a feature of dairy industries in many countries including the US (Butler and Smith, 1989; Beam and Butler, 1999), UK (Royal et al., 2000, Wall et al., 2003) and Norway (Ranberg et al., 2003). This decline in fertility is often ascribed to increased negative energy balance in early lactation, brought about by higher milk yields (Butler and Smith, 1989). Several studies have attempted to demonstrate the link between characteristics of energy balance in early lactation and different measures of fertility (see for example Butler et al., 1981; De Vries et al., 1999). Despite reporting several significant associations, the predictive ability of these relationships has proved to be low with R^2 values of 0.36 (Butler et al., 1981) and 0.22 (De Vries et al., 1999).

Many of the measures of fertility used to study the relationship between energy balance and fertility rely on farm observation which may not always reflect the real underlying physiological state of the animal. The traditional on-farm measurement of fertility serves a useful purpose but is vulnerable to missing key events through silent heats and poor observation. Progesterone profiling provides a more objective method for tracking reproductive events in dairy cows (Lamming and Bulman, 1976) and not only has it become the method of choice in fertility research, it also is being developed for on-farm use.

The Langhill herd maintained by The Scottish Agricultural College (SAC) at its Crichton Royal Farm (CRF) provides the opportunity to study the relationship between energy balance and fertility. This herd is monitored for feed intake, condition scored and weighed regularly, and feeds sampled and analysed routinely. In addition milk progesterone profiles have been monitored, and thus provide a more objective view of a cow's reproductive physiology, in addition to displays of oestrus. This paper explores the link between energy balance and fertility using data from the CRF herd.

Materials and methods

The experimental herd

The data used in this study were collected from cows on SAC's Dairy Research Centre at CRF, Dumfries, Scotland (55° 02' N, 3° 34' W; ~40 m above sea level). The cows had previously comprised the Langhill Herd, Edinburgh (Veerkamp *et al.*, 1995; Langhill, 1999; Pryce *et al.*, 1999) and transferred to CRF in September 2001. The herd was part of a long-term breeding experiment comparing two genetic lines on two systems of production. The herd consisted of approximately 200 milking cows divided evenly between four groups made up from all possible combinations of the two genetic lines and two production systems, described by Pollott and Coffey (2008). The various combinations of high and low genetic merit cows kept on high-concentrate and high-forage diets resulted in a wide range of energy balance profiles being found in the herd. This set of energy balance profiles was used to investigate the link between energy balance and fertility in this study.

Traits and recording

Fertility in dairy cows is commonly measured as a series of time related events and/or intervals between them. However, the use of milk progesterone profiles allows an insight into one set of key underlying mechanisms. A typical normal progesterone profile is illustrated in Figure 1. This cow first exceeded the threshold level of progesterone of 3 ng/ml (Lamming and Bulman, 1976) on day 35, had a cycle with no service, had another cycle starting on day 53, was served and had a positive pregnancy diagnosis on day 109. Milk samples were taken on Mondays, Wednesdays and

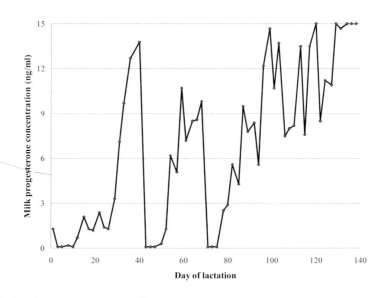

Figure 1. A typical progesterone profile.

Fridays each week from cows in the herd calving from September 2003 to August 2005, during the first 140 days of the subsequent lactation. The milk samples were analysed by enzyme-linked immunoabsorbant assay (Ridgeway Science Ltd.) and milk progesterone levels recorded (ng/ml). At least two consecutive readings of ≥3 ng/ml were taken as evidence of luteal activity. The trait data were calculated from the progesterone readings using the methods outlined by Royal *et al.* (2000). These included interval from calving to first luteal activity (CLA; days) and average progesterone level (ng/ml) during the oestrus cycle. In addition the number of cycles per lactation was counted.

Cows were milked three times per day and mostly fed and maintained indoors. This provided many opportunities for farm observation for signs of breeding activity. The days of heats, services and calvings were recorded. This provided data from which to calculate the day of first observed heat (DFH), the day of first service and, in combination with subsequent calving information, the day of successful service. One further trait, the incidence of silent heats, was calculated by combining the farm and cycle data. In this case a silent heat was assumed to have occurred when a cycle, identified from the progesterone profiles, was not preceded by an observed heat.

Cows were weighed at every milking and their liveweights recorded. They were also condition scored weekly using the six-point scale of Lowman *et al.*(1976). Seventh order polynomials were fitted to each condition score profile to obtain an estimate of condition score on each day of lactation for every cow. In addition the occurrence of any health events was recorded daily for each cow. In these analyses calving difficulty, metritis, mastitis, cystic ovaries, milk fever, non-feeding behaviour, locomotion and retained placentae were the main health events used in the analyses. Calving difficulty was scored on a five-point scale (normal, assisted normal, malpresentation, caesarean and abortion) and locomotion was classed as good or poor using the scale of Manson and Leaver (1988).

Energy balance

The mean daily liveweight and the estimated daily condition score were used to estimate the daily energy content of each cow using the EB2 methodology of Coffey *et al.* (2001). Energy balance was calculated as the difference in energy content between any given day and the previous day. A typical energy balance profile is shown in Figure 2. A number of traits were calculated from the energy balance curve; day of return to positive energy balance (PEB), day of energy balance nadir (lowest point in the energy balance profile), level of energy balance nadir, mean energy content of the animal over the first 25 days of lactation and the mean energy balance over the first 25 days of lactation.

Inspection of individual cow energy balance profiles throughout lactation indicated that there was a variety of patterns found in the herd. These are described in Table 1 and shown in Figure 3. The energy balance profiles were categorised into the six pattern types shown in Figure 3 and Table 1.

Statistical analyses

In exploring the link between energy balance and fertility the question arose about the effect of health events on either or both of the variables being explored. As a first analysis the distribution of health events between various fixed effects classes was explored using chi-squared tests. Seven health conditions recorded at CRF were investigated to see if they could be associated with genetic line, production system, lactation number and energy balance profile type. Two different sets of analyses were undertaken; the first looked at health events prior to the commencement of luteal activity and the second prior to day of first heat. Each health condition/fixed effect combination was analysed using a chi-squared test to see if there was a higher incidence than expected in any particular sub-class. Where there was a significant deviation from expectation the class responsible for the deviation was noted.

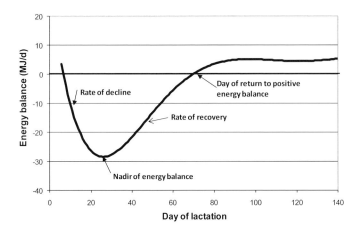

Figure 2. A typical energy balance profile.

Table 1. Type of energy balance patterns shown in Figure 3.

Type	Description
1	Day 1 at 0 balance; rapid drop in early lactation; recovery to, and maintained in, positive energy balance thereafter.
2	Day 1 in deep negative energy balance; recovery to, and maintained in, positive energy balance thereafter.
3	Day 1 in deep negative energy balance; never recovers to positive energy balance.
4	Day 1 in large positive energy balance; falls into negative energy balance and then recovers later to positive energy balance.
5	Day 1 in large positive energy balance; falls but maintained around 0.
6	Very little movement about 0 throughout lactation.

The relationship between health events and fertility was investigated by using a mixed model in a REML analysis (SAS MIXED Procedure; SAS, 2004). Both day of commencement of luteal activity and day of first heat were analysed using the mixed model shown in Equation 1.

$$Y_{ijklmnopqrst} = \mu + G_i + S_j + LN_k + P_l + CD_m + MT_n + MS_o + CO_p + LC_q + R_r + A_{s(ij)} + e_{ijklmnopqrst} \quad (1)$$

where:
$Y_{ijklmnopqrst}$ was either the day of commencement of luteal activity or day of first heat;
μ the overall mean;
G_i the effect of the ith genetic line (i = 1, 2);
S_j the effect of the jth production system (j = 1, 2);
LN_k the effect of the kth lactation number (k = 1 to 3);
P_l the effect of the lth energy balance profile (l = 1 to 6);
CD_m the effect of the mth calving difficulty type (m = 1 to 5);
MT_n the effect of a metritis occurrence (n = 1, 2);
MS_o the effect of the oth mastitis occurrence (o = 1, 2);
CO_p the effect of the pth occurrence of cyctic ovaries (p = 1 to 4);
LC_q the effect of the qth locomotion condition (q = 1, 2);

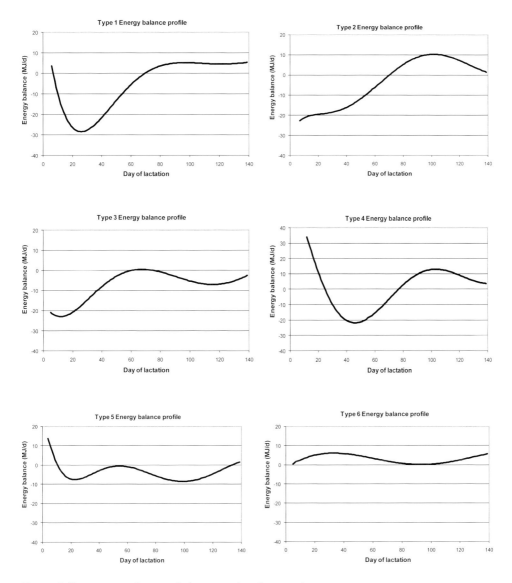

Figure 3. Six patterns of energy balance used in these analyses.

R_r the effect of the rth retained placenta (r = 1, 2);

$A_{s(ij)}$ the effect of the sth cow within genetic line/production system group distributed with variance σ_a^2; and

$e_{ijklmnopqrst}$ the residual term with variance σ_e^2.

Aligning states approach

The general failure of approaches using energy balance characteristics to predict measures of fertility may be due to the nature of the two traits involved. Energy balance follows one of the patterns shown in Figure 3 whilst heat opportunities occur at particular points in the series of oestrus cycles, as shown in Figure 1. If the energy balance pattern is superimposed on the luteal cycles then it becomes clear why a trait such as day of first heat is difficult to predict from characteristics of the energy balance curve. If a cow has 21-day oestrus cycles then heat can only be observed every 21 days. However, energy balance may change dramatically during the 21-day oestrus cycle. It may be better to look at the prevailing states of both curves and see which characteristics are aligned when oestrus occurs. In order to do this each, heat opportunities up to the first observed heat was considered from the perspective of the energy balance state at the time.

Inspection of the correspondence between the energy balance and luteal cycle curves indicated a number of possible links between the two. The first cycle was rarely preceded by an observed heat. Heat was observed when the cow was in positive energy balance at the start of the preceding luteal cycle. Heat was observed following a luteal cycle with high levels of progesterone (mean progesterone level >10 ng/ml). Heat was observed when the cow was in positive energy balance. Such relationships showed less concordance in heifers. Cows which had a high nadir of energy balance or a fast rate of recovery tended to come into heat before a return to positive energy balance. These observations were put together and each lactation was categorised as to whether first heat occurred under a specific set of conditions. This summary was carried out for healthy cows and 'unhealthy' cows separately and their results compared using a chi-squared test.

Results

The distribution of the 305 lactations between health status and energy balance profile type is shown in Table 2. Only one lactation was classified in energy balance profile Type 5 and the largest groups were Types 1, 2 and 4. Cows with a health condition had reduced fertility characteristics, and lactations with energy balance profiles 3 and 6 had longer intervals between parturition and the commencement of luteal activity.

The relationship between health conditions and various environmental factors – *Chi* **squared analyses**

A series of chi-squared analyses was carried out on the incidence of various diseases and conditions occurring before the commencement of luteal activity and before the day of first heat by genetic line,

Table 2. The means and distribution of lactations by health status and energy balance profile type for day of commencement of luteal activity (CLA) and day of first heat (DFH).

Energy balance type	No.	Mean CLA (d)	No.	Mean DFH
Healthy cows				
1	33	29.3	29	59.4
2	79	29.3	76	57.4
3	3	38.7	3	99.0
4	31	29.9	27	60.1
6	9	37.2	9	46.7
Total	155	30.1	144	58.5
Unhealthy cows				
1	16	41.0	16	81.4
2	90	37.0	90	79.0
3	4	70.5	4	96.8
4	36	34.9	36	89.2
5	1	76.0	1	71.0
6	3	55.7	3	104.3
Total	150	38.5	150	82.6

production system, energy balance profile type and lactation number. These are summarised in Table 3, except for energy balance profile type which was not found to contain any significant differences.

Up to the commencement of luteal activity, there were differences between the genetic lines in the incidence of mastitis and cows being off their feed. In both cases the Control Line had fewer occurrences of the condition than the Selection Line cows. Comparing the production system results, there was a higher incidence of locomotion problems in High-Concentrate fed cows. The incidence of several conditions was related to lactation number. Calving difficulty, metritis, and retained placentae were more common in heifers than older cows, whilst locomotion problems and milk fever was less common in heifers. Up to the day of first heat, the results were similar to those found up to the commencement of luteal activity.

Health and fertility – mixed model

A mixed model was constructed to investigate the effect of various health conditions on fertility, as measured by the commencement of luteal activity and the day of first heat. An ANOVA summary is shown in Table 4 for these two traits. The least-squares means of the significant health effects are summarised in Table 5.

The commencement of luteal activity was affected by four health conditions; calving difficulty, metritis, locomotion problems and retained placentae, but not by energy balance profile type. The day of first heat was affected by metritis, cystic ovaries and locomotion problems.

Table 3. The distribution of health conditions between genetic line, production system and lactation number up to commencement of luteal activity and day of first heat from cows at Crichton Royal Farm.

	Calving difficulty	Mastitis	Metritis	Locomotion	Milk fever	Off feed	Retained placenta
Up to CLA							
Genetic line	ns	**	ns	ns	ns	*	ns
Control		less				less	
Production system	ns	ns	ns	*	ns	ns	ns
High concentrate				more			
Lactation number	***	ns	*	***	**	ns	**
1	more		more	less	less		more
2	less		less		less		less
3+	less			more	more		less
Up to DFH							
Genetic line	ns	**	ns	ns	ns	*	ns
Control		less				less	
Production system	ns	ns	ns	*	ns	ns	ns
High concentrate				more			
Lactation number	***	ns	*	***	**	ns	**
1	more		more	less	less		more
2	less		less	less	less		less
3+	less			more	more		less

$* = P < 0.05$; $** = P < 0.01$; $*** = P < 0.001$; ns $= P > 0.05$.

Breeding for robustness in cattle

Table 4. ANOVA summary of fitting a mixed model to investigate the effect of health traits on the commencement of luteal activity and day of first heat in dairy cows at Crichton Royal Farm.

	Commencement of luteal activity	Day of first heat
Genetic line	**	**
Production system	ns	ns
Lactation number	**	ns
Energy balance profile type	ns	ns
Calving difficulty	*	ns
Metritis	***	***
Mastitis	ns	ns
Cystic ovaries	ns	**
Locomotion	***	**
Retained placenta	*	ns

$* = P < 0.05$; $** = P < 0.01$; $*** = P < 0.001$; ns $= P > 0.05$.

Table 5. Least-squares means for the commencement of luteal activity and day of first heat as affected by various health traits.

		Commencement of luteal activity (d)	Day of first heat
Calving difficulty	normal	42.7[a]	ns
	assisted normal	39.1[ab]	
	malpresentation	47.1[abc]	
	caesarean	16.5[abcd]	
	abortion	27.9[d]	
Metritis	no	28.1	84.2
	yes	37.6	108.2
Cystic ovaries	none	ns	67.7[a]
	1 occurrence		109.6[b]
	2 occurrences		123.9[bc]
	3 occurrences		90.3[abc]
Locomotion	good	27.3	85.1
	poor	38.7	107.1
Retained placenta	none	29.4	ns
	retained	36.6	

[a-d] Means with the same superscripts within a health category were not significantly different.

The extent of the health effects on the commencement of luteal activity varied. Normal calvings and malpresentations did not affect the start of luteal activity differently but cows which had aborted fetuses started cycling much earlier. Animals suffering from metritis, poor locomotion and retained placenta all had a delayed commencement of luteal activity by about 10 days compared to unaffected cows. Calving difficulty and retained placentae did not affect day of first heat but cows with metritis or poor locomotion delayed it by about 23 days. Cycles resulting in cystic ovaries delayed the occurrence of first heat by at least 30 days, the more cystic ovaries occurring before heat the more delay there was.

A heat opportunities approach

Every cycle, as indicated by the progesterone profile of the lactations under study, should be preceded by an ovulation and a heat. The absence of an observed heat before a cycle is referred to as a silent heat and has been shown to occur in this dataset in 60.9% of the cycles in the CRF herd (Pollott and Coffey, 2008). Every cycle can be considered to be preceded by a 'heat opportunity' and the outcome of a heat opportunity can be investigated. This will either be an observed heat or not. The conditions that are prevalent at the time of each heat opportunity can be analysed to see what conditions are associated with an observed heat at each heat opportunity.

The distribution of heat opportunities and observed heats by cycle number are shown in Table 6. The percentage of heat opportunities that resulted in observed heats is also shown in Table 6. Overall, 34.2% of all heat opportunities resulted in an observed heat. However, very few first cycles were preceded by an observed heat and these cycles will be ignored in the following analyses.

Table 7 summarises the conditions under which the first heat in a lactation was observed. The majority of cows exhibited their first heat at the first opportunity following the specified sequence of events; the cow must have been in positive balance by the start of the preceding luteal cycle, that cycle must have had a high level of progesterone and the cow must have been in positive energy balance at the

Table 6. The number of heat opportunities and observed heats by cycle number.

Cycle number	Number of heat opportunities	Number of observed heats	% observed heats
1	327	4	1.2
2	327	112	34.3
3	300	124	41.3
4	216	107	49.5
5	146	86	58.9
6	65	39	60.0
7	9	3	33.3
Total	1,390	475	34.2

Table 7. Summary of prevailing characteristics at the time of first observed heat for both healthy cows and those with a health condition prior to first heat; number of cows and percentage of health group (in parentheses).

	Healthy cows (n=142)	Unhealthy cows (n=144)
Met criteria[1]	89 (63)	76 (53)
Observed heat 1 cycle later	3 (2)	20 (14)
Heat immediately before return to PEB – low nadir	16 (11)	10 (7)
Heat immediately before return to PEB – fast rate of recovery	9 (6)	2 (1)
Later first heat - unexplainable	25 (18)	36 (25)

[1] Criteria were: cow in positive energy balance at the start of the preceding luteal cycle; mean progesterone level >10 ng/ml during preceding luteal cycle; cow in positive energy balance at the time of heat.

time of oestrus. Either the method of measuring energy balance did not give a true reflection of the actual day of return to positive energy balance or else cows were able to exhibit oestrus before they were exactly back to PEB. Seventeen healthy cows and 8 cows with a health condition had their first heat prior to the return to positive energy balance; most of these had a very high nadir of energy balance and so were not in deep negative energy deficit, whilst a small number were recovering rapidly at the time of first heat. Eighteen percent of healthy cows and 25% of cows with a health condition had their first heat later than expected with no apparently explainable reason. A chi-squared test was carried out on the data in Table 7 and there was a significant deviation of observed results from expectation ($P<0.001$). There was no difference between the two health status groups for cows meeting the stated criteria but more non-healthy cows had their first heat during the cycle after expected and more healthy cows had their first heat just before reaching positive energy balance. Inspection of the data by energy balance profile type showed that more cows with an unexplainably late first heat were found to be Type 2 cows (see Figure 3) and several of these were heifers.

Discussion

This paper describes work which has attempted to link various aspects of energy balance with fertility in post-partum dairy cattle. A number of factors relating to the recording and calculation of both energy balance and fertility may have affected the results described. The point at which a cow moves from being in negative energy balance to being in positive energy balance may be very precise but the method used here to define that point could be considered crude and therefore may be less accurate than is desirable. Also much of this work relies on the observation of oestrus which could be missed for a number of practical reasons; occurrence at night, faint signs of oestrus etc.

Given the difficulties outlined above the approach used in this paper has been quite successful at predicting day of first observed heat given information on both the energy balance of the cow and its underlying hormonal profile during early lactation. There is clearly still work to be done refining the method, particularly with respect to the exact point when a cow signals that it is in positive energy balance, and hence ready to express heat. This was clearly demonstrated by the number of cows with a high nadir of energy balance or which approached positive energy balance very rapidly which also expressed oestrus behaviour just prior to a return to positive energy balance. It is difficult to compare this method with the regression approach of Butler *et al.* (1981) in terms of success. Not all first heats were predicted correctly here but the characteristics of energy balance which influence fertility have been investigated and the relationship between them and fertility has been highlighted.

The link between energy balance and cow health has not been investigated here but the effect on fertility of different health conditions has been shown. Calving difficulties and retained placentae influence the onset of luteal activity but not day of first heat. Other conditions, such as metritis and poor locomotion affected both fertility traits. Further work is needed to link energy balance to health conditions but these results suggest that some health problems which may be linked to energy balance also affect fertility.

The effect of energy balance profile on fertility indicated that cows which did not get back into positive energy balance had delayed luteal activity and first heat. The numbers of cows in this category was small so further results are needed in order to see if this is a general result.

Conclusions

The heat opportunities approach used in this work has highlighted how different aspects of a cow's energy balance affect fertility, as measured by the commencement of luteal activity and day of first heat. Clearly the return to positive energy balance is a critical feature but the level of nadir and the rate of return to positive energy balance are also important. Health events also affect fertility and if these can also be linked to energy balance then there may be further links between energy balance and fertility to explore.

Acknowledgements

The authors acknowledge the funding and support of the UK Department for Environment, Food and Rural Affairs, National Milk Records, Cattle Information Services, Genus, Cogent, Holstein UK, BOCM Pauls, Dartington Cattle Breeding Trust, and RSPCA through the LINK Sustainable Livestock Production Programme. The Scottish Agricultural College receives financial support from the Scottish Executive Environment and Rural Affairs Department.

References

Beam, S.W. and W.R. Butler, 1999. Effects of energy balance on follicular development and first ovulation in postpartum dairy cows. J. Repr. Fert. (Suppl.), 54: 411-424.

Butler, W.R. and R.D. Smith, 1989. Interrelationships between energy balance and postpartum reproductive function in dairy cattle. J. Dairy Sci., 72: 767-783.

Butler, W.R., R.W. Everett and C.E. Coppock, 1981. The relationships between energy balance, milk production and ovulation in postpartum Holstein cows. J. Animal Sci., 53: 742-748.

Coffey, M.P., G.C. Emmans and S. Brotherstone, 2001. Genetic evaluation of dairy bulls for energy balance traits using random regression. Animal Sci., 73: 29-40.

De Vries, M.J., S. van der Beek, L.M.T.E. Kaal-Lansbergen, W. Ouweltjes and J.B.M. Wilmink, 1999. Modeling of energy balance in early lactation and the effect of energy deficits in early lactation on first detected estrus postpartum in dairy cows. J. Dairy Sci., 82: 1927–1934.

Lamming, G.E. and D.C. Bulman, 1976. The use of milk progesterone radioimmunoassy in the diagnosis and treatment of subfertility in dairy cows. British Veterinary Journal, 132: 289.

Langhill, 1999. Report from the Langhill Dairy Cattle Research Centre, Roslin, UK.

Lowman, B.G., N.A. Scott and S.H. Somerville, 1976. Condition Scoring of Cattle. Edinburgh School of Agriculture, UK.

Manson, F.J. and J.D. Leaver, 1988. The influence of concentrate amount on locomotion and clinical lameness in dairy cattle. Animal Production, 47: 185-190.

Pollott, G.E. and M.P. Coffey, 2008. The effect of genetic merit and production system on dairy cow fertility, measured using progesterone profiles and on-farm recording. J. Dairy Sci., 91: 3649-3660.

Pryce, J.E., B.L. Neilson, R.F. Veerkamp and G. Simm, G. 1999. Genotype and feeding system effects and interactions for health and fertility traits in dairy cattle. Livest. Prod. Sci., 57: 193-201.

Ranberg, I.M.A., B. Heringstad, G. Klemetsdal, M. Svendsen and T. Steine, 2003. Heifer fertility in Norwegian dairy cattle: variance components and genetic change. J. Dairy Sci., 86: 2706-2714.

Royal, M.D., A.O. Darwash, A.P.F. Flint, R. Webb, J.A. Wooliams and G.E. Lamming, 2000. Delcining fertility in dairy cattle: changes in traditional and endocrine parameters of fertility. Animal Science, 70: 487-501.

SAS, 2004. The Statistical Analysis Software. SAS Institute; Cary, North Carolina, USA.

Veerkamp, R.F., G. Simm and J.D. Oldham, 1995. Genotype by environment interactions: experience from Langhill. In: Breeding and feeding the high genetic merit dairy cow, T.L.J. Lawrence, F.J. Gordon and A. Carson (eds.), British Society of Animal Science Occasional Publication. No. 19, pp. 59-77.

Wall, E., S. Brotherstone, J.A. Wooliams, G. Banos and M.P. Coffey, 2003. Genetic evaluation of fertility using direct and correlated traits. J. Dairy Sci., 86: 4093-4102.

Potential to genetically alter intake and energy balance in grass fed dairy cows

D.P. Berry, M. O'Donovan and P. Dillon
Teagasc, Dairy Production Research Center, Fermoy, Co. Cork, Ireland

Abstract

There is currently a large gap in knowledge relating to estimates of genetic parameters in dairy cows fed predominantly grazed grass for two traits related to robustness, dry matter intake (DMI) and energy balance (EB). The objective of this study was to estimate, using data from 2 research farms, variance components for DMI and EB in multiparous Holstein-Friesian dairy cows. Data included in the present study consisted of 5,050 test-day records for DMI and 5,017 test-day records for EB, from 1,588 lactations on 755 cows. Variance components were estimated using random regression methodology and eigenvalues and eigenvectors were calculated from the additive genetic covariance matrix which were subsequently used to calculate the associated eigenfunctions and the genetic gain attributable to selection on the eigenvector of the additive genetic covariance matrix. Heritability for DMI and EB across days post-calving varied from 0.10 (8 days post-calving) to 0.30 (169 days post-calving), and, respectively, from 0.06 (29 days post-calving) to 0.29 (305 days post-calving). Genetic correlations between DMI across days post-calving were all ≥0.10; genetic correlations between EB across days post-calving varied from -0.36 to 1.00. Analysis of the expected responses to selection on the eigenvectors of additive genetic covariance matrix indicate large potential to genetically alter the shape of the lactation curves for both DMI and EB, thereby facilitating selection on the ability of the animal to remain close to nutritional homeostasis. This is seen as a characteristic of robustness.

Keywords: genetics, energy balance, intake, grass, dairy

Introduction

Several definitions of robustness in dairy cattle are possible varying from the ability of an animal to have a long productive life as well as the ability to perform in different environments (i.e. a generalist). In the present study dairy cow robustness is defined as the ability of the animal to remain close to nutritional homeostasis, or in other words minimise the extent and duration of negative energy balance (EB). Few studies have attempted to quantify the genetic variation in EB in Holstein-Friesian dairy cows (Svendsen *et al.*, 1994; Veerkamp *et al.*, 2000) although others (Van Arendonk *et al.*, 1991; Veerkamp *et al.*, 1995) have reported significant heritability estimates for a related trait, residual feed intake. Heritability estimates for EB (0.06 to 0.33; Svendsen *et al.*, 1994; Veerkamp *et al.*, 2000) and residual feed intake (0.14 to 0.38; Van Arendonk *et al.*, 1991; Veerkamp *et al.*, 1995) suggest that up to 38% of the variation in robustness, as defined here, may be attributable to genetic differences among animals. Nonetheless, all the aforementioned studies were undertaken on cows fed predominantly ensiled forages and concentrates in confinement production systems. Systems of milk production based on grazed grass predominate in Ireland and New Zealand, although the reduced costs associated with such systems have increased awareness in other countries such as Australia, Argentina and the UK and USA.

There is currently a gap in knowledge relating to estimates of genetic parameters for dry matter intake and energy balance in cows fed grazed grass. Therefore, the objective of this study was to estimate genetic parameters for dry matter intake (DMI) and EB in Holstein-Friesian multiparous cows fed predominantly grazed grass under controlled experimental conditions.

Materials and methods

The data used in the present study originated from several experiments (Buckley *et al.*, 2000; Kennedy *et al.*, 2003, 2006; O'Donovan and Delaby, 2005; Horan *et al.*, 2006; McCarthy *et al.*, 2007; McEvoy *et al.*, 2007) on 2 research farms comparing alternative genotypes of Holstein-Friesian dairy cows on different production systems or alternative grazing experiments run over the years 1996 to 2006. All cows grazed perennial ryegrass (*Lolium perenne*) pastures in a rotational grazing system. Concentrate fed varied from 113 to 1,452 kg per cow per year. Individual cow milk yield was recorded daily and milk fat and protein concentration was recorded weekly. Body weight was also recorded weekly.

Individual cow DMI was measured up to 4 times per lactation when cows were fed grass only or grass plus concentrates. Cow DMI was estimated using the n-alkane technique (Mayes *et al.*, 1986) as modified by Dillon (1993). A total of 5,118 records for DMI were available between calving and 305 days post-calving. Energy balance was estimated for all these test-days based on the Net Energy System as outlined by O'Mara (2000) using the data on DMI, body weight, milk production and stage of gestation. Animals calving greater than 200 days from the median age at calving within parity were discarded as were animals calving after the 20th week of the year. Following the retention of animals with a known sire and dam, 5,050 test-day records for DMI and 5,017 test-day records for EB, from 1,588 lactations on 755 cows remained. Pedigree information 4 generations deep was collated and consisted of 1,622 non-founder animals.

Additive genetic and within lactation permanent environmental variance components for DMI and EB were estimated using random regression methodology with Legrende polynomials in ASREML (Gilmour *et al.*, 2006). Residual variances were estimated within 6 stages of lactation: 8 to 50 days post-calving, 51 to 100 days post-calving, 101 to 150 days post-calving, 151 to 200 days post-calving, 201 to 250 days post-calving, and 251 to 305 days post-calving. A permanent environmental effect was also fitted across lactation. Fixed effects included in the model of analysis were treatment-test-date, fortnight of the year at calving, age at calving nested within parity and a cubic regression on days post-calving at the time of DMI and EB measurement interacting with parity. A linear regression on concentrate feeding level at the respective test-day was also included when the dependent variable was DMI.

The log-likelihood ratio test was used to determine the significance of higher order random regression coefficients. However, if the variance of the higher order regression coefficients was bound at zero then the immediately lower order polynomial was chosen. Eigenvalues and eigenvectors were calculated from the additive genetic covariance matrix and used to calculate the eigenfunctions. The genetic gain (ΔG_i) attributable to the ith eigenvector index was defined as:

$$\Delta G_i = \Phi \cdot K \cdot e_i \cdot \left(\frac{i}{\sqrt{\lambda_i}} \right)$$

where:
Φ is the matrix of Legrende regression coefficients;
K is the additive genetic covariance matrix estimated from the random regression model;
e_i is the ith normalised eigenvector;
i is the selection intensity (assumed to be one in the present study); and
λ_i is the eigenvalue associated with the ith eigenvector.

Results and discussion

A quadratic random regression was optimal to model both the additive genetic and within lactation permanent environment (co)variance components for both DMI and EB. Heritability for both DMI and EB changed during lactation (Figure 1). Heritability estimates for DMI varied from 0.10 at 8 days to 0.30 at 169 days post-calving while the heritability estimates for EB tended to increase with days post-calving, varying from 0.06 at 29 days to 0.29 at 305 days post-calving. The lower heritability estimates at the start of lactation in the present study were mainly attributable to greater residual variances for both traits and to a lesser extent lower genetic variance for DMI. Heritability estimates from the present study where grazed grass constituted the majority, if not all, of the cow's diet at the time of measurement were similar to previous international estimates from confinement systems of milk production for DMI (Koenen and Veerkamp, 1998; Veerkamp and Thompson, 1999) and EB (Svendsen *et al.*, 1994; Veerkamp *et al.*, 2000). Heritability estimates reported suggest that genetic selection for changes in DMI and EB may be fruitful.

Genetic correlations between DMI across days post-calving and between EB across days post-calving are described in Figure 2. Genetic correlations between days close to each other were strong as represented by the unshaded area in the center of Figure 2. Genetic correlations between DMI across days post-calving decreased in strength as the interval between days compared lengthened as represented by decrease in genetic correlations going from the center of Figure 2 to the top left (EB) and bottom right (DMI). The phenotypic correlations between DMI at different days post-calving varied from 0.05 (day 8 to day 246 post-calving) to 1.00 and between EB at different days post-calving varied from -0.06 (day 8 to day 218 post-calving) to 1.00. All genetic correlations between DMI at different days post-calving were positive with the weakest correlation (0.10) being between 8 days and 305 days post-calving. Genetic correlation between EB across days post-calving varied from -0.36 (between 8 days and 305 days post-calving) to 1.00. Low genetic correlations between days post-calving for DMI have also been reported in previous studies in cows fed total mixed rations (Koenen and Veerkamp, 1998; Veerkamp and Thompson, 1999). This suggests the influence of different genes on both DMI and EB at different stages of lactation. The low correlations also suggest a poor predictive ability of genetic merit for DMI or EB in early lactation from data generated in late lactation.

The ratio of the 3 eigenvalues of the additive genetic co-variance matrix to the sum of the 3 eigenvalues was 0.88, 0.11 and 0.001 for DMI and 0.67, 0.30 and 0.03 for EB. The shape of the eigenfunctions associated with each of the eigenvalues for DMI and EB are illustrated in Figures 3

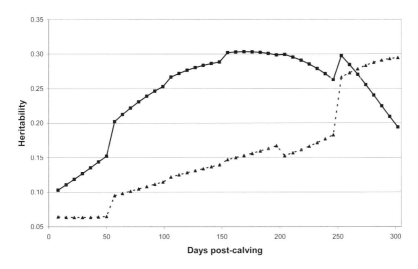

Figure 1. Heritability estimates across days post-calving for dry matter intake (squares and continuous line) and energy balance (triangles and broken line).

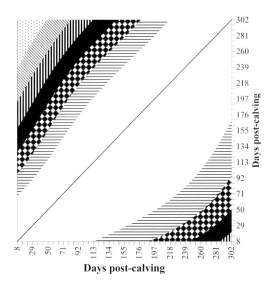

Figure 2. Genetic correlations among different days post-calving for energy balance (top left) and dry matter intake (bottom right). Range of genetic correlations represented are 0.80 to 1.00 (unshaded), 0.60 to 0.79 (horizontal lines), 0.40 to 0.59 (checked boxes), 0.20 to 0.39 (dark shaded), 0.00 to 0.19 (vertical lines), -0.20 to -0.01 (diagonal lines), -0.40 to -0.21 (dots).

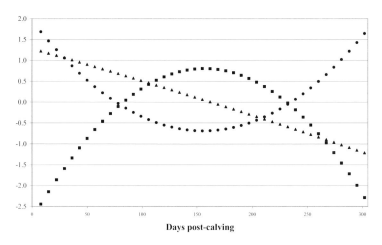

Figure 3. Eigenfunctions associated with the largest (■; 0.88), middle (▲; 0.11) and smallest (●; 0.001) eigenvalues for dry matter intake.

and 4, respectively. The eigenvalues and eigenfunctions of the additive genetic covariance matrix provide an insight into how the lactation profile of the trait under investigation is likely to change with genetic selection (Kirkpatrick and Heckman, 1989). The larger the eigenvalue the greater the expected rate of change in the shape of the lactation profile depicted by the associated eigenfunction. The largest eigenfunction for DMI changed sign during lactation and was negative early and late in lactation and positive in mid lactation. This suggests that selection on this eigenfunction will increase (decrease) DMI in early and late lactation but decrease (increase) DMI in mid-lactation. The eigenfunction associated with the largest eigenvalue for EB was positive throughout lactation

Breeding for robustness in cattle

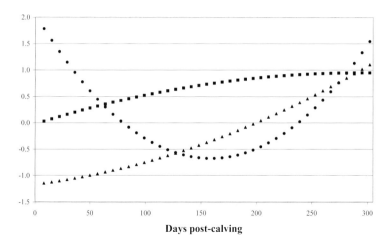

Figure 4. Eigenfunctions associated with the largest (■; 0.67), middle (▲; 0.30) and smallest (●; 0.03) eigenvalues for energy balance.

suggesting that the majority of genetic variance in EB is explained by a principal component acting equally throughout lactation.

Figure 5 shows the expected response in EB lactation profiles to selection using different relative weighting factors on the 3 eigenfunctions of the additive genetic covariance matrix. There is clearly a large potential to alter the shape of the lactation profile for EB using different relative weightings on the eigenfunctions. The optimal weighting factors may be derived using multiple trait selection index theory based on either economic impact of different lactation profiles or desired gains in areas of the lactation profile of most importance, for example those most strongly correlated with reproductive performance and health.

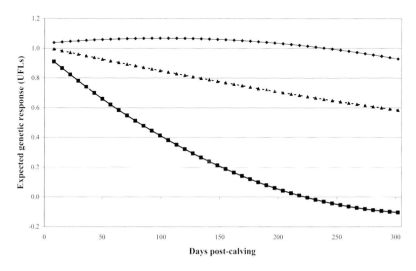

Figure 5. Expected genetic response to selection on the eigenfunctions associated with the largest, middle and smallest eigenvalue with respective weighting of 1:1:1 (squares with thick continuous line), 2:1:1 (triangles with broken line) and 2.5:1:1 (diamonds with thin continuous line).

Conclusions

There is currently a gap in knowledge relating to genetic parameters for DMI and EB, two traits associated with robustness, in dairy cows fed a basal grazed grass based diet. Results from the present study indicate significant genetic variation in DMI and EB with heritability estimates consistent with previous estimates from cows fed total mixed rations. Analysis of the eigenvalues and eigenfunctions of the additive genetic covariance matrix also indicate the ability to genetically alter the profile of these lactation curves to potentially improve dairy cow robustness.

References

Buckley, F., P. Dillon, M. Rath and R.F. Veerkamp, 2000. The relationship between genetic merit for yield and liveweight, condition score and energy balance of spring calving Holstein-Friesian dairy cows on grass based systems of milk production. J. Dairy Sci., 83: 1878-1886.

Dillon, P., 1993. The use of n-alkanes as markers to determine intake, botanical composition of available or consumed herbage in studies of digesta kinetics with dairy cows. Ph. D. Thesis, National University Ireland, Dublin, Ireland.

Gilmour, A.R., B.R. Cullis, S.J. Welham and R. Thompson, 2006. ASREML Reference Manual. New South Wales Agriculture, Orange Agricultural Institute, Orange, NSW, Australia.

Horan, B., P. Faverdin, L. Delaby, M. Rath and P. Dillon, 2006. The effect of strain of Holstein-Friesian dairy cows and pasture-based system on grass intake and milk production. Anim. Sci., 82: 435-444.

Kennedy, J., P. Dillon, P. Faverdin, L. Delaby, G. Stakelum and M. Rath, 2003. Effect of genetic merit and concentrate supplementation on grass intake and milk production with Holstein-Friesian dairy cows. J. Dairy Sci., 86: 610-621.

Kennedy, E., M. O'Donovan, J.P. Murphy, F.P. O'Mara and L. Delaby, 2006. The effect of initial grazing date and subsequent stocking rate on the grazing management, grass dry matter intake and milk production of dairy cows in summer. Grass Forage Sci., 61: 375-384.

Kirkpatrick, M. and N. Heckman, 1989. A quantitative genetic model for growth, shape and other infinite-dimensional characters. J. Math. Biol., 27: 429-450.

Koenen, E.P.C. and R.F. Veerkamp, 1998. Genetic covariance functions for live weight, condition score, and dry-matter intake measured at different lactation stages of Holstein-Friesian heifers. Livest. Prod. Sci., 57: 67-77.

Mayes, R.W., C.S. Lamb and P.M. Colgrove, 1986. The use of dosed and herbage n-alkanes as markers for the determination of herbage intake. J. Agric. Sci., Camb., 107: 161-170.

McCarthy, S., D.P. Berry, P. Dillon, M. Rath and B. Horan, 2007. Effect of strain of Holstein-Friesian and feed system on udder health and milking characteristics. Livest. Sci., 107: 1-28.

McEvoy, M., E. Kennedy, J.P. Murphy, T.M. Boland, L. Delaby and M. O'Donovan, 2008. The effect of herbage allowance and concentrate supplementation on milk production performance and dry matter intake of spring-calving dairy cows in early lactation. J. Dairy Sci., 91: 1258-1269.

O'Donovan, M. and L. Delaby, 2005. A comparison of perennial ryegrass cultivars differing in heading date and grass ploidy with spring calving dairy cows grazed at two different stocking rates. Anim. Res., 54: 337-350.

O'Mara, F., 2000. A net energy system for cattle and sheep. University College Dublin. Version 1.2.

Svendsen, M., P. Skipenes and I.L. Mao, 1994. Genetic correlations in the feed conversion complex of primiparous cows at a recommended and reduced plane of nutrition. J. Anim. Sci., 72: 1441-1449.

Van Arendonk, J.A., G.J. Nieuwhof, H. Vos and S. Korver, 1991. Genetic aspects of feed intake and efficiency in lactating dairy heifers. Livestock Prod. Sci., 29: 263-275.

Veerkamp, R.F., J.K. Oldenbroek, H.J. van der Gaast and J.H.J. van der Werf, 2000. Genetic correlations between days until start of luteal activity and milk yield, energy balance, and live weights. J. Dairy Sci., 83: 577-583.

Veerkamp, R.F., G.C. Emmans, A.R. Cromie and G. Simm, 1995. Variance components for residual feed intake in dairy cows. Livest. Prod. Sci., 41: 111-120.

Veerkamp, R.F. and R. Thompson, 1999. A covariance function for feed intake, live weight, and milk yield estimated using a random regression model. J. Dairy Sci., 82: 1565-1573.

Part 5:

Hot climate conditions

Profitable dairy cow traits for hot climatic conditions

A. De Vries[1] and J.B. Cole[2]
[1]*Department of Animal Sciences, University of Florida, Gainesville, FL 32611, USA*
[2]*Animal Improvement Programs Laboratory, Agricultural Research Service, USDA, Beltsville, MD 20705, USA*

Abstract

Permanent differences in environment have led to the distinct cattle races that are currently present. *Bos indicus* cattle evolved in the tropical areas of southern Asia. These cattle are heat tolerant through a sleek coat, loose skin, a high sweating capacity, long ears, resistance to ticks and reduced metabolic heat production. The major dairy breeds in the USA are *Bos taurus* cattle which evolved in temperate environments. They are less heat tolerant than *Bos indicus* cattle, but generally have higher reproductive rates, higher milk production, and better meat quality attributes. Differences in milk production have made the Holstein the preferred breed in much of the USA, including the South where cows may be subjected to heat stress during much of the year. The relative economic value of changes in various dairy traits may be different in hot Southern climates compared to temperate Northern climates in the USA. Major determining factors are the market prices of milk, feed, and replacement animals. In the southeastern USA with its fluid milk markets, the value of protein is nil while the value of skim milk is greater than in the North. The biological efficiency of milk production may be less under heat stress. The lack of quality premiums in much of the South reduces the value of changes in somatic cell counts. On the other hand, changes in fertility, productive life, body size, and calving ability are likely worth more in the southern than the northern USA. A light coat colour and slick hair would also have some value in hot climates. Direct selection for heat tolerance is possible within breeds, but could compromise overall performance. There is also variation within breeds to reduce the negative effects of heat stress, for example by putting more emphasis on functional traits. Crossbreeding may also overcome some of the negative effects. But the associated reduction in emphasis on milk production should be carefully considered, especially in Southeastern markets with higher milk prices. Dairy farmers in hot climates whose dairy cattle experience heat stress should value traits differently than dairy farmers in temperate climates.

Keywords: robustness, hot climate, traits, profitability

Introduction

Some dairy cows are more desirable than others. For dairy farmers, this typically means that more desirable cows are more profitable while being easy to handle. This contribution to the topic of robustness of dairy cattle discusses dairy cow traits that may be relatively more important (profitable) in hot climates than in more temperate climates in the USA. Historically, permanent differences in environment have led to distinct genetic adaptations to those environments. Adaptation can be defined as an animal's ability to survive, grow, produce milk, and reproduce in the presence of endemic stressors of its environment (e.g. parasites, diseases, climates, availability of nutrition). Genetic adaptation to an environment implies that animals have evolved in that environment over many generations and carry genes that allow them to survive and thrive there (Bourdon, 2000). Cattle genetically adapted to tropical environments with hot climates are physically different from cattle that are genetically adapted to temperate environments. First, we briefly discuss in our contribution differences in traits between cattle that evolved in these two environments.

The major breeds in the USA all evolved in temperate climates. Holstein is the dominant dairy breed (>90% of all dairy cattle in the USA (Jordan, 2003)) because of its high milk production. Other major breeds in the USA are Jersey, Ayrshire, Brown Swiss, Guernsey and Milking Shorthorn. These breeds are also kept in hot and (sub)tropical environments. The climate in the southern USA is hot enough to cause at least several months of heat stress, and there are concerns that global warming may further accentuate this problem (West, 2003). St-Pierre *et al.* (2003) estimated that the economic loss caused by heat stress was $897 million ($1 ≈ €0.64 in July, 2008) annually for the dairy industry in the USA. This equals almost $100 per adult dairy cow per year. Losses are primarily caused by decreased dry matter intake, decreased milk production, impaired fertility, and an increased death rate (St-Pierre *et al.,* 2003). Therefore, the second focus of this contribution is the relative importance of traits of the dominant Western breeds exposed to long periods of heat stress in the southern USA.

Thirdly, we briefly discuss within-breed selection and crossbreeding as options to overcome effects of heat stress. In our concluding remarks, we also address the phenomenon of genotype-by-environment interaction.

Dairy cattle traits developed in temperate and hot environments

Differences in permanent environments led cattle to evolve into two geographic races over 600,000 years ago (MacHugh *et al.,* 1997). Both were domesticated approximately 10,000 years ago (Loftus *et al.,* 1994). One group of breeds adapted mostly to temperate environments in Europe and the Near East (the non-humped taurine breeds, *Bos taurus)*. The other group of breeds adapted to more tropical environments in southern Asia (the humped Zebu breeds, *Bos indicus)*. A third distinct group evolved more recently in tropical environments. They are true *Bos taurus* that retain some of the productive attributes of *Bos taurus* but are better adapted to tropical environments. They include the southern African Sanga breeds, West African humpless breeds and Criollo breeds of Latin America and the Caribbean, e.g. Romosinuano (Prayaga *et al.,* 2006). The major dairy breeds in the USA are *Bos taurus* cattle with little influence from *Bos indicus*. Some beef cattle in the southeastern USA have a considerably amount of *Bos indicus* influence, such as the widespread Brangus cattle, a crossbred between the Brahman *(Bos indicus)* and Angus *(Bos taurus)* (Cartwright, 1980; Turner, 1980).

Genetic adaptation to one type of environment may decrease genetic adaption to other environments (Gotthard and Nylin, 1995; De Jong and Bijma, 2002). For example, *Bos indicus* cattle are well known for their ability to regulate their body temperature under heat stress conditions. Manifestations of effective thermoregulation include a sleek coat, high sweating capacity, loose skin, long ears, and reduced metabolic heat production. *Bos indicus* cattle are also resistant to the cattle tick, *Boophilus microplus*, which is prevalent in tropical climates worldwide. However, *Bos indicus* breeds have lower reproductive rates, lower milk production, and poorer meat quality attributes than *Bos taurus* breeds that are less adapted to the stressors of tropical climates. *Bos indicus* cattle may also be more difficult to handle under extensive management conditions (Prayaga *et al.,* 2006). *Bos taurus* cattle possess the ability to fatten on forage, providing insulation and energy reserves to help them survive cold winters. Tropical climates usually have marginal nutrition and income from both milk and meat is desired.

In general, tropically adapted genotypes are more robust than temperately adapted breeds if the environment changes from temperate to tropical or *vice versa*. A cow may be said to be robust if she is capable of coping well with variations (sometimes unpredictable variations) in her operating environment with minimal damage, alteration or loss of functionality (after www.wikipedia.org). In other words, robustness is the ability of an animal to adjust (adapt) to (short-term) changes in the environment. Adaptation implies here that the cow adjusts to changing environments with minimal effects on production. Therefore, more robustness means less variation in production when the environment changes. It usually follows that a robust cow performs less well in an optimal environment, but better under challenging conditions such as heat stress. This principle will be discussed for temperately adapted breeds (genotypes) and tropically adapted breeds (genotypes).

Temperately adapted breeds outperform tropically adapted breeds in temperate environments, but both genotypes perform quite well. In tropical environments, both genotypes perform at a lower level because the environment is more stressful (more heat, humidity, insects and parasites), but the loss in productivity is less for the tropically adapted breed. These genotype-by-environment interactions occur when the relative difference in performance between two genotypes depends on the environment they are in (Bourdon, 2000). Figure 1 shows the interaction.

In many countries with tropical climates, the preferred dairy cow is probably a cross with both *Bos indicus* and *Bos taurus* genes. Although the temperate breeds have potentially greater milk production, their lack of robustness results in significant production losses, higher mortality, and greater treatment cost than in temperate climates. Some *Bos indicus* influence is desired to overcome the severity of these negative effects (McDowell *et al.*, 1996). In tropical environments, the economic importance of cow and calf survival and calf growth is increased relative to milk yield and milk components (Madelena, 1986).

One of the sources of robustness is phenotypic plasticity which is the ability of a genotype to express different phenotypes in response to varying environments (Hoffman and Parsons, 1994; Strandberg, 2008). When traits are under directional selection (such as selection for increased milk yield) alleles affecting that trait are fixed (the gene frequency goes to 1). As alleles become fixed the additive genetic variance decreases, and there is little value in alternative alleles. Genes that affect the trait in stressful environments are probably not under consistent directional selection, resulting in an increase in additive genetic variance (and heritability) under stressful conditions. When the environmental differences between populations are small, as may be the case with intensively-managed livestock in controlled environments, then there is probably little value to robustness. The best genetically adapted breed to that environment is preferred. It is still possible that the more robust cow also performs the best in that environment, but the fact that she is more robust is not relevant. Robustness may be

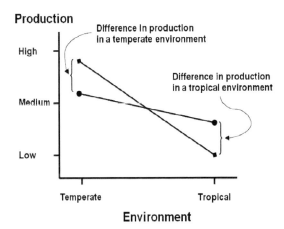

Figure 1. Example of a genotype-by-environment interaction. Tropically adapted breeds are generally more robust for changes in the environment. After Bourdon (2000).

Breeding for robustness in cattle

important in management systems that allow for significant variation in the environment because it increases the ability of the cow to respond to changes in weather, feed quality, and other factors. The optimal level of robustness depends on the loss under more favourable conditions and the gain under more challenging conditions.

Dairy cattle traits currently selected for in the USA

In the USA, the major dairy breeds (*Bos taurus*) are also used in the South with its hot summer climate. Southern climates vary from very hot during the day and cool at night with low humidity in Southwestern states such as Arizona, to hot and humid during the day with some cooling and high humidity at night in Southeastern states such as Florida. Cow performance is compromised under heat stress (Jordan, 2003; West, 2003). The question then is if the marginal values of important traits are different under environments causing heat stress than in temperate environments. For example, would improvement in fertility be worth more in regions with heat stress than in temperate regions?

A useful start to compare the economic value of dairy cow traits is a selection index. A selection index is a prediction of an animal's breeding merits for total economic merit (Shook, 2006). The objective of a selection index is to breed the most desirable (profitable) cow. Breeding goals, selection index theory, and applications were reviewed by Hazel *et al.* (1994), Groen *et al.* (1997), Goddard (1998), Dekkers and Gibson (1998) and VanRaden (2004).

Total economic merit is a linear combination of an animal's predicted transmitting abilities (PTA) for all economically important traits, with each trait weighted by its net economic value. Predicted transmitting ability measures the level of genetic superiority or inferiority an animal is expected to transmit to its offspring for a given trait. The economic value of a trait is the change in profit per unit change of the trait, given no change in other traits in the selection index. The economic values determine how selection pressure and selection response are distributed among traits (Shook, 2006). Total economic merit then is the expected economic value, or relative profit, of a particular animal's offspring compared to an average animal's offspring, or the difference in lifetime profit between a particular cow and an average cow of the same breed.

In the USA, PTAs for 27 traits are calculated quarterly by the USDA-Animal Improvement Programs Laboratory (AIPL) for bulls and cows (VanRaden and Multi-State Project S-1008, 2006; Cassell, 2007). Predicted transmitting abilities for protein, fat, and milk are expressed in pounds (0.454 kg) per 305-day lactation on a mature equivalent basis assuming two milkings per day. Somatic cell score (SCS) measures udder health and is a log2-transformation of somatic cell counts (SCC) (Schutz, 1994). Predicted transmitting abilities for SCS need to be reduced by 3 before they are used in the indices. Productive life (PL) measures the time between first calving and culling (VanRaden *et al.*, 2006). Seventeen linear type traits are evaluated for breeds other than Holstein (Gengler *et al.*, 1998); Holstein conformation evaluations are produced quarterly by the Holstein Association USA (Brattleboro, VT). For all breeds, udder, feet and legs, and body size are composite traits that are calculated from the linear traits, which are scored by trained evaluators from the individual breed associations. Calving ability (CA$) is an index that combines direct and maternal calving ease (dystocia) and stillbirths, and is expressed in dollars (Cole *et al.*, 2007a). Daughter pregnancy rate (DPR) measures the speed at which cows become pregnant, calculated as the probability of conception per 21 eligible days open (VanRaden *et al.*, 2004). Negative responses to SCS and body size are desired. Positive responses are desired for the other traits.

Ten traits are included in the net merit (NM$), cheese merit (CM$) and fluid merit (FM$) selection indices: milk, fat, and protein yield, SCS, PL, CA$, the udder, feet and legs, and body size composites, and DPR. The three indices provide estimates of lifetime profit under different assumptions about milk and components utilisation (VanRaden and Multi-State Project S-1008, 2006). Economic values for these traits depend on the prices for milk components, replacement heifers, cull cows, calves, milk quality premiums, feed cost, labour, veterinary, and breeding cost. The economic values of the traits do not include any double accounting. For example, a lower pregnancy rate will decrease productive life. The economic value of DPR therefore does not include the cost associated with culling because cull cost is already included in the economic value of PL.

The economic values are expressed as lifetime values. Lifetime is approximately three years or 2.57 lactations for Holsteins. For example, one extra kg of fat per 305 days is worth $5.95 over the animal's lifetime. The relative value for each trait is calculated by multiplying its economic value by its standard deviation of true transmitting abilities and then dividing the individual value by the sum of the absolute values. The relative value is expressed as a percentage of the total selection emphasis. Table 1 lists the current ten traits included in NM$, CM$, and FM$ with their economic values. Predicted transmitting abilities for these traits are calculated for Ayrshire, Brown Swiss, Guernsey, Holstein, and Jersey cattle, except that the PTA for calving ability are only available for Holsteins and Brown Swiss. For the remaining breeds, the relative values of the other nine traits in the NM$ and FM$ each increase by a factor 1.06 because the 6% of emphasis on CA$ is excluded. The relative values increase by 1.04 for the CM$. The NM$, CM$ and FM$ of an animal is then simply the sum of the PTAs for each trait multiplied by their economic values.

Lifetime net merit is a general selection index, and one of the most widely used. The economic values used in the CM$ are more appropriate in markets where milk is used principally for making cheese. The economic values used in the FM$ are more appropriate in markets were milk is primarily

Table 1. The ten traits used in the net merit (NM$), cheese merit (CM$) and fluid merit (FM$) selection indices in the USA (2006 revision). Sources: VanRaden and Multi-State Project S-1008 (2006) and Cassell (2007).

Trait[1]	PTA units	Standard deviation (SD)	Economic value ($/PTA unit)			Relative value (%)		
			NM$	CM$	FM$	NM$	CM$	FM$
Protein	kg[2]	10.0	7.82	12.62	0	23	28	0
Fat	kg[2]	13.6	5.95	5.95	5.95	23	18	23
Milk	kg[2]	354.1	0	-0.15	0.23	0	-12	24
PL	months[3]	2.1	29	29	29	17	13	17
SCS	log[4]	0.2	-150	-150	-150	-9	-7	-9
Udder	composite[5]	0.78	28	28	28	6	5	6
Feet/legs	composite[5]	0.88	13	13	13	3	3	3
Body size	composite[5]	0.94	-14	-14	-14	-4	-3	-4
DPR	percent[6]	1.4	21	21	21	9	7	8
CA$	dollars[7]	20	1	1	1	6	4	6

[1] PL = productive life; SCS = somatic cell score; DPR = daughter pregnancy rate; CA$ = calving ability.
[2] Kg per 305 days (1 lactation).
[3] Months in the milking herd, weighted.
[4] Log scores (0-9) from lactation average somatic cell counts.
[5] Function of the original data on a 50-point scale.
[6] Percent new pregnancies in a 21-day period.
[7] A composite of calving difficulty and stillbirths, expressed in dollars.

processed for fluid consumption and no premium is paid for protein. This is the case in the southeastern USA and Arizona which also have hot summer climates.

Many of the breed associations have their own selection indices. For example the Total Performance IndexTM calculated by the Holstein Association USA (Brattleboro, VT) is very popular in that breed. The correlation between NM$ and the Total Performance Index is approximately 95%.

The relative importance of traits can be easily calculated by dividing the economic value of one trait by the economic value of another trait. For example in a NM$ market, the value of one kg of protein equals 1.32 kg of fat (7.82/5.95). The value of one month longer PL equals 1.38 percentage points higher DPR (28/21). One kg more fat per 305 days equals 0.53 months (16.03 days) longer PL ($2.57 \times 5.95 / 29$).

The economic values of the various traits have changed over time as Table 2 illustrates for the USA (VanRaden, 2004). In the earliest selection index, only milk and fat were included. Consequently, much emphasis was placed on these two traits. Since 1994, more emphasis has been placed on functional traits that include conformation, fertility, longevity and health traits. Their incorporation became possible because accurate genetic evaluations became available for lower-heritability traits (VanRaden, 2004). Increased emphasis on functional traits was also a reaction to concerns from dairy producers that too many cows had to be culled because they failed to get pregnant or had health problems. Incorporation of these new traits in the selection indices has improved the progress in total economic merit of animals.

Table 2. Relative economic values (%) for selected USDA selection indices. Source: VanRaden (2004) and VanRaden et al. (2006).

Trait[6]	Index (year of introduction)						
	MF$[1] 1971	MFP$[2] 1976	CY$[3] 1984	NM$[4] 1994	NM$[5] 2000	NM$[5] 2003	NM$[5] 2006
Milk	52	27	-2	6	5	0	0
Fat	48	46	45	25	21	22	23
Protein		27	53	43	36	33	23
PL				20	14	11	17
SCS				-6	-9	-9	-9
DPR						7	9
SCE						-2	
DCE						-2	
CA$							6
Udder					7	7	6
Feed/legs					4	4	3
Body size					-4	-3	-4

[1] Milk-fat dollars.

[2] Milk-fat-protein dollars.

[3] Cheese yield dollars.

[4] Net merit dollars.

[5] Lifetime net merit.

[6] PL = productive life; SCS = somatic cell score; DPR = daughter pregnancy rate; SCE = service sire calving ease; DCE = daughter calving ease, CA$ = calving ability.

Breeding for robustness in cattle

Currently, records from all breeds, including crossbreds, are combined and analysed together in one animal model for genetic evaluations in the USA (VanRaden *et al.,* 2007). Evaluations are calculated initially on an all-breed base and then are converted to traditional within-breed genetic bases for publication. Trait means for base cows are shown in Table 3. Holsteins produce on average more milk, fat, and protein than the other five breeds but PL, SCS, and DPR are more favourable in at least one of the other breeds. Jersey bulls have higher breeding values for PL than Holsteins, with daughters remaining in the herd an average of 5.7 months longer; Guernseys have the shortest average herd lives, averaging 1.0 month shorter PL than Holsteins. Guernseys are, on average, the poorest-performing breed. Jersey and Milking Shorthorn cows have dramatically better fertility than Holstein cows. However, actual pregnancy rates observed by dairy producers (DRMS, 2008) are usually several percentage points lower than predicted by PTA for DPR. Calves sired by Brown Swiss and Jersey bulls are born with fewer calving difficulties than those sired by Holsteins (Cole *et al.,* 2005). The relative profitability of different traits, and therefore of the various breeds, may change under different conditions. The following discussion focuses on the impact of heat stress on the economic value of these and other traits.

Effects of heat stress on economic value of traits

Cows have optimal temperature zones within which no additional energy above maintenance is expended to heat or cool the body (West, 2003). This thermo-neutral zone for dairy cattle is estimated to be from -0.5 to 20 °C with an upper critical air temperature of approximately 25 °C (West, 2003). Dairy animals become heat stressed when the effective temperature conditions venture outside their zone of thermal comfort.

High environmental temperatures reduce productive and reproductive efficiency in dairy cattle (Fuquay, 1981; West, 2003; Jordan, 2003; Collier *et al.,* 2006). Cows gain heat through metabolism and from their environment (Fuquay, 1981). The basic thermoregulatory strategy of dairy cows is to maintain a core body temperature higher than the ambient temperature to allow heat to flow out via conduction, convection, radiation, and evaporation (Collier *et al.,* 2006). When the ambient temperature rises, the primary non-evaporative means of cooling for the cow (radiation, conduction, convection) become less effective and she relies more on evaporative means (panting and sweating) (West, 2003). Evaporative cooling requires additional energy above the needs for maintenance.

Table 3. Trait means for base cows (born in 2000) of the six breeds used in the USDA genetic evaluations (April 2008). Source: USDA-AIPL, http://aipl.arsusda.gov/eval/summary/Bmean_bases_het.cfm.

Breed	Trait[1]					
	Milk (kg/305 days)	Fat (kg/305 days)	Protein (kg/305 days)	PL (months)	SCS	DPR (%)
Ayrshire	8,232	317	257	31.8	2.96	21.8
Brown Swiss	9,677	390	320	30.4	2.92	20.4
Guernsey	7,541	334	246	26.5	3.29	19.9
Holstein	11,536	420	346	27.5	3.07	21.0
Jersey	8,102	371	287	33.2	3.33	26.0
Milking Shorthorn	7,683	274	237	29.8	3.10	24.0

[1] PL = productive life; SCS = somatic cell score; DPR = daughter pregnancy rate.

A high relative humidity compromises evaporative cooling. Cows may not be able to dissipate sufficient body heat to prevent a rise in body temperate. Consequently, heat stress occurs when the amount of heat energy produced by the animal is greater than the amount of energy flowing from the animal to its environment. This imbalance occurs when the combination of environmental factors (e.g. air temperature, sunlight), animal properties (e.g. rate of metabolism, moisture loss), and thermoregulatory mechanisms (conduction, radiation, convection, and evaporation) changes. However, given sufficient night cooling, cows can tolerate a relatively high daytime air temperature (West, 2003).

Temperature-humidity indices (THI) are frequently-used measures of heat stress in dairy cattle. The THI is calculated by combining temperature and humidity into one value; a commonly-used formulation (NOAA, 1976) is:

$$THI = (9/5 \times \text{temperature } ^{\circ}C + 32) - (11/2 - 11/2 \times \text{humidity}) \times (9/5 \times \text{temperature } ^{\circ}C - 26) \quad (1)$$

Heat stress in dairy cattle starts at a THI of 72, which corresponds to 22 °C at 100% humidity, 25 °C at 50% humidity, or 28 °C at 20% humidity (Ravagnolo *et al.,* 2000; Jordan, 2003). Thus, heat stress starts at lower temperatures when humidity increases. Bohmanova *et al.* (2007) have suggested that different THI may be appropriate in different conditions, for example one index may be more appropriate in a hot, dry climate and another on a hot, humid climate.

Due to the hot and often humid climates in the southern USA, heat stress abatement methods are widespread to cool cows. West (2003) and Collier *et al.* (2006) reviewed cooling methods. Most dairy producers make use of fans, sprinklers, misters, or shade cloth, and in many areas free stall barns have been built. Heat stress abatement methods are still changing, given for example the recent interest in tunnel barns that are modeled after housing in the poultry industry. Tunnel barns use evaporative cooling by sucking air though barns, thus creating a horizontal air flow. The air is typically cooled with high pressure mist. The amount of heat stress abatement is one of economics, not technology, and therefore few applications cool cows enough to avoid heat stress altogether.

Dairy cows of better production genotypes produce more body heat as a result to their greater metabolic activity. These animals are more susceptible to heat stress. Selection for increased production in the USA may have led to reduced heat tolerance, and bulls differ in the ability of their daughters to resist heat stress (Misztal *et al.,* 2006). This means that across the range of temperatures in the USA, from cold in the North to hot and humid in the Southeast, cows are becoming less robust for milk production.

The NM$, FM$, and CM$ selection indices are linear combinations of traits. The marginal value of a change in one unit of a trait is assumed to be constant. This implies that the marginal value is independent of the level of the trait. However, in many cases the total merit may be a non-linear function of these traits. This means that the marginal value of a change in one unit depends on the level of the unit. Nevertheless, Dekkers and Gibson (1998) concluded that unless profit functions are extremely nonlinear and cannot be approximated by quadratic functions, linear breeding goals and linear selection indexes can still be used for the selection of sires and dams to breed the next generation. However, Gibson (1989) earlier showed that economic responses to selection vary dramatically when plausible alternative values are used, underscoring the need to derive accurate economic values for traits. The use of a single selection index for various environments can lead to losses in revenue for individual producers because there is considerable variation among countries and regions within countries with respect to production circumstances (Groen *et al.,* 1997). This problem is ameliorated somewhat in the USA by the availability of the three merit indices from a market perspective, but given the considerable variation among production conditions in different geographic areas of the country more specific indices may be needed to accommodate local conditions.

The following discussion explores how heat stress may impact the marginal value of individual traits and also discusses the impact of market pricing on possible changes in marginal values of individual traits We first discuss traits that are directly included in the current USDA selection indices: milk, fat, protein, somatic cell counts, reproduction, productive life, body size, calving ability, udder, feed, and legs. Secondly, we discuss some traits not included in the current USDA selection indices (feed intake, coat colour, slick hair, animal wellbeing and heat tolerance) but that have a direct relation with heat stress.

Milk, fat, and protein

The effects of heat stress on milk production have been reviewed by Fuquay (1981) and West (2003) among many others. Numerous physiologic changes occur as a result of heat stress. Milk production loss is partly the result of deceases in dry matter intake (West, 2003).

According to West (2003), earlier work showed that annual milk production decreased up to 500 kg/cow per year in the southern USA under heat stress conditions. Estimates varied widely. Losses were greater for higher producing cows and cows in mid lactation. Similarly, higher producing cows were more prone to heat stress because of the extra heat produced as a result of the increased milk yield. When heat stress ends, milk production does not quite return to normal because the energy deficiency cannot be fully compensated in high-producing cows. The permanent drop in the current lactation is proportional to the length of the heat stress period.

West (2003) quotes earlier work by Bianca on the effects of heat stress on milk production of Holstein, Jersey and Brown Swiss cows. At a temperature of 29 °C and 40% relative humidity, the milk production of the three breeds was 97, 93, and 98% of normal, respectively. When relative humidity was increased to 90% (increasing heat stress), the yields of Holstein, Jersey and Brown Swiss cows were 69, 75, and 83% of normal, respectively. Thus, Holsteins appeared to be less robust and Brown Swiss cows more robust to heat stress for milk production.

Ravagnolo *et al.* (2000) found that milk production decreased at about 0.2 kg per unit increase in the THI when THI $\geq$72 in Georgia (southeast USA). Fat and protein declined at a rate of 0.012 and 0.009 kg per unit THI $\geq$72, respectively.

St-Pierre *et al.* (2003) modelled the reduction in milk for dairy cows as a non-linear function of the THI as follows:

$$\text{milk}_{loss} \text{ (kg/cow/day)} = 0.0695 \times (\text{THI}_{max} - \text{THI}_{threshold})^2 \times D \quad (2)$$

where:
THI_{max} is the maximum THI during the day,
$\text{THI}_{threshold}$ was set at 70, and
D is the proportion of the day where THI > $\text{THI}_{threshold}$, for example 0.33.

Thus in this model heat stress had a quadratic effect on reduction in milk loss.

Although milk production is compromised under heat stress, the marginal value of milk yield changes only if the milk price varies or if the cost to produce the milk increases. For the majority ($\approx$75%) of farm milk produced in the USA, the price is calculated following the rules in the Federal Milk Marketing Orders (FMO). The FMO pricing system is cumbersome and relatively complex (USDA-ARS, 2008). In the FMO, supply and demand of milk products play an important role in the price dairy farmers receive.

There are 11 FMO in the USA. Seven FMO use a multiple component pricing scheme to pay producers. Payment depends on the amounts of protein, butterfat, and other solids in the milk the dairy farmer sells. In general, most of the milk in these areas is used for cheese and butter and a smaller fraction is used for fluid milk. Four FMO use a skim milk/butterfat scheme to pay dairy farmers. Farmers in these FMO are paid based on the amount of skim milk and butterfat.

The four FMO that use the skim milk/butterfat scheme are all in the southern part of the USA where climates are hot during the summer. In general, the states in the Southeast are in the so called Florida, Southeast, or Appalachian FMO. These areas are generally milk deficient and a large percentage of the milk is processed into fluid milk. The fourth marketing area is Arizona, a state with a surplus of milk.

Whereas the NM$ selection index is based on national (USA) conditions, the economic values in the FM$ are more appropriate in markets were milk is primarily processed for fluid consumption and nothing is paid for protein. This is the case in the southeastern USA and Arizona. Therefore, milk protein is not considered a profitable trait in most of the areas with the most heat stress. Skim milk (volume) on the other hand is worth more than in the temperate North. Thus, profitable milk production traits in these hot climates are a large milk volume with a high fat content. The economic values for the traits in the FM$ are the more appropriate for cows in these regions than the NM$ or CM$.

The Southwest FMO, which covers Texas and New Mexico, uses the multiple component pricing scheme. Texas and New Mexico have hot climates that vary between hot and humid climates in the Southeast and the hot and dry climate of Arizona. Protein and somatic cell counts are directly rewarded in this marketing area. In conclusion, the milk pricing scheme determines to a large extend the marginal value of milk components in areas with heat stress.

Somatic cell counts

Heat stress is generally considered to increase the risk of mastitis because the udder's defense mechanisms become deficient and greater exposure to bacteria that thrive in hot and humid weather. Mastitis increases SCC. Consequently, average SCC in the hot and humid Southeast is generally higher than in the temperate northern USA (Norman *et al.*, 2000). For example, a recent comparison between thousands of Holstein herds in the Mid-west and South showed a difference of 32,000 cells (339,000 SCC vs. 371,000 SCC) (DRMS, 2008). The legal limit for SCC in the USA is 750,000 cells/ml.

Some of these regional differences in SCC are due to the quality premiums that are paid in Northern markets because of the emphasis on cheese manufacturing. Milk processors in Southeastern markets produce primarily fluid milk where the level of SCC has been less of interest. For example, a lower SCC is not directly rewarded in the Florida, Southeast, and Appalachian FMOs. As a result many of the dairy farmers in these regions have had less incentive to reduce their SCC than dairy farmers in Northern cheese markets.

The economic value of SCS (-$150/log unit) is the same in the NM$, CM$, and FM$ selection indices. It includes losses due to labour, drugs, discarded milk and milk shipments lost because of antibiotic residue. The reduction in milk production is not included in this value because that is already accounted for in the PTA for milk. About 72% of the $150 ($109) is allocated for premiums. The actual marginal value of a reduction in SCS is higher for herds with more mastitis (higher SCC) and lower for herds with less mastitis because premiums are linear with SCC rather than with SCS. Nevertheless, the value of a marginal reduction in SCS is much lower than $150/log unit when no premiums are paid as is the case in many regions with heat stress and primarily fluid milk processing.

Reproduction

The negative impact of heat stress or season on reproductive efficiency has been a topic of many studies (Jordan, 2003; Huang *et al.,* 2008). In particular, estrous behavior, follicular development, growth and function of the dominant follicle and oocyte competence are compromised (Jordan, 2003; Collier *et al.,* 2006). Rates of embryonic mortality also increase with temperature (Hansen, 2002). The negative effects of heat stress on reproductive efficiency start at a lower THI than on milk production. Pregnancy rates in the summer in the southeastern USA may be half of their values in the winter, e.g. 10 vs. 20% (De Vries and Risco, 2005). High production resulted in a faster decline of conception rates in Georgia (Huang *et al.,* 2008) and Florida (Al-Katanani *et al.,* 1999) under heat stress.

The negative effects are difficult to quantify because there is a concurrent and a delayed effect of heat stress on reproduction (Jordan, 2003; Collier *et al.,* 2006). Negative effects of heat stress have been identified from 42 days before to 40 days after insemination (Jordan, 2003). St-Pierre *et al.* (2003) modelled the effect of heat stress on reproduction through effects on pregnancy rate, days open, and reproductive culling:

$$\text{pregnancy rate} = 0.20 - 0.0009 \times \text{THI}_{\text{load}}. \tag{3}$$

Then using the change in pregnancy rate, they calculated the change in the average days open per month as:

$$\Delta \text{ days open} = 164.5 - (184.5 \times \text{pregnancy rate}) + (29.38 \times \text{pregnancy rate}^2) - 128.8 \tag{4}$$

The annual loss in days open is calculated by adding the 12 monthly losses.

Dairy producers make increased use of timed artificial insemination, without the need for estrus detection, in the summer (Collier *et al.,* 2006). However, conception rates with timed artificial insemination programs are still reduced in the summer (Jordan, 2003). Thus, without the seasonal variation in estrus detection rate, pregnancy rates are less variable throughout the year.

The marginal increase in pregnancy rate is more valuable when pregnancy rate is low (say 14%) than when it is high (say 25%) (Boichard, 1990; De Vries, 2007, 2008). For example, assuming an average herd in the USA, the economic values of a 1-percentage point increase in pregnancy rate around 7%, 14%, 18%, 23%, and 34% were $32.04, $14.49, $9.92, $6.67, and $3.31 per cow per year (De Vries, 2007).

The economic value of DPR in the NM$, FM$, and CM$ indexes is $21 per lifetime, or approximately $7 per cow per year. This value does not include the cost of culling and replacement which are included in PL. Daughter pregnancy rate is clearly a non-linear trait and is worth more when pregnancy rate is low, as is typically the case when cows are heat stressed.

In seasonal herds with summer heat stress, the marginal value of improving pregnancy rate may be higher in the winter than in the summer, however (De Vries, 2008). Much of this result is attributed to the reduction of milk production in the summer and fall. Cows that get pregnant in the summer have lower peak milk production and lower fertility. Consequently, many dairy farmers opt for somewhat seasonal production where cows calve in the fall and winter and breeding in the summer is avoided. Many dairy farmers in Florida that do not significantly cool their cows avoid inseminating cows in the summer. The use of cooling systems has reduced the impact of hot climates on fertility (Jordan, 2003).

Thus, increased fertility may be worth more in hot climates than in temperate climates because overall fertility is lower. However, in herds with seasonal milk production and fertility, the marginal value could be worth more in the cooler season.

Productive life

West (2002) suggested that severe heat waves could increase the mortality of cattle. Indeed, in the Southeast the risk of death is greater in the hot summer than in the cooler winter (DRMS, 2008). St-Pierre *et al.* (2003) modelled the effect of heat stress on the increase in monthly death rate as:

$$\Delta \text{ death rate} = 0.000855 \times \exp(0.00981 \times \text{THI}_{\text{load}}) \tag{5}$$

They (N.R. St-Pierre, personal communication) also modelled the increase in monthly reproductive cull rate as:

$$\Delta \text{ reproductive cull rate} = (100 - 102.7 \times (1 - 1.101 \times \exp(-10.19 \times \text{pregnancy rate})))/100 \tag{6}$$

Both formulas lead to shorter productive life as a result of heat stress. However, regional differences in productive life in the USA are small and not obvious (DRMS, 2008).

The value of PL is primarily determined by the revenues from a cull cow, the cost of a replacement heifer, and the average length of productive life (VanRaden *et al.*, 2006). Anecdotal evidence suggests that the cost of raising replacement heifers in Florida may be greater than in the more temperate northern USA because of higher feed cost. Further, a greater percentage of Florida producers purchase heifers instead of raising them themselves. Typically, purchase costs are somewhat greater than raising costs. However, these trends may not hold in other Southern states. Higher replacement costs increase the marginal value of PL. Cull cow prices are not known to differ from elsewhere. A greater mortality rate in heat stressed areas means that more cows leave the herd without salvage income. Furthermore, shorter average PL also increases the marginal value of PL. It is clear then that the marginal value of PL is likely greater in hot climates than in temperate climates in the USA.

Body size

Mature body size of dairy cattle has a negative economic value because the cost to maintain body size is greater than the added revenue when culling heavier cows (Groen *et al.*, 1997; VanRaden *et al.*, 2006). Heat stress results in less efficient maintenance. Therefore the marginal value of a reduction in body size is probably greater in hot climates than in temperate climates.

Calving ability

There is some evidence that calves delivered to heat stressed mothers are smaller at birth and less likely to be vigorous and thrive after birth (West, 2003). This is partly caused by shorter gestation lengths of heat stressed cattle. Smaller calves at birth are also associated with less dystocia.

Calving ability (CA$) is an index consisting of the calving ease and stillbirth measures from both the service sire (paternal effect) and the dam (maternal effect) (Cole *et al.*,2007b). Difficult calvings and still birth may result in more calf death losses, reduced milk production, lower fertility, and shorter longevity. The maternal costs are already included in the economic values for yield, DPR and PL, but the paternal costs are included in the marginal value for CA$.

High summer temperatures were associated with shorter gestation lengths, lower calf birth weights, and reduced incidence of dystocia in Australian Holsteins (McClintock *et al.*, 2003), and calf mortality was higher in the winter than the summer. These findings are consistent with those of Meijering (1984), who found that dystocia is less frequent in the summer than in the winter (level of heat stress not known). Results are less clear for stillbirths, with Meyer *et al.* (2000) reporting higher stillbirth rates for US Holsteins in the summer than the winter, and that the difference has increased over time. In his review article, Meijering (1984) reported contradictory findings on the relationship between stillbirth and season of calving. The calving traits evaluations in the USA include the fixed effect of year-season of calving (Cole *et al.*, 2007a), and the magnitude of the herd-year effects is similar for calving ease and stillbirth. While there is a clear relationship between season of calving and dystocia, there is not a similar relationship between season of calving and stillbirth. Greater values for DPR and perhaps PL in heat stressed cattle environments indicate that the marginal value of CA$ in hot climates may be greater than in temperate climates in the USA.

Udder, feet and legs

There are no clear effects of heat stress on the economic value of improved conformation for udder and feet and legs. While there are a number of ways in which heat stress can affect an animal physiologically, there is no documented mechanism by which conformation can be altered. Misztal *et al.* (2006) reported differences in conformation among daughters of the least and most heat tolerant Holstein sires, but those differences are probably due to the differences in type among bulls with high and low milk and components yields. Bulls with higher heat tolerance have lower yields than bulls with lower heat tolerance, and correspondingly differ in conformation.

Feed intake

Heat stressed dairy cattle experience an increase in respiration rate, rectal temperature, water intake and overall maintenance requirements. This leads to a decrease in ruminal contraction, a decrease in gut mobility, dry matter intake (DMI), and a decrease in the rate of food passage. The effects are greater in cows with multiple lactations than in cows in their first lactation. In the Nutrient Requirements for Dairy Cattle guidelines (NRC, 2001), DMI is modeled as a linear function of fat corrected milk production (FCM) with an adjustment for body weight (BW) and week of lactation (WOL):

$$\text{DMI (kg/d)} = (0.372 \times \text{FCM} + 0.0968 \times \text{BW}^{0.75}) \times (1 - \exp(-0.192 \times (\text{WOL} + 3.67))) \quad (7)$$

Thus a reduction in milk production due to heat stress would in this formula lead to a linear decrease in DMI and therefore not affect the efficiency of DMI conversion into milk. The NRC authors did not include a temperature or humidity adjustment factor in the equation because of insufficient DMI data outside the thermal-neutral zone.

Not considering the effect of reduced milk production, Eastridge *et al.* (1998) suggested that temperature >20 °C would reduce DMI by $(1 - ((°C - 20) \times 0.005922) \times 100\%)$ (NRC, 2001). St-Pierre *et al.* (2003) modelled the reduction in DMI for dairy cows as a non-linear function of the THI and calculated the loss in DMI as:

$$\text{DMI}_{\text{loss}} \text{ (kg/d)} = 0.0345 \times (\text{THI}_{\text{max}} - \text{THI}_{\text{threshold}})^2 \times D \quad (8)$$

Thus in this model heat stress had a quadratic effect on DMI loss but the ratio of milk loss and DMI loss remained a constant of 2.01 regardless of level of heat stress. There may be a delayed effect of THI on DMI loss (West, 2003).

Not only feed intake is reduced under heat stress, but several studies also reported a reduction in the efficiency of converting feed energy units to milk production energy units during heat stress

(Fuquay, 1981). The reduced efficiency is probably due to energy expended in getting rid of the excess heat load by way of increased respiration (panting and sweating) and other related activities (Fuquay, 1981). Under heat stress, more nutrients go to maintenance of body functions and less is available for production (West, 2003).

Because of lower DMI, rations in the summer are often formulated to be more nutrient dense (West, 2003). Dietary heat production and avoiding nutrient excesses are also considered when rations are reformulated. A typical approach is to reduce forage intake and increase concentrate content of the diet. However, such changes may lead to more rumen acidosis (Collier *et al.*, 2006).

In conclusion, heat stress reduces DMI. There is some evidence to suggest that the efficiency of milk production (Δ milk/Δ DMI) is reduced under heat stress conditions. Not considering the value of milk, this effect would lower the economic values of milk, fat and protein.

Feed cost in the Southeast are typically greater than in other parts of the USA due to the difficulty of growing high quality forages in subtropical areas. However, this is not the case in hot but dry climates such as in Arizona where irrigation can produce high quality and affordable forages. Consequently, the marginal value of an increase in DMI under conditions of heat stress could be greater than in temperate areas because the greater return from increased milk yield typically more than offsets the increase in feed cost. Feed intake is a trait that is not directly selected for with the current selection indices.

Coat colour

Coat colour is another trait that is not directly selected for with the current selection indices. There is evidence that hair colour influences the susceptibility to heat stress because a lighter coat can absorb more heat from solar radiation (West, 2003). Indeed, *Bos taurus* beef cattle with dark coats had greater heat transfer to the skin, higher body temperatures, and reduced weight gains compared to those with white coats (Finch, 1986). Increased woolliness of the coat increased the colour effect. In dairy cows, studies showed that cows with more white colour might be more fertile and have greater milk production than dark coloured cows (West, 2003). Although coat colour is heritable, it is not clear if it is useful to select for coat colour (West, 2003), especially because most cows in regions with heat stress have access to shade.

Slick hair

Olson *et al.* (2003) described evidence for the existence of a major dominant gene (designated as the slick hair gene) that is responsible for producing a very short, sleek hair coat. The gene is found in Senepol cattle and criollo breeds (both *Bos taurus*) in Central and South America. Purebred and crossbred cattle with slick hair were observed to maintain lower rectal temperatures and be more heat tolerant (Olson *et al.*, 2003). In Carora × Holstein crossbred cows there was a positive effect of slick hair on milk yield under dry, tropical conditions. The performance and profitability of crossbreds that are mostly Holsteins and carry the slick hair gene are currently being evaluated at the University of Florida.

Animal wellbeing

An animal's wellbeing is compromised when it experiences heat stress. Wellbeing is not included the current selection indices in the USA but interest in such a trait is growing. The merit of traits is not always measureable in monetary units (Sölkner *et al.*, 2008). Olesen *et al.* (2000) described a method for constructing selection objectives that include both market values and non-market values, which are values partially or completely unrecognised by current markets, such as animal wellbeing.

Nielsen *et al.* (2006) extended that method to the case of multiple traits with non-market values, and demonstrated the use of such indices in dairy cattle breeding. Domestic milk markets do not currently assign any direct value to the ability of a cow to resist the effects of heat stress, and thereby improving its wellbeing, making it a possible target for the approaches of Olesen *et al.* (2000) and Nielsen *et al.* (2006). An index constructed using non-market values would, for example, sacrifice some selection response for milk yield in order to increase the non-market value associated with resistance to heat stress. Consumer preferences also are increasingly important, particularly with respect to perceptions of animal wellbeing. Non-market values can also be used to account for producer frustration at dealing with animals that are often sick or die on the farm, or in organic farming.

Heat tolerance

Work by Misztal *et al.* (2006) showed that direct selection for heat tolerance is possible. Heat tolerance is the ability to remain productive under hot conditions. Heat tolerant sires have daughters that vary less in milk production over a range of temperatures than the daughters of less heat tolerant sires. The genetic correlation between heat tolerance and production was approximately -0.3 (West, 2003). Continued selection for greater performance in the absence of consideration of heat tolerance will result in greater susceptibility to heat stress (West, 2003) and thus lead to reduced robustness for variations in climate. Finch (1986) pointed out that within *Bos taurus* beef cattle an increased capacity for thermoregulation is accompanied by a reduction in energy metabolism which is associated with lower performance. Thus, if performance is not considered during selection, direct selection for increased heat tolerance would likely result in lower overall performance of the cow (West, 2003). However, because the genetic correlation is small, a combined selection for production and heat tolerance is possible (Collier *et al.*, 2006). It is clear that the trait 'heat tolerance' is more profitable in hot climates than in temperate climates.

Within breed selection to overcome effects of heat stress

VanRaden (2004) suggested that lines of cows selected for regional environmental or market conditions may be more profitable than those from a global population selected for some global objective. However, except for niche markets this may not prove to be the case in the USA. While Holstein cows in the South are affected substantially by heat stress, they still produce so much more milk and components that they are still preferred to other breeds; the costs of decreased fertility and increased morbidity and mortality do not yet offset the value of the greater yield. In general, dairy farmers in the USA are not constrained by milk quotas, which would almost certainly favour breeds such as the Jersey and Brown Swiss, which have lower milk yields but favourable components yields, much lower rates of dystocia, and longer productive lives than Holsteins. Still, there is room within each breed to improve more valuable functional traits under heat stress (Table 1).

Crossbreeding to overcome effects of heat stress

There is an increased interest in crossbreeding in the USA to overcome concerns about fertility, cow health, and calf survival with primarily the Holstein cow (Jordan, 2003; Heins *et al.*, 2006). Jordan (2003) summarised the effects of crossbreeding on fertility and found that crossbreds had typically better reproductive performance. Much of the value of crossbreeding comes from heterosis.

In a study of seven California herds, where heat stress may be observed part of the year, first-service conception rates were 22% for Holsteins, 35% for Normande/Holstein crosses, 31% for Montbeliarde/Holstein crosses, and 30% for Scandinavian Red/Holstein crossbreds. Days open were 150 for pure Holsteins, 123 for Normande/Holstein, 131 for Montbeliarde/Holstein, and 129 for Scandinavian Red/Holstein crossbreds, and all three crossbred groups had significantly fewer days open than pure Holsteins (Heins *et al.*, 2006).

Because of improved fitness traits compared to purebreds, *Bos taurus* crossbreds may be more valuable in the hot southern USA than in the more temperate North. Differences in milk production need to be strongly considered, especially because milk is typically more valuable in regions with heat stress.

The advantages of *Bos indicus* cattle compared to *Bos taurus* cattle to deal with heat stress were mentioned previously. In an Australian study, Brahman (*Bos indicus*) × Friesian (*Bos taurus*) crosses gained weight faster under heat stress than their purebred contemporaries (Colditz and Kellaway, 1972). Earlier, many studies investigated crossbreds between *Bos indicus* breeds and *Bos taurus* breeds (Hayman, 1972). However, West (2003) and Jordan (2003) believed it is questionable that crosses between *Bos taurus* and *Bos indicus* breeds could be sufficiently productive to meet the needs of dairy farmers in the USA. The gains in coping with heat stress do not weigh up against the loss in production.

Therefore, for the last several decades the focus in the USA has been on improving the environment and nutrition of dairy cows while selecting for production, rather than improving resistance to stressors (Collier *et al.*, 2006). However, as energy costs of cooling are rising and animal welfare becomes a more important issue, there is renewed interest in identifying genes that could improve resistance to heat stress without adversely affecting productivity (Collier *et al.*, 2006).

Concluding remarks

The principle of genotype-by-environment interaction was illustrated (Figure 1) and explained earlier in this contribution. It was concluded that tropically adapted breeds are generally more robust for changes in thermal environments than temperately adapted breeds. Collier *et al.* (2006) reported that genotype-by-environment interactions in dairy cattle are larger than originally thought. Bryant *et al.* (2007) showed that in New Zealand Jerseys were able to sustain high levels of milk production over a wider range of thermal environments than Holstein-Friesians. This implies that Jerseys are more robust across a range of temperatures than Holstein Friesians. However, these cows did not experience significant heat stress.

The ranking of sires for daughters' performance in temperate and hot climates appears to be similar. Bohmanova *et al.* (2008) showed that differences among bulls for heat tolerance explained only a small amount of the differences between regions in the USA. However, the perception among many dairy farmers in Florida is that variation in genotype is not being expressed as well in hot climates as it is in temperate climates. Thus the variation among sires for daughters' performance in hot climates is thought to be less than for daughters' in temperate climates.

Misztal *et al.* (2006) reported that the most heat tolerant Holstein sires transmitted lower milk yields with higher components than did the least heat tolerant sires. The daughters of the most heat tolerant sires had also more desirable udder and body size composites, longer PL, and higher DPR than did daughters of the least heat tolerant sires. They noted further that the sires most widely used in the southeastern USA were less heat tolerant than average, which may be a consequence of the fluid milk market in that part of the USA. If the marginal value of improved fertility is substantially higher than the marginal value of increased milk yield then the use of heat tolerant bulls seems more justified. The difference in Total Performance Index™ between the most and least heat tolerant sires was small (T.J. Lawlor, personal communication).

Barlow (1981) in a literature review concluded that there is evidence that an interaction of heterosis-by-environment is the rule rather than the exception. Ruvuna *et al.* (1983) evaluated the performance of first-lactation purebred and crossbred Holsteins, Jerseys, and Brown Swiss based on the season of calving (cool vs. warm). Differences in milk production were more favourable for crossbreds in the warm season than in the cool season compared to purebred Holsteins. They also concluded that for reproduction there may be even more favourable heterosis in the warm season, especially for Holstein crosses.

Genotype-by-environment interactions have been reported between and within countries, although most within-country interactions are not significant (Weigel *et al.,* 2001; Rekaya *et al.,* 2003). Ravagnolo and Misztal (2000) found that the heritability of milk yield decreases with increasing THI because the additive variance of heat tolerance increased to a magnitude similar to that of milk yield. These results were confirmed by Zwald *et al.* (2003), who reported that heritabilities calculated using herds from hot climates were lower than those using herds from cold climates. Records from cows milked in hot environments thus appear to be affected more by environmental factors than those in cold or temperate climates.

The relative value of traits may differ between hot and temperate environments because variances as well as the economic values, as explained before, differ between these environments. Phenotypic and genetic (co) variances are used in selection index calculations, so any change in those values changes the weights in the final index. If the change in (co)variance components is relatively small between environments then the index is affected more by changes in the (marginal) economic values of the traits. Dairy farmers in hot climates whose dairy cattle experience heat stress should value traits differently than dairy farmers in temperate climates.

References

Al-Katanani, Y.M., D.W. Webb and P.J. Hansen, 1999. Factors affecting seasonal variation in 90-day nonreturn rate to first service in lactating Holstein cows in a hot climate. J. Dairy Sci., 82: 2611-2616.

Barlow, R., 1981. Experimental evidence for interaction between heterosis and environment in animals. Anim. Breed. Abstr., 49: 715.

Bohmanova, J., I. Misztal and J.B. Cole, 2007. Temperature-humidity indices as indicators of milk production losses due to heat stress. J. Dairy Sci., 90: 1947-1956.

Bohmanova, J., I. Misztal, S. Tsuruta, H.D. Norman and T.J. Lawlot, 2008. Short Communication: Genotype by environment interaction due to heat stress. J. Dairy Sci., 91: 840-846.

Boichard, D., 1990. Estimation of the economic value of conception rate in dairy cattle. Livest. Prod. Sci., 24: 187-204.

Bourdon, R.M., 2000. Understanding animal breeding. 2nd Ed. Prentice Hall, Upper Saddle River, NJ.

Bryant, J.R., N. López-Villalobos, J.E. Pryce, C.W. Holmes, D.L. Johnson and D.J. Garrick, 2007. Environmental sensitivity in New Zealand dairy cattle. J. Dairy Sci., 90: 1538-1547.

Cartwright, T.C., 1980. Prognosis of Zebu cattle: Research and application. J. Anim. Sci., 50: 1221-1226.

Cassell, B., 2007. Sire evaluations for health and fitness traits. Virgina Cooperative Extension Service. Publication 404-087. Available at: http://www.extension.org/pages/Sire_Evaluations_for_Health_and_Fitness_Traits (Accessed: 29 September 2008).

Colditz, P.J. and R.C. Kellaway, 1972. The effect of diet and heat stress on feed intake, growth, nitrogen metabolism, in Friesian, F1 Brahman x Friesian, and Brahman heifers. Aust. J. Agric. Res., 23: 717-725.

Cole, J.B., R.C. Goodling, Jr., G.R. Wiggans and P.M. VanRaden. 2005. Genetic evaluation of calving ease for Brown Swiss and Jersey bulls from purebred and crossbred calvings. J. Dairy Sci., 88: 1529-1539.

Cole, J.B., G.R. Wiggans and P.M. VanRaden, 2007a. Genetic evaluation of stillbirth in United States Holsteins using a sire-maternal grandsire threshold model. J. Dairy Sci., 90: 2480-2488.

Cole, J.B., G.R. Wiggans, P.M. VanRaden and R.H. Miller, 2007b. Stillbirth (co)variance components for a sire-maternal grandsire threshold model and development of a calving ability index for sire selection. J. Dairy Sci., 90: 2489-2496.

Collier, R.J., G.E. Dahl and M.J. VanBaale, 2006. Major advances associated with environmental effects on dairy cattle. J. Dairy Sci., 89: 1244-1253.

De Jong, G. and P. Bijma, 2002. Selection and phenotypic plasticity in evolutionary biology and animal breeding. Livest. Prod. Sci., 78: 195-214.

De Vries, A. and C.A. Risco, 2005. Trends and seasonality of reproductive performance in Florida and Georgia dairy herds from 1976 to 2002. J. Dairy Sci., 88: 3155-3165.

De Vries, A., 2007. Economic value of a marginal increase in pregnancy rate in dairy cattle. J. Dairy Sci., 90 (Suppl. 1): 423.

De Vries, A., 2008. What is improved dairy cattle reproductive performance worth? In: Proc. 13th Intern. Congr. ANEMBE. Salamanca, Spain, pp. 145-154.

Dekkers, J.C.M. and J.P. Gibson, 1998. Applying breeding objectives to dairy cattle improvement. J. Dairy Sci., 81: 19-35.

DRMS (Dairy Record Management Systems), 2008. DairyMetrics – selected statistics. Available at: http://www.drms. org (Accessed July 15, 2008).

Eastridge, M.L., H.F. Bucholtz, A.L. Slater and C.S. Hall, 1998. Nutrient requirements for dairy cattle of the National Research Council versus some commonly used ration software. J. Dairy Sci., 81: 3049-3062.

Finch, V.A., 1986. Body temperature in beef cattle: its control and relevance to production in the tropics. J. Animal Sci., 62: 531-542.

Fuquay, J.W., 1981. Heat stress as it affects animal production. J. Dairy Sci., 52: 164-174.

Gengler, N., G.R. Wiggans, C.W. Wolfe and J.R. Wright, 1998. Genetic evaluations for type for breeds other than Holstein. Available at: http://aipl.arsusda.gov/reference/type/typedef.htm (Accessed: 25 July 2008).

Gibson, J.P., 1989. The effect of pricing systems, economic weights, and population parameters on economic response to selection on milk components. J. Dairy Sci., 72: 3314-3326.

Goddard, M.E., 1998. Consensus and debate in the definition of breeding objectives. J. Dairy Sci., 81 (Suppl. 2): 6-18.

Gotthard, K. and S. Nylin, 1995. Adaptive plasticity and plasticity as an adaptation: a selective review of plasticity in animal morphology and life history. OIKOS, 74: 3-17.

Groen, A.F., T. Steine, J-J. Colleau, J. Pedersen, J. Pribyl and N. Reinsch, 1997. Economic values in dairy cattle breeding, with special reference to functional traits. Report of an EAAP-working group. Livest. Prod. Sci., 49: 1-21.

Hansen, P.J., 2002. Embryonic mortality in cattle from the embryo's perspective. J. Anim. Sci., 80: E33-E44.

Hayman, R.H., 1972. Bos indicus and Bos taurus crossbred dairy cattle in Australia. I. Crossbreeding with selection among filial generations. Austr. J. Agric. Res., 23: 519-532.

Hazel, L.N., G.E. Dickerson and A.E. Freeman, 1994. The selection index – then, now, and for the future. J. Dairy Sci., 77: 3236-3251.

Heins, B.J., L.B. Hansen and A.J. Seykora, 2006. Fertility and survival of pure Holsteins versus crossbreds of Holstein with Normande, Montbeliarde, and Scandinavian Red. J. Dairy Sci., 89: 4944-4951.

Hoffman, A.A. and P.A. Parsons, 1994. Evolutionary genetics and environmental stress. Oxford University Press, New York, NY.

Huang, C., S. Tsuruta, J.K. Bertrand, I. Misztal, T.J. Lawlor and J.S. Clay, 2008. Environmental effects on conception rates of Holsteins in New York and Georgia. J. Dairy Sci., 91: 818–825.

Jordan, E.R., 2003. Effects of heat stress on reproduction. J. Dairy Sci., 86 (E. Suppl.): E104-E114.

Loftus, R.T., D.E. MacHugh, D.G. Bradley, P.M. Sharp and P. Cunningham, 1994. Evidence for two independent domestications of cattle. PNAS, 91: 2757-2761.

MacHugh, D.E., M.D. Shriver, R.T. Loftus, P. Cunningham and D.G. Bradley, 1997. Microsatellite DNA variation and the evolution, domestication and phylogeography of Taurine and Zebu cattle (Bos taurus and Bos indicus). Genetics, 146: 1071-1086.

Madelena, F.E., 1986. Economic evaluation of breeding objectives for milk and beef production in tropical environments. Proc. 3rd World Congress on Genetics Appl. to Livest. Prod., Lincoln, NE. XII, p. 33.

McClintock, S., K. Beard, A. Gilmour and M. Goddard, 2003. Relationships between calving traits in heifers and mature cows in Australia. Interbull Bull., 31: 102-106.

McDowell, R.E., J.C. Wilk and C.W. Talbott, 1996. Economic viability of crosses of Bos taurus and Bos indicus for dairying in warm climates. J. Dairy Sci., 79: 1292-1303.

Meijering, A., 1984. Dystocia and stillbirth in cattle – a review of causes, relations and implications. Livest. Prod. Sci., 11: 143-177.

Meyer, C.L., P.J. Berger and K.J. Koehler, 2000. Interactions among factors affecting stillbirths in Holstein cattle in the United States. J. Dairy Sci., 83: 2657-2663.

Misztal, I., J. Bohmanova, M. Freitas, S. Tsuratu, H.D. Norman and T.J. Lawlor, 2006. Issues in genetic evaluation of dairy cattle for heat tolerance. Proc. 8th World Congress on Genetics Appl. to Livest. Prod., August 13-18, Belo Horizonte, MG, Brazil.

Nielsen, H.M., L.G. Christensen and J. Ødegard, 2006. A method to define breeding goals for sustainable dairy cattle production. J. Dairy Sci., 89: 3615-3625.

NOAA (National Oceanic and Atmospheric Administration), 1976. Livestock hot weather stress. Regional Operations Manual Letter C-31-76. US Dep. Commerce, Natl. Oceanic and Atmospheric Admin., Natl. Weather Service Central Region, Kansas City, MO.

Norman, H.D., R.H. Miller, J.R. Wright and G.R. Wiggans, 2000. Herd and state means for somatic cell count from Dairy Herd Improvement. J. Dairy Sci., 83: 2782-2788.

NRC (National Research Council), 2001. Nutrient requirements of dairy cattle. 7th revised ed. National Academy Press, Washington, DC.

Olesen, I., A.F. Groen and B. Gjende, 2000. Definition of animal breeding goals for sustainable production systems. J. Anim. Sci., 78: 570-582.

Olson, T.A., C. Lucena, C.C. Chase, Jr. and A.C. Hammond, 2003. Evidence of a major gene influencing hair length and heat tolerance in *Bos taurus* cattle. J. Animal Sci., 81: 80-90.

Prayaga, K.C., W. Barendse and H.M. Burrow, 2006. Genetics of tropical adaptation. Proc.8th World Congress on Genetics Appl. to Livest. Prod., August 13-18, Belo Horizonte, MG, Brazil.

Ravagnolo, O. and I. Misztal, 2000. Genetic component of heat stress in dairy cattle, parameter estimation. J. Dairy Sci., 83: 2126-2130.

Ravagnolo, O., I. Misztal and G. Hoogenboom, 2000. Genetic component of heat stress in dairy cattle, development of heat index function. J. Dairy Sci., 83: 2120-2125.

Rekaya, R., K.A. Weigel and D. Gianola, 2003. Bayesian estimation of parameters of a structural model for genetic covariances between milk yield in five regions of the United States. J. Dairy Sci., 86: 1837-1844.

Ruvuna, F., B.T. McDaniel, R.E. McDowell, J.C. Johnson, Jr., B.T. Hollon and G.W. Brandt, 1983. Crossbred and purebred dairy cattle in warm and cool seasons. J. Dairy Sci., 66: 2408-2417.

Schutz, M.M., 1994. Genetic evaluation of somatic cell scores for United States dairy cattle. J. Dairy Sci., 77: 2113-2129.

Shook, G.E., 2006. Major advances in determining appropriate selection goals. J. Dairy Sci., 89: 1349-1361.

Sölkner, J., H. Grausgruber, A.M. Okeyo, P. Ruckenbauer and M. Wurzinger, 2008. Breeding objectives and the relative importance of traits in plant and animal breeding: a comparative review. Euphytica, 161: 273-282.

Strandberg, E. 2008. The role of environmental sensitivity and plasticity: lessons from evolutionary genetics. In: Breeding for robustness in cattle. Klopcic, M., R. Reents, J. Philipsson and A. Kuipers (eds.) EAAP Scientific Book No. 126, Wageningen Academic Publishers, The Netherlands, pp. 17-33.

St-Pierre, N.R., B. Cobanov and G. Schnitkey, 2003. Economic losses from heat stress by US livestock industries. J. Dairy Sci., 86 (E. Suppl.): E52-E77.

Turner, J.W., 1980. Genetic and biological aspects of Zebu adaptability. J. Anim. Sci., 50: 1201-1205.

USDA-ARS, 2008. Federal Order Reform – Final Order. Available at: http://www.ams.usda.gov/AMSv1.0 (Accessed: 31 July 2008).

VanRaden, P.M., 2004. Invited review: Selection on net merit to improve lifetime profit. J. Dairy Sci., 87: 3125-3131.

VanRaden, P.M., A.H. Sanders, M.E. Tooker, R.H. Miller, H.D. Norman, M.T. Kuhn and G.R. Wiggans, 2004. Development of a national genetic evaluation for cow fertility. J. Dairy Sci., 87: 2285–2292.

VanRaden, P.M., C.M.B. Dematawewa, R.E. Pearson and M.E. Tooker, 2006. Productive life including all lactations and longer lactations with diminishing credits. J. Dairy Sci., 89: 3213-3220.

VanRaden, P.M. and Multi-State Project S-1008, 2006. Net merit as a measure of lifetime profit: 2006 revision. Available at: http://aipl.arsusda.gov/reference/nmcalc-2006.htm (Accessed 23 July 2008).

VanRaden, P.M., M.E. Tooker, J.B. Cole, G.R. Wiggans and J.H. Megonigal, 2007. Genetic evaluations for mixed-breed populations. J. Dairy Sci., 90: 2434-2441.

Weigel, K.A., R. Rekaya, N.R. Zwald and W.F. Fikse, 2001. International genetic evaluation of dairy sires using a multiple-trait model with individual animal performance records. J. Dairy Sci., 84: 2789-2795.

West, J.W., 2003. Effects of heat-stress on production of dairy cattle. J. Dairy Sci., 86: 2131-2144.

Zwald, N.R., K.A. Weigel, W.F. Fikse and R. Rekaya, 2003. Identification of factors that cause genotype by environment interaction between herds of Holstein cattle in seventeen countries. J. Dairy Sci., 86: 1009-1018.

Part 6:
Application and attitude

Perception of robustness traits in breeding goal for dairy cattle in a new EU country

M. Klopčič[1] and A. Kuipers[2]
[1]*University of Ljubljana, Biotechnical Faculty, Department of Animal Science, Domžale, Slovenia*
[2]*Expertisecentre for Farm Management and Knowledge Transfer, Wageningen University and Research Centre, Wageningen, the Netherlands*

Abstract

The objective of this study was to examine the farmers' perception of robustness (sustainable) traits in the breeding goal, and any associations of this perception with the farm's and farmers' characteristics and plans for the future in countries in transition to the EU environment. As a case, the new member state country of Slovenia was taken. As a tool, a questionnaire was used: 1,114 questionnaires, about 20% of the distributed ones, were anonymously returned. This implies that 12% of the dairy farmers' population is part of the analysis. The farmers were asked if they desired more emphasis on a variety of traits, e.g. production and robustness traits. The robustness trait was constructed afterwards from the fertility, health and longevity traits and simply calculated as the sum of the answers on the question whether (1) or not (0) more emphasis should be placed on the trait in the breeding goal. It was analysed whether the perception of these traits was associated with farm and farmers' characteristics, and interest in different aspects of farming. Also the relations with breed of herd and plans for the future of farmers were studied. Principal components, characterised as 'age of farmer', 'size of farm' and 'number of activities other than dairy farming' were indicated as main factors characterising the farm and farmer. A total of 25% of the farmers did not see a need to change the emphasis in the existing breeding goal. A majority of the farmers would like more emphasis to be placed on health traits (62%), fertility traits (55%) and protein content (56%). The perception of farmers was significantly dependent on farm size, breed of herd and plans for the future. In the farmers' population in transition, a positive attitude towards sustainable traits was found. A general interest for animal breeding work was not associated with a desired increase in emphasis on a specific trait. This is different for economically oriented farmers and farmers who are interested in feeding or calf rearing practices. These farmers focus more on, respectively, longevity, health and fertility traits. This indicates that it may be worthwhile to think further about the composition of the working group(s) to be involved in the preparation of breeding goals to obtain a well balanced sustainable breeding program. Moreover, it would provide valuable insight in addition to model calculations and the results of this study, when trying out a Profile experiment in combination with other techniques in some farmers' populations in the transition countries.

Keywords: production and robustness traits, selection focus, questionnaire, farmers' attitudes and input

Introduction

Breeding programmes are constructed to improve each future generation. What constitutes 'improvement', however, is a 'political' decision and various basic principles exist to define the desired improvement. For each basic principle, there are various scientific methods to translate 'improvement' into index weights for the traits in the breeding goal. Examples of such basic principles are:
1. Maximise the total economic response for a dairy herd (Hazel, 1943; Perez-Cabal and Alenda, 2003; Dekkers *et al.,* 2004; VanRaden, 2004; Nielsen *et al.,* 2005; Mulder *et al.,* 2006; Vangen, 2003, 2008).

2. Correct weaknesses relative to the competition of other breeds or populations (Rauw *et al.,* 1998; Dillon and Veerkamp, 2001).
3. Work on trouble-free production (Hamoen *et al.,* 2008; Pryce *et al.,* 2008; Wall *et al.,* 2008).
4. Look for desired gains according to the preferences of dairy farmers (Veerkamp, 1998; Tozer and Stokes, 2002; Nielsen and Amer, 2007).
5. Look for desired gains based on preferences in society (Oleson *et al.* 2000; Nielsen *et al.,* 2006).

Combinations of the above basic principles are also possible (Amer, 2006).

Various methods exist to identify the 'best' future parents of the next generation. BLUP methods are most commonly used for estimating breeding values in dairy cattle breeding. Each estimated breeding value (EBV) of an individual is weighted with the index weight of the trait. For optimisation of the index weights a merit equation is used including the economic values of the traits (Kuipers and Shook, 1980; Mulder and Jansen, 2001; Sölkner and Fürst, 2002; Pedersen *et al.,* 2002; VanRaden and Seykora, 2003). De Vries and Cole (2008), discussed the economic weights of traits for hot climatic conditions. The gene-flow method by McClintock and Cunningham (1974) has widely been used to calculate the economic values, both considering the long term effects of selection and to get the values of all traits of the sexes at a given point in time, e.g. at insemination. The selection index combines the weighted EBV's into a single figure for each selection candidate. However, concerning the reasons behind the choice of a certain combination of traits limited information is available (Pearson, 1986; Philipsson *et al.,* 1994; Philipsson and Lindhé, 2003; Berry *et al.,* 2005; Sölkner *et al.,* 2008).

The basic principle for defining 'improvement', as described above, is often chosen implicitly. Concerning the impact of the main target group, the farmers on the process of composing breeding goals and indices is hardly any literature found. Nevertheless, the opinions and experiences of farmers or groups of farmers as main users of the products of animal breeding do play a role. The Farm Animal Breeding and Reproduction Technology Platform (FABRE, 2006) even states: 'Changes in the breeding goal are always decided upon by the members of breeding organisations – by the dairy farmers. Breeding organisations ensure the health and welfare of the animals they keep and select. They are engaged in the search for selectable traits that are indicative of species-specific animal welfare'. Perhaps it would be closer to practice to say 'Changes in breeding programmes *should always be* decided upon in cooperation with the members of breeding organisations – the dairy farmers'. Amer *et al.* (1998) explained that 'a well-defined breeding objective is the first requirement of any genetic improvement program and comprise those traits, which one attempts to improve genetically because they influence returns and costs to the producer. The breeding objective should closely align with the overall objectives of the target groups in the livestock business, who are the critical link in the use of genetically improved animals'. Nielsen *et al.* (2006) add that 'the objectives of any breeding program are to achieve the goals of the breeder, and some of the goals of the breeder may not be economic or may include goals that are not directly measurable using some economic gauge'. Veerkamp (1998) and Tozer and Stokes (2002) state that 'it is possible to incorporate non-economic as well as economic objectives of dairy producers into a multiple-objective breeding model'.

The Research Institute for Organic Agriculture (FiBL) in Switzerland carried out a survey under 1000 organic dairy farmers to obtain information on the state of breeding affairs. The response of 608 out of 1000 returned questionnaires was high. The farmers weighted functional traits to be very important in the breeding program. The most important criteria for selection appeared to be fertility (84% of farmers ranked this as first priority), low somatic cell count (81%), longevity (78%), good milk production from forage (77%) and milk quality, especially protein content (72%). On farms with high milk yield (>7,000 kg milk per cow/lactation), protein content was considered to be most important. Fleckvieh holders emphasised in particular the selection criterion of milking speed. Milk yield was important on farms in the valleys, whereas meat production was of great concern to Fleckvieh holders, farms with low average milk yield and farms in the mountain areas. Persistence of yield over lactation was found very important in the French speaking part of Switzerland and on farms with high milk yield (Haas and Bapst, 2004; Bapst *et al.*, 2005). This shift of emphasis from production traits to functional traits was also observed in Austria (Schwarzenbacher, 2001; Schwarzenbacher *et al.*, 2003).

A survey among 132 Dutch organic dairy farmers revealed that 55% of the farmers were specialised in milk production and 45% were running a multi-functional farm. Farmers from both strategies were asked to value different breeding aspects of the animals. In general, the two groups of farmers valued the various aspects more or less the same: they wanted a robust, long living cow, with good udder health and fertility (Nauta *et al.*, 2006). A survey involving 18 organic dairy farms in Ontario, Canada, was carried out to collect data on their production systems, breeding policies and concerns. An organic index was constructed based on farmers' preferences. The relative weight of production to functional traits (28% vs. 72%) was substantially different from those in the Canadian Lifetime Profit Index (54% vs. 46%), but similar to those used in conventional indices in Sweden and Denmark and in the Swiss organic index (Rozzi *et al.*, 2007).

When studying literature the impression arises that the 'young organic/ecological sector' gives, relatively to the traditional sector, quite some attention to the opinions of farmers in formulating the breeding goals, whereas it may not be as well documented in the traditional sector. Sölkner and Fürst (2002), who compared index methods across countries, 'found it very difficult to find details on the rationale for choosing traits included in the index and the methodology used for derivation of the index weights'. VanRaden (2004) believes that 'trait values often are assigned by committee and consensus rather than by strict economic or mathematical models'. He adds: 'Some difficulties Sölkner and Fürst (2002) encountered may be caused by economic goals being debated informally in local languages and not translated into published scientific documents'. Madalena (2008) states that 'In Brazil, the increased popularity of more fertile and adapted breeds indicates that farmers are at least aware of the economic set back associated to high yield genetics, although nonetheless more research on the better alternatives is needed. Unfortunately, research and information are too often directed more towards the vendors' interest than the farmers' needs'.

A practical way of customising the Total Merit Indexes according to different needs of various farmers' groups has been applied for a longer time in Sweden (Philipsson and Lindhé, 2003; Philipsson *et al.*, 2005). Hamoen *et al.* (2008) indicate that inclusion of a robustness score in the overall score for conformation was discussed in member (farmers') meetings of this Dairy Herd Improvement Organisation. However, the impact of these discussions on the decision making is not described quantitatively. Bebe *et al.* (2003) studied the breed preferences and breeding practices in small holder dairy systems in the Kenya highlands. They looked at traits, like hardiness, high milk yield, traction ability, high butterfat and attractive looks, but included in the same question practices like availability of semen of choice and input of the extension service.

A wide variety of indices exist (Miglior *et al.*, 2005; Shook, 2006; Miglior and Sewalem, 2008). Sustainability or robustness of animals becomes a more and more important characteristic. In fact, robustness is often seen as a combination of functional traits. However, in the 'new' EU Central and Eastern European countries and adjacent non-EU countries to the East virtually all emphasis in the selection process is still on the production and type traits.

Pärna *et al.* (2003) and Wolfova *et al.* (2007) described the merits of including various functional traits in the 'total merit index', using both the same bio-economic model adapted to respectively the Estonian and Czech Republic circumstances. On basis of their study, Wolfova *et al.* (2007) advise that 'somatic cell count should be introduced in the breeding goal, and in the future also emphasis should be given to length of productive life and the reproductive performance of cows'.

It can be observed that these 'new' EU Central and Eastern European countries are in a transition phase (Kuipers *et al.*, 2006). The country reports show that each is adapting in their own way to the EU policy and environment, which also concerns developments in the cattle sectors and breeding goals. In this context several questions can be raised:
- How are farmers in transition countries looking at the sustainability of their cows and do they want to change the currently used indices to more sustainable indices? Assessing the attitudes of farmers towards sustainability in breeding programs may help in the decision to adapt the indices and in choosing the time frame to do this.
- Is it possible to identify different groups of farmers with specific strategies and/or interests that show a different attitude towards the breeding goal? If so, this may help to formulate a strategy to adapt the breeding program in the most successful way.

These questions are addressed in this paper.

As a case, the 'new' member state of Slovenia was taken. The farmers in Slovenia represent a community in transition (Osterc *et al.*, 2003) with a small herd size and good opportunities for diversification of the farm business, because of their positioning in the outskirts of the mountains and large numbers of tourists visiting (Klopcic and Osterc, 2005). In these aspects, Slovenia is similar to regions in transition in Poland, Czech Republic, Romania and Bulgaria, etc. In Slovenia a revised total merit index for sires and cows has been introduced in 2005. The TMI for market orientation on milk (Holstein-Friesian breed) has the following relative index weights decided in small committee: production traits 40%, type traits 45%, age at first calving and calving interval 10%, and calving ease 5%. TMI for dual-purpose breeds (Simmental and Brown) include milk production traits (9%), type traits (52%), age at first calving and calving interval (10%), calving ease (9%) and daily gain (20%). As can be seen some functional traits have recently been included in breeding index. However, it is experienced that by far the most selection emphasis in the field is on the production and type traits (Klopcic and Osterc, 2005).

In summary, the objective of this study is to examine farmers' perceptions towards robustness (sustainable) traits, and associations with farm and farmers' characteristics and future plans in countries in transition to the EU environment.

Material and methods

Data

In year 2005/2006 questionnaires were sent to dairy farmers in Slovenia. Questions were asked about:
- Characteristics of the farm and farmers.
- Interests of the farmers in various aspects of farming.
- The farmer's plans for the future, e.g. specialisation in dairy farming or diversification of the farm or both.
- The farmer's preferences for changes in emphasis on traits in breeding goal under the new EU policies. Seven traits were presented as well as the option 'I want to keep the breeding program the same'. As production traits were listed protein and butterfat content, milk yield and beef characteristics and as robustness traits were considered health, fertility and longevity. The farmers were requested to tick the traits of their choice or tick the option 'keep breeding goal as it is'.

The questionnaires were distributed to 5,000 dairy farmers out of a total of 10,000 dairy farmers in Slovenia: milk haulers distributed the questionnaires to farmers in the cooperatives and the researchers to farmers present at organised meetings. 1,114 questionnaires were returned anonymously in a closed envelope resulting in a response of 22%. This group of farmers represented 11% of the total dairy farm population. The response was very satisfactory. Nevertheless, we have to realise that the returned questionnaires are not fully a representative sample of the complete Slovenian dairy farm population. That is one of the reasons that we include in the results a detailed description of the farm and farmers' characteristics of the sample.

By interpretation of the data in Results chapter, we must also realise that the breeding program in Slovenia is focusing on the production traits. In other words, more emphasis on a production trait is additional to the weight already given to this trait in the current breeding index, while more emphasis on functional (say robustness) traits is a signal to enter these traits (or some of those) into the breeding program.

Variables

Some continuous variables were included in the survey as a number of classes with a range. The farmer was asked to mark the applicable class. For the analysis, the central value of each class was used to reconstruct the continuous variable again. This was done for instance for the variables Quota size and Farm size. If questions in the questionnaire were not answered, the value was indicated as a missing value and not included in the analysis. In cases where options for answers were 'yes', 'no', 'perhaps' or 'don't know', the values for this variable were reduced to a binomial variable: 1 is 'yes' and 0 is 'not yes'.

The composite trait 'robustness' was introduced. This artificial trait is derived adding the answers to the question about preferred emphasis on the fertility, health and longevity traits together: robustness (0, 1, 2, 3) = fertility (0, 1) + health (0, 1) + longevity (0, 1).

Statistical methods

In order to check the answers in the questionnaire to be associated with the different types of farms and farmers, some 'characterising variables' were selected to represent the types of farms and farmers. In a preliminary analysis using Principal Component Analysis (PCA) (STATISTIX 7, 2000) it was found that three variables highly determine the type of farm and farmer. Two 'characterising variables' were related to the farm: (1) 'farm size (ha of agricultural land)' and (2) 'number of other activities than dairy'. The third 'characterising variable' was for the farmer: (3) 'age of farmer (years)'. This offers the opportunity to identify groups of farms: farms with young or old farmers, small or large farms, and very specialised farms versus farms with more activities on the farm. To apply a PCA, categorical variables, like breed of herd were transformed to binominal variables (for example: only Holstein-Friesian versus other breeds). Breed of the herd, however, is in this study also used as independent variable to see whether the preferences for a change of emphasis in the breeding goal were breed-dependent. In this case the original categorical variable with more classes was used.

The STATISTIX 7 statistical program (2000) was used to analyse the data. For each question of the questionnaire the answers were summarised in terms of the mean and standard deviation. With multiple regressions it was analysed whether answers were associated with the 'characterising variables'. R^2 is used to indicate the fraction sums of squares explained by the 'characterising variables'. Significance is indicated by * if $P<0.05$, by ** if $P<0.01$ and by *** if $P<0.001$.

Results

Farm and farmers' characteristics

The average milk quota of the farms in this sample is 108 tons (Table 1), which is about twice the average amount of all dairy farms in Slovenia (Klopcic and Huba, 2006). The average farm size is 17.1 ha, which is high because this is 5.9 ha for all agricultural farms in Slovenia (SORS, 2002). A total of 77% of the farmers in this sample participate in milk recording with an average production of 5,473 kg, while in practice 54% of farmers record the milk production of their herds with an average production of 4,896 kg (SORS, 2007). The percentage of 69% of farmers that expect to have a successor is very high and without doubt higher than in the total population. This description of farmer and farms in Table 1 indeed illustrates that the sample of farmers in this study is not representative for all Slovenian dairy farms, but represents the larger farms with a higher average production and relatively often with a known successor. It reflects farmers who opt for continuity.

Emphasis on traits

The majority of farmers (see Table 2) want more emphasis on health traits (62% of farmers), protein yield (56%) and fertility (55%). Lowest interest is in putting more emphasis on butterfat (28%) and beef characteristics (only14%). 25% of farmers appeared to be content with the current breeding program and see no reason for change.

Table 1. Mean and standard deviations (SD) of characteristics of farms and farmers.

Nr	Variable (answer)	N	Mean	SD
	Characteristics of the farm			
	Milk quota for processing plant (1000 kg)	1,098	108.0	109.9
	Number of dairy cows	1,101	19.0	15.5
	Number of young stock (calves and heifers)	1,114	16.7	13.1
	Milk quota for direct sales (1000 kg)	1,114	3.2	7.2
	Average milk production per cow (kg/year)	1,059	5,473	1,504
	Agricultural land in use (ha)	1,114	17.1	10.6
	Farms with hilly or mountainous land (0); farms with flat or less favourable land (1)	1,114	0.67	0.47
	Farms with only Holstein Friesian cows (1); farms with other breeds or a mixture of breeds (0)	1,109	0.13	0.34
	Milk recording (no=0, yes=1)	1,067	0.77	0.42
	Number of fattening bulls	428	6.1	5.85
	Number of pigs	420	14.3	44.5
	Land for grain and maize (ha)	888	7.0	7.1
	Forestry on the farm (no=0, yes=1)	1,114	0.25	0.43
	Number of other activities on the farm than dairy [1]	1,114	2.0	1.5
	Characteristics of the farmer			
	Non agricultural employment of farmer/wife (no=0, yes=1)	1,062	0.32	0.47
	Successor on farm (no=0, yes=1)	1,092	0.69	0.46
	Age of farmer (years)	1,100	51.5	12.7
	Farmers with education at public school level (0); education higher than public school (1)	1,103	0.60	0.49

[1] In total there was a choice of 22 different activities. Choices related to dairy activities (calves, heifers, land for grain and maize, maize for silage) were not counted in this variable.

Associations with farm and farmer characteristics

More emphasis on protein yield is positively associated with farm size, while emphasis on butterfat, although limited (Table 2), is mostly favoured on the smaller farms. Surprisingly, a higher emphasis on milk yield is not associated with the Variables that characterise farm and farmer. A higher emphasis on the robustness traits is positively associated with farm size. More emphasis on longevity is associated with the somewhat younger farmer, as is the composite robustness trait. More emphasis on beef characteristics is expressed by farmers with a higher number of other activities than dairy on the farm. Farmers who want the breeding program to stay the same are older farmers on smaller farms.

Associations with fields of interest of farmer

The desire for more emphasis on milk yield is positively associated with interest in grassland management and milking and milk quality, but negatively with the interest in working environmentally friendly (Table 3). More emphasis on fertility is positively associated with interest in calf rearing, while more emphasis on both fertility and health is related to interest in care for these traits, as can be expected. A higher emphasis on health traits was associated with interest in feeding practices. Farmers focussing on farm economics desire strongly more attention for longevity. The composite robustness trait shows a clear association with farmers' interest in farm economics, care for health & fertility and feeding. It accumulates the associations of the underlying traits. Emphasis on beef traits is more often expressed by farmers who find calf rearing interesting than by farmers who do not think so. Remarkably, interest in animal breeding work is not significantly associated with more focus on either production or robustness or beef traits. Most positive to a change in breeding program appear to be farmers with interest in feeding and calf rearing. However, the analysis is somewhat sensitive to some of the tasks included. A clear change occurs, when deleting the interest field 'milking and milk quality' from analysis, because focus on milk protein becomes positively

Table 2. Mean and standard deviations (SD) of milk, beef and robustness traits and associations with the variables characterising farm and farmer.

More emphasis on traits or keep program the same (yes/no)	N	Mean	SD	Variables characterising farm and farmer[1]			
				Farm size	No. of other activities	Age of farmer	R^2
Protein %	1114	0.56	0.50	+***			2.10
Butterfat %	1114	0.28	0.45	-***			1.82
Milk yield	1114	0.43	0.50				0.08
Fertility	1114	0.55	0.50	+***			2.32
Health	1114	0.62	0.49	+**			1.56
Longevity	1114	0.39	0.49	+***		-**	9.76
Robustness composite	1114	1.57	1.13	+***		-*	6.21
Beef characteristics	1114	0.14	0.35		+***		4.69
Keep breeding program the same	1114	0.25	0.43	-***		+*	3.09

[1] Associations are tested by a linear regression model: variable = constant + $b_1F + b_2O + b_3A$. Constant is not presented. F is farm size; O is number of other activities, and A is age of farmer. Significance of b's is indicated by: * ($P<0.05$); ** ($P<0.01$); *** ($P<0.001$). The sign of b is indicated by - in case of negative association and + for a positive association.

Table 3 Associations of 'more emphasis' on milk, beef and robustness traits with interest in different tasks of farming[1].

More emphasis on trait/ keep program the same (yes/no)	Interest in aspects of dairy farming (low (1), average (3), high (5))								
	Grassland management	Breeding work	Farm economics	Working environmentally friendly	Rearing of calves	Care for health & fertility	Feeding	Milking & milk quality	R^2
Protein %									3.48
Butterfat %									0.89
Milk yield	+*			-**				+*	2.85
Fertility					+*	+**			4.68
Health						+*	+*		3.64
Longevity			+***						6.08
Robustness composite			+**			+**	+*		6.76
Beef characteristics					+***				3.20
To keep breeding program the same					-*		-*		2.88

[1] Associations are tested by a linear regression model: trait = constant + sum (b_i*interest$_i$), with $_i$ is 1 to 8. Constant is not presented. Significance of b's is indicated by: * ($P<0.05$); ** ($P<0.01$); *** ($P<0.001$). The sign of b is indicated by - in case of negative association and + for a positive association.

associated with interest in grassland management, economical farming and good feeding practices, while leaving most other associations the same.

Association with breed of herd

In general, when analysing the focus on traits for the three breed groups, herds with crossbreds or multiple breeds (Mixed herd) are between Holstein-Friesian herds (HF) and Simmental or Brown herds (Table 4). The wish for more emphasis on protein yield is significantly higher for farmers with Holstein-Friesian and Mixed herds than for farmers with Simmental and Brown (S&B) herds. On the contrary, HF farmers have less interest in increasing butterfat yield than S&B and mixed herd farmers. HF farmers showed a highly significantly higher interest in focus on robustness traits then S&B farmers. This is opposite for the beef characteristics. In addition, more S&B farmers than HF farmers chose for keeping the breeding goal as it is.

Associations with plans for the future

The desired emphasis on various traits appeared to depend also on the future orientation of the farmer (Table 5). Farmers who do express plans for further development of the farm business want a higher focus on protein content and robustness traits than farmers who intend to keep the farm the same. However the emphasis on robustness traits does not differ significantly between farmers who want

Table 4. Associations of 'more emphasis' on milk, beef and robustness traits with breed of farm herd (% yes).

More emphasis on traits or keep program the same (yes/no)	No. of farmers who say yes	Breed of farm herd (% yes)[1]			Total (% yes)
		Holstein-Friesian	Simmental and/or Brown	Mixed herd	
Protein %	619	65.5[a]	49.6[b]	58.6[a]	55.8
Butterfat %	311	19.3[a]	29.3[b]	29.3[b]	28.0
Milk yield	475	35.9	42.8	44.8	42.8
Fertility	615	61.4[a]	50.4[b]	58.2[a]	55.4
Health	691	75.2[a]	56.3[b]	64.0[c]	62.3
Longevity	439	52.4[a]	30.9[b]	43.8[a]	39.6
Beef characteristics	155	4.8[a]	18.3[b]	12.7[c]	14.0
Keep breeding program the same	273	15.2[a]	28.9[b]	23.4[c]	24.6
Total of farmers	1,110	145	460	505	1,110

[1] Percentages within rows with different superscripts are significantly different with $P<0.05$, using Bonferroni t-test.

Table 5. Associations of 'more emphasis' on milk, beef and robustness traits with future plans of farmers (% yes).

More emphasis on traits or keep program the same (yes/no)	No. of farmers who say yes	Plans for the future (% yes)[1]			
		Keep farm the same	Specia-lisation[3]	Diversi-fication[4]	Both[5]
Protein %	544	52.4[a]	64.4[b]	54.4[a]	68.9[b]
Butterfat %	269	30.7	25.7	25.0	34.4
Milk yield	431	40.2[a]	55.7[b]	37.8[a]	62.2[c]
Fertility	544	47.6[a]	63.5[b]	62.2[b]	71.1[b]
Health	614	60.6[a]	64.8[b]	71.7[b]	66.7[b]
Longevity	406	27.9[a]	52.2[b]	52.8[b]	62.2[b]
Beef characteristics	141	13.0[a]	10.4[a]	22.8[b]	14.4[a]
Keep breeding program the same	225	32.2[a]	18.7[b]	13.3[b]	18.9[b]
Total of farmers	901[2]	401	230	180	90

[1] Percentages within rows with different superscripts are significantly different with $P<0.05$, using Bonferroni t test.
[2] 901 farmers = total of 1,114 farmers minus 213 farmers who intend to stop farming or continue as hobby or who did not make a future choice.
[3] Specialisation by increase in number of dairy cows.
[4] Diversification by start of new branch(es) or increasing this branch(es).
[5] Both: increase in number of dairy cows and start of new branch(es).

to specialise or to diversify the farm business or to develop in both directions. As can be expected, farmers who choose for diversification are less interested in increasing the milk yield level and want more emphasis on beef characteristics than the other two groups of farmers.

Discussion

Farm and farmers' characteristics

In general, the sample of farmers in this study reflects farmers who opt for continuity, as described in Farm and Farmers' Characteristics. This implies that the means and SD's are not representative for all farmers in this case country, but the calculation of associations and relationships, on which most of the analysis in this study relies, is usually assumed to be less sensitive to such a kind of sample. Moreover, this sample still showed a substantial variation in attitudes: farmers who want to keep the breeding program the same or to adapt, and farmers who want to specialise or to diversify, and farmers with different interest in the various aspects of farming. This sample allows to analyse the posed research questions.

Attitude towards sustainable traits

The first research question raised was *how farmers in transition countries look at the sustainability of their cows and if they want to change the currently used indices to more sustainable indices?* This study gives indeed an impression of the attitude of farmers towards the individual traits as part of the breeding program. A majority of farmers express that they like more focus on the health and fertility traits. This is most distinctive when compared with the interest in more emphasis on butterfat content. A lower emphasis on longevity as trait is signalled than on fertility and health. However, farmers who are economical oriented express relatively more interest in having 'longevity' in the breeding goal. In general, this analysis shows a substantial interest for including 'robustness traits' into the breeding indices, which is a good base for authorities and animal breeding associations in the 'new' Central and Eastern European countries to do so. The 'robustness traits' can be specified and measured in a variety of ways. Many countries participate in Interbull evaluations with longevity, defined usually as productive life or as surviving 1, 2 or more lactations (J. Philipsson, 2008, personal communication). Longevity can be also derived from the type classification data, as explained by Hamoen *et al.* (2008).

However, the economic weights that farmers may wish to assign to the various traits cannot be derived from the questionnaire data of this study. Nielsen and Amer (2007) explain that enabling 'the estimation of economic weights from questionnaires demands the relative importance of the traits and the trade-off or marginal rates of substitution between the traits that farmers are willing to take to be established'. Partial profile experiments can be used to achieve this goal. In a choice experiment, a set of alternatives (the choice set), that are prespecified in terms of levels of attributes, are incorporated into a questionnaire (Nielsen and Amer, 2007). Respondents are then asked to view various alternative descriptions of a good, differentiated by their attributes and levels, and asked to choose their most preferred alternative in a given choice set. In this study, the alternatives could be breed of herd (3) or future plans (4), while the attributes in each alternative are represented with the traits (7) each expressed at different performance levels.

Jabbar *et al.* (1999) and Bebe *et al.* (2003) took a simpler approach applied in developing countries. Bebe *et al.* (2003) asked farmers to give their primary preference to the attributes (traits and practices) for keeping the breed. The odds ratio was presented as a measure of the relative performance for an attribute in a given breed. One of the breeds was chosen as reference (base) to compare with. Rozzi *et al.* (2007) based an organic breeding index also on farmers' preferences, but used a score of 0-5 for each trait. Scores were averaged across farmers to determine the relative (subjective) weight for each trait.

As discussed by Nielsen *et al.* (2005, 2006), a breeding goal including economic value (EV) may be 'too narrow minded', because the EV may represent only economic aspects of the current market opportunity. In contrast, the nonmarket value (NV) represents a wider perspective, like the value of improved animal welfare or other societal influences on animal production. Among others, NV is a desired gain based on consumers' or societies' willingness to pay for a certain product. In other words, Nielsen *at al.* (2006) argue that farmers are not the only stakeholders to be heard when establishing really sustainable breeding goals.

Wolfova *et al.* (2007), applying a bio-economic model to the Czech Republic circumstances, calculated somatic cell count as a measure for cow health and longevity to get the highest economic weights relative to milk yield (range of 30-40%), while various fertility traits received low weights. Thus, these model results are (only) partly in agreement with the impressions obtained from measuring farmers' attitudes towards the traits in this study.

Sölkner *et al.* (1999) discussed plus and minuses of including type traits in the index. In a questionnaire Austrian Simmental farmers were asked to give subjective weights of dairy vs. beef vs. functional traits vs. conformation traits. The average ratio's expressed by 7,137 breeders were 44:22:19:15. Sölkner *et al.* (1999) state that, 'Although this is definitely not comparable with an economic weight, it gives some indication of the importance of conformation to farmers. However it is arguable whether the farmer thinks about the part of conformation related to fitness or about the beauty of the cow when he is placing this subjective weight'. In this study conformation traits were not included in the questionnaire. This may have been an omission, but at the same time it made the comparison between production and functional traits probably more straightforward for the farmer.

We think that it would provide valuable insight in addition to model calculations and the results of this study, when trying out a Profile experiment in combination with other techniques described above in some farmers' populations in the transition countries.

Differentiation of farmers towards breeding goals

The second research question raised was *if it is possible to identify different groups of farmers with specific strategies and/or interests that show a different attitude towards the breeding goal?* This study indeed showed groups of farmers who reacted quite differently towards the preference for the series of traits proposed to them. The attitude towards the traits was affected by the size of the farm business as well as the interest in various management tasks: a larger farm as well as interest in economics, calf rearing, and care expressed for animal health and fertility correspond with the wish for a greater emphasis on robustness traits. Different breeds on the farm as well as different future plans also lead to a variety of wishes concerning the breeding goal. A practical way of customising the Total Merit Indexes according to different needs of various farmers' groups is practiced in Sweden: all sub-indexes with their weights are published together with the TMI so that farmers may apply other weights than those practised for the population as a whole (Philipsson and Lindhé, 2003; Philipsson *et al.*, 2005).

Because of lack of similar research, the results of this study cannot be validated easily. Often, breeding goals for different breeds are prepared separately. This study underlines this choice, because significant differences in attitudes towards some traits are noticed between the breeds. Nauta *et al.* (2005) found that organic farmers who followed different strategies on their farm comparable to this study, i.e. specialisation or diversification, did value the various breeding aspects about the same. This is the same in this study for the robustness traits, but specialised farmers did react differently towards the production traits than farmers who (want to) diversify. Especially the higher emphasis expressed by the specialised farmers on protein content is notable. This is similar to results of Huba *et al.* (2006), who concluded that the transition of Slovakia to the EU requires updating of the selection index by favouring protein to fat yield. We would also have expected the specialised dairy farmers to prefer healthy, fertile and easy to handle cows even more than the farmers who diversify. This makes work more efficient in a larger herd. The attitude may be different towards longevity. Farmers with smaller herds and diversified farms may be more interested in long living cows, than the specialist dairy farm manager who likes the young healthy cow. This is one of the reasons that the average culling age has decreased for many years in the past. Hamoen *et al.* (2008) stated that farmers in the Netherlands 'want to give more attention to feet and legs, while cows should not become taller, but do need adequate body condition and weight to handle the milk production'.

Sometimes it is argued that perceptions of people may change rapidly over time. In other words, results of a questionnaire like this present just a momentarily impression. In this reasoning profit equations or bio-economic models are favoured to compute economic weights. Indeed, the answers of farmers are expected to be influenced somewhat by the news of the day. But the preferences and associations found in this study show a high degree of logic. It is postulated that the tendencies in attitudes registered would not dramatically differ when the experiment would be repeated in the same country. That no association is found between general interest in breeding work and focus on certain groups of traits, however, is curious. But we must realise that the associations are calculated between groups of farmers with certain interests. It is possible that within the group of farmers with interest in animal breeding work variation exists in focus on specific traits. Anyway, this may be a signal for some extra thought about the composition of the working group(s) that are involved in preparing breeding goals for the future. It is possible that a group of farmers with a variety of practical know-how, including economical and environmental insight, may contribute the best in making a well balanced and sustainable breeding plan. Moreover, when choosing farmers to be involved, it can be advised on base of this study to look for farmers who opt for further development of the farm business, because they show a more sustainable attitude towards the breeding goal than the farmer who wants to keep the farm the same.

We believe that this analysis presents useful information for the authorities responsible for the breeding goals in the 'new' Central and Eastern European countries, but more in particular for somewhat similar regions as Slovenia. Also Animal Breeding Companies in Western Europe and elsewhere may be interested in the perceptions of farmers in Central and Eastern Europe towards breeding goals and the emphasis on robustness as part of this.

Conclusions

- 75% of farmers in the case country in transition like to adapt the breeding program; a majority of farmers in this transition country desire more emphasis on the health, fertility and protein traits
- The reaction of farmers was highly influenced by the size of farm, the breed of herd and by the future strategic plans the farmer has in mind
- Farmers who intend to develop the farm further have a significant higher wish in adapting the breeding program; the choice for specialisation or diversification influences the emphasis wanted on milk yield and beef characteristics significantly, while it does not affect the emphasis desired on the robustness traits. The impression is that the variation in attitudes of these two groups of farmers towards the production traits is largely similar to the preference of farmers with different breeds.
- Longevity as indicator provides similar associations with farm and farmers' characteristics, breed of herd and future strategic plans as the composite robustness trait.

- Interest for animal breeding work, as expressed by a group of farmers, appears to be not associated with the wish for more focus on specific traits. This is different for the group of economical oriented farmers and farmers who have interest in feeding and/or calf rearing practices. These farmers focus respectively more on the longevity, fertility and health traits. This may indicate that it is beneficial to choose a diverse composition of the working group(s) to be involved in the preparation of the breeding goals to obtain a well balanced sustainable breeding program. Moreover, future oriented farmers may have the better input.
- It is advised and planned to check the repeatability of these results in some other countries in Eastern Europe to make the conclusions more generally applicable. Use of 'Profile experiments' in combination with other techniques may be considered.

Acknowledgements

The authors gratefully acknowledge all comments and suggestions of dr. Jan ten Napel, Animal Sciences Group, Wageningen UR, The Netherlands, professor Jan Philipsson, Dept. of Animal Breeding and Genetics, Swedish University of Agricultural Sciences, Uppsala, Sweden, dr. Wiebe Koops, Animal Production Systems, Wageningen University, The Netherlands, dr. Peter Amer, Abacus Bio, New Zealand, and prof. Jože Osterc, Dept. of Animal Science, Biotechnical Faculty, University of Ljubljana, Slovenia.

References

Amer, P.R., N. Mpofu and O. Bondoc, 1998. Definition of breeding objectives for sustainable production systems. In: 6th World Congr. Genet. Appl. Livest. Prod., 28: 97-104.

Amer, P.R., 2006. Approaches to formulating breeding objectives. In: 8th World Congr. Genet. Appl. Livest. Prod., August 13-18, 2006, Belo Horizonte, Brazil, p. 7.

Bapst, B., A. Bieber and E. Haas, 2005. Untersuchungen zur Zuchtstrategie in Schweizer Bio-Braunviehbetrieben - Analysis of breeding strategies on Swiss organic Braunvieh-farms. In: Ende der Nische. Heß, J. and G. Rahmann (eds.) Wissenschaftstagung Ökologischer Landbau, Kassel, 1-4 March 2005. Available at: http://orgprints.org/3612.

Bebe, B.O., H.M.J. Udo, G.J. Rowlands and W. Thorpe, 2003. Smallholder dairy systems in the Kenya highlands: breed preferences and breeding practices. Livest. Prod. Sci., 82: 117-127.

Berry, D.P., F. Buckley, P. Dillon and R.F. Veerkamp, 2005. Dairy cattle breeding objectives combining production and non-production traits for pasture based systems in Ireland. In: Final Report of Project No. 5066, Teagasc, Ireland, p 12.

Dekkers, J.C.M., J.P. Gibson, P. Bijma and J.A.M. van Arendonk, 2004. Design and optimisation of animal breeding programmes. In: Proceeding of Animal breeding strategies course, Wageningen, 2004, p 16.

DeVries, A. and J.B. Cole, 2008. Profitable dairy cow traits for hot climatic conditions. In: Breeding for robustness in cattle. Klopcic M., R. Reents, J. Philipsson and A. Kuipers (eds.) EAAP Scientific Series No. 126, Wageningen Academic Publishers, the Netherlands, pp. 227-245.

Dillon, P. and R.F. Veerkamp, 2001. Breeding strategies. In: Proceedings of the National Dairy Conference, Tralee, Ireland, 16-Nov-2001, pp. 41-54.

FABRE, 2006. Sustainable farm animal breeding and reproduction. A vision for 2025. Available at: http://www.euroqualityfiles.net/vision_pdf/vision_fabre.pdf

Haas, E. and B. Bapst, 2004. Swiss organic dairy farmer survey: Which path for the organic cow in the future? In: Organic livestock farming: potential and limitations of husbandry practice to secure animal health and welfare and food quality. Hovi, M., A. Sundrum and S. Padel (eds.). Proceedings of the 2nd SAFO Workshop, 25-27 March 2004, Witzenhausen, Germany, pp. 35-41.

Hamoen, A., G. de Jong and M. van Pelt, 2008. The role of robustness in type classification. In: Breeding for robustness in cattle. Klopcic M., R. Reents, J. Philipsson and A. Kuipers (eds.) EAAP Scientific Series No. 126, Wageningen Academic Publishers, the Netherlands, pp. 265-270.

Hazel, L.N., 1943. The genetic basis for constructing selection indices. Genetics, 28: 476-490.

Huba, J., S. Mihina, M. Stefanikova, M. Zahumensky and J. Brocko, 2006. Structural and farm development as consequence of the milk quota introduction in Slovakia. In: Farm management and extension needs in Central and Eastern European countries under the EU milk quota system. Kuipers A., M. Klopcic and A. Svitojus (eds.). EAAP Technical Series No. 8, Wageningen Academic Publishers, the Netherlands, pp. 119-130.

Jabbar, A.M., B.M. Swallow and J.E.O. Rege, 1999. Incorporation of farmer knowledge and preferences in designing breeding and conservation strategy for domestic animals. Outlook Agric., 28: 239-243.

Klopcic, M. and J. Osterc, 2005. Extension work in milk and beef production in Slovenia. In: Knowledge transfer in cattle husbandry: new management practices, attitudes and adaptation. Kuipers A., M. Klopcic and C. Thomas (eds.). EAAP Scientific Series No. 117, Wageningen Academic Publishers, the Netherlands, pp. 63-76.

Klopcic, M. and J. Huba, 2006. Farm management under quota in small and large herd CEE countries. In: Farm management and extension needs in Central and Eastern European countries under the EU milk quota system. Kuipers A., M. Klopcic and A. Svitojus (eds.). EAAP Technical Series No. 8, Wageningen Academic Publishers, the Netherlands, pp. 237-251.

Kuipers, A. and G.E. Shook, 1980. Net returns from selection under various component testing plans and milk pricing schemes. J. Dairy Sci., 63: 1006-1018.

Kuipers, A., M. Klopcic and A. Svitojus (eds.), 2006. Farm management and extension needs in Central and Eastern European countries under the EU milk quota system. EAAP Technical Series No. 8, Wageningen Academic Publishers, the Netherlands.

Madalena, F.E., 2008. How sustainable are the breeding programs of the global stream dairy breeds? The Latin-American situation. Livestock Research for Rural Development, 20: Art. 19. Available at: http://www.cipav.org.co/lrrd/lrrd20/2/mada20019.htm

McClintock, A.E. and E.P. Cunningham, 1974. Selection in dual purpose cattle populations: defining the breeding objective. Animal Production, 18: 237-247.

Miglior, F., B.L. Muir and B.J. van Doormaal, 2005. Selection indices in Holstein Cattle of various countries. J. Dairy Sci., 88: 1255-1263.

Miglior, F. and A. Sewalem, 2008. A Review on breeding for functional longevity of dairy cow. In: Breeding for robustness in cattle. Klopcic M., R. Reents, J. Philipsson and A. Kuipers (eds.) EAAP Scientific Series No. 126, Wageningen Academic Publishers, the Netherlands, pp. 113-122.

Mulder, H. and G. Jansen, 2001. Derivation of economic values using lifetime profitability of Canadian Holstein cows. Available at: www.cdn.ca/committees/Sept2001/MulderJansen.pdf.

Mulder, H.A., R.F. Veerkamp, B.J. Ducro, J.A.M. van Arendonk and P. Bijma, 2006. Optimization of dairy cattle breeding programs for different environments with genotype by environment interaction. J. Dairy Sci., 89: 1740-1752.

Nauta, W.J., A.F. Groen, R.F. Veerkamp, D. Roep and T. Baars, 2005. Animal breeding in organic dairy farming: an inventory of farmers' views and difficulties to overcome. NJAS, 53: 19-45. Available at: http://library.wur.nl/ojs/index.php/njas/article/viewFile/329/48.

Nauta, W.J., H. Saatkamp, T. Baars and D. Roep, 2006. Breeding in organic farming: different strategies, different demands. In: Paper presented at Joint Organic Congress, Odense, Denmark, May 30-31, 2006. Available at: http://orgprints.org/7506/01/OrganicbreedNauta.doc.

Nielsen, H.M., L.G. Christensen and A.F. Groen, 2005. Derivation of sustainable breeding goals for dairy cattle using selection index theory. J. Dairy Sci., 88: 1882-1890.

Nielsen, H.M., L.G. Christensen and J. Ødegard, 2006. A Method to define breeding goals for sustainable dairy cattle production. J. Dairy Sci., 89: 3615-3625.

Nielsen, H.M. and P.R. Amer, 2007. An approach to derive economic weights in breeding objectives using partial profile choice experiments. Animal, 1: 1254-1262.

Olesen, I., B. Gjerde and A.F. Groen, 1999. Methodology for deriving non-market trait values in animal breeding goals for sustainable production systems. In: Breeding goals and selection schemes. Proceeding of International workshop on EU concerted action Genetic improvement of functional traits in Cattle (GIFT); 7-9th November 1999, Wageningen, The Netherlands. Interbull Bulletin, 23: 13-23.

Osterc, J., S. Čepin, M. Klopcic, I. Štuhec, A. Holcman and A. Komprej, 2003. Competitiveness of livestock production in Slovenia during the process of association to the EU. ACS, Agric. Conspec. Sci. (Tisak). 68: 55-63.

Pärna, E., K. Pärna and I.A. Dewi, 2003. Economic value of milk production and functional traits in the Estonian Holstein population. In: EFITA 2003 Conference, Debrecen, Hungary, pp. 352-359.

Pearson, R.E., 1986. Economic evaluation of breeding objectives in dairy cattle: Intensive specialised milk production in temperate zones. In: Proc. 3rd World Cong. Genet. Appl. Livest. Prod. IX, pp. 11-17.

Pedersen J., U.S. Nielsen and G.P. Aamand, 2002. Economic values in the Danish Total Merit Index. In: Proceeding of the 2002 Interbull Meeting, 26-27 May 2002, Interlaken, Switzerland. Interbull Bulletin, 29: 150-154.

Perez-Cabal, M.A. and R. Alenda, 2003. Lifetime profit as an individual trait and prediction of its breeding values in Spanish Holstein cows. J. Dairy Sci., 86: 4115-4122.

Philipsson, J., G. Banos and T. Arnason, 1994. Present and future uses of selection index methodology in dairy cattle. J. Dairy Sci., 77: 3252-3261.

Philipsson, J. and B. Lindhé, 2003. Experiences of including reproduction and health traits in Scandinavian dairy cattle-breeding programs. Livest. Prod. Sci., 83: 99-112.

Philipsson, J., J.Å. Eriksson and H. Stålhammar, 2005. Know-how transfer in animal breeding - the power of integrated cow databases for farmer's selection of bulls to improve functional traits in dairy cows. In: Knowledge transfer in cattle husbandry: new management practices, attitudes and adaptation. Kuipers A., M. Klopcic and C. Thomas (eds.). EAAP Scientific Series No. 117, Wageningen Academic Publishers, the Netherlands, pp. 85-95.

Pryce, J., B.L. Harris and W.A. Montgomerie, 2008. Do 'robust' dairy cows already exist? In: Breeding for robustness in cattle. Klopcic M., R. Reents, J. Philipsson and A. Kuipers (eds.) EAAP Scientific Series No. 126, Wageningen Academic Publishers, the Netherlands, pp. 99-109.

Rauw, W.M., E. Kanis, E.N. Noordhuizen-Stassen and F.J. Grommers, 1998. Undesirable side effects of selection for high production efficiency in farm animals: a review. Livest. Prod. Sci., 56: 15-33

Rozzi, P., F. Miglior and K.J. Hand, 2007. A total merit selection index for Ontario organic dairy farmers. J. Dairy Sci., 90: 1584-1593.

Schwarzenbacher, H., 2001. Vergleich von biologischen mit konventionellen Milchviehbetrieben in Niederösterreich. Diplomarbeit. Institut für Nutztierwissenschaften, Universitat für Bodenkultur, Wien: 89 p.

Schwarzenbacher, H., J. Sölkner and C. Fürst, 2003. Stand der Züchtung auf biologischen Milchviehbetrieben in Österreich. In: Ökologischer Landbau der Zukunft. Freyyer B. (ed.) Beiträge zur 7. Wissenschaftstagung zum Ökologischen Landbau, Februar 2003, Wien, pp. 249-252.

Shook, G.E., 2006. Major advances in determining appropriate selection goals. J. Dairy Sci., 89: 1349-1361.

Sölkner, J., A. Willam, E. Gierzinger and C. Egger-Danner, 1999. Effects of including conformation in total merit indices of cattle. In: Breeding goals and selection schemes. Proceeding of International workshop on EU concerted action Genetic improvement of functional traits in Cattle (GIFT); 7-9th November 1999, Wageningen, the Netherlands. Interbull Bulletin 23: 143-150.

Sölkner, J. and C. Fürst, 2002. Breeding for functional traits in high yielding dairy cows. In: Proc. 7th World Cong. Genet. Appl. Livest. Prod., 29: 107-114.

Sölkner, J., H. Grausgruber, A. Mwai Okeyo, P. Ruckenbauer and M. Wurzinger, 2008. Breeding objectives and the relative importance of traits in plant and animal breeding: a comparative review. Euphytica, 161: 273-282.

SORS, 2002. Agricultural Census, Slovenia, 2000. Results of surveys. Statistical Office of the Republic of Slovenia. No. 777, pp. 56-64.

SORS, 2007. Statistical Yearbook of the Republic Slovenia, 2006. Statistical Office of the Republic of Slovenia, pp. 299-304.

STATISTIX 7, 2000. Statistical Analysis Software. http://www.statistix.com/

Tozer, P.R. and J.R. Stokes, 2002. Producer breeding objectives and optimal sire selection. J. Dairy Sci., 85: 3518-3525.

Vangen, O., 2003. Modern breeding programmes. Available at: http://www.nordgen.org/ngh/download/bokartikkel-odd.doc

Vangen, O., 2008. Norwegian breeding strategies – A success story of long-term benefits. Available at: www.umb.no (Accessed: 29 June 2008).

VanRaden, P.M., 2004. Invited review: Selection on net merit to improve lifetime profit. J. Dairy Sci., 87: 3125-3131.

VanRaden, P.M. and A.J. Seykora, 2003. Net merit as a measure of lifetime profit: 2003 revision. AIPL research report. Available at: http://aipl.arsusda.gov/reference/nmcalc.htm

Veerkamp, R.F., 1998. Selection for economic efficiency of dairy cattle using information on live weight and feed. intake: a review. J. Dairy Sci., 81: 1109-1119.

Wall, E., M.P. Coffey and P.R. Amer, 2008. Derivation of direct economic values for body tissue mobilisation in dairy cows. In: Breeding for robustness in cattle. Klopcic M., R. Reents, J. Philipsson and A. Kuipers (eds.) EAAP Scientific Series No. 126, Wageningen Academic Publishers, the Netherlands, pp. 201-205,

Wolfova, M., J. Wolf, J. Kvapilik and J. Kica, 2007. Selection for profit in cattle: I. Economic weights for purebred dairy cattle in the Czech Republic. J. Dairy Sci., 90: 2442-2445.

The role of robustness in type classification

A. Hamoen, G. de Jong and M. van Pelt
CRV, Wassenaarweg 20, 6843 NW Arnhem, the Netherlands

Abstract

Robustness was introduced as general characteristic in the type classification system in the Netherlands and Flanders. The trait robustness is directly related to Longevity and is composed out of four linear conformation traits: rump width, chest width, body depth and body condition score. All four traits have an optimum with longevity and therefore also with robustness. The most important trait appears to be body condition score, followed by body depth, chest width and rump width. The classifier uses a module that computes a score for robustness based on the scores for rump width, chest width, body depth and body condition score. Robustness is weighted with 10% in the overall score for conformation.

Keywords: robustness, longevity, conformation

Introduction

In the type classification system in the Netherlands and Flanders four general characteristic traits are used to give an overall judgement on the quality of the cow. The four traits are frame, type, udder and feet & legs. These four traits combined determine the overall score for the cow. The general characteristics are scored on a scale of 71 to 89. Besides the general characteristic traits 18 linear traits are scored on each cow in the conformation classification system, using a scale of 1 to 9.

In 2007 the general characteristic trait type was replaced by 'robustness'. This new robustness trait is the way the Netherlands and Flanders have defined now the international trait dairy strength.

In 2005 the members of NRS herdbook organisation discussed in their annual member meetings how they would like to change the cow in the near future. They talked about which conformation traits need more attention and which traits less. The outcome was that farmers made clear that the quality of the udders was good enough but feet and legs should have more attention. Further the farmers wanted to have cows which do not get taller, but which have adequate body condition and body weight to be able to handle the milk production. This means having healthier and more robust cows with a higher milk production. These qualities were not expressed sufficiently in one of the general characteristic traits.

The wish of the farmers was to breed a durable cow and this is now becoming true with the new conformation trait robustness or dairy strength. But what exactly does this conformation trait robustness mean? This paper describes the development and background of robustness.

Relation with survival

When discussing the trait or term robustness with farmers they have the following interpretation: a robust cow is a cow which looks strong, which has enough body condition and shows enough power in her body when judging her body width, body depth and rump width. At the same time a cow is robust if she is able to stay long in the herd. So the goal when defining robustness as a new trait for the type classification was to combine these four body traits with longevity.

Based on data of more than 600,000 cows the relation was determined between the scores for the linear traits rump width, chest width, body depth, body condition score and the percentage of cows which were able to start their third lactation, i.e. the survival percentage (Figures 1 to 4). From these figures it is clear that all four traits can be considered as optimum traits in their relation with survival or longevity. This means that cows scoring in the range of 4-6 for a linear trait stay longer in the herd than cows scoring 1 or 9. Further body condition Score shows larger difference in survival between the optimum score and the extreme scores 1 and 9, than the other traits. The relationship of the four linear traits with survival is the basis for the trait robustness. A transformation has been developed using multiple regression to convert the relationship between the four linear traits with survival to a score for robustness on a scale of 71 to 89. At the same time the score for the linear traits are corrected for lactation stage and age at calving. The derived robustness scores show a normal distribution and have a positive relationship with longevity (Figure 5).

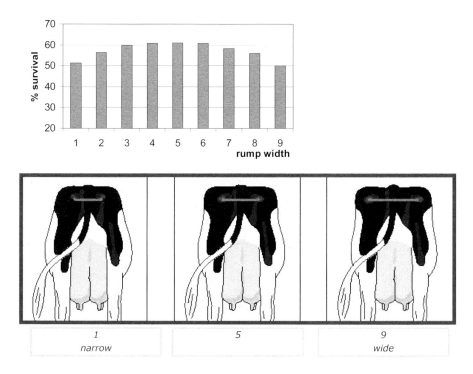

Figure 1. Relation rump width with survival percentage, being the percentage of cows that is able to start the third lactation (Dutch average 62%).

Body condition score has the biggest influence on the number of cows starting their third lactation and therefore on the score for robustness. The order of importance of the four linear traits for the robustness score is body condition score, followed by body depth, chest width and rump width (Figure 6). This order decides actually the weighing of the traits in the trait robustness.

Figure 6 shows the average score for robustness of a heifer for the different linear trait scores. Heifers with much or little condition will be 'punished' more severely in their robustness score than heifers with a wide or narrow rump. Heifers with an optimum condition will be rewarded more than heifers with an optimum rump width.

Breeding for robustness in cattle

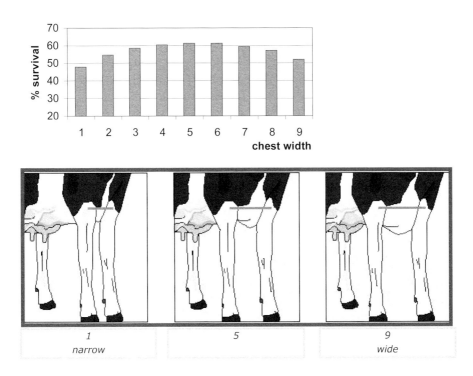

Figure 2. Relation Chest Width with survival percentage at third lactation (Dutch average 62%).

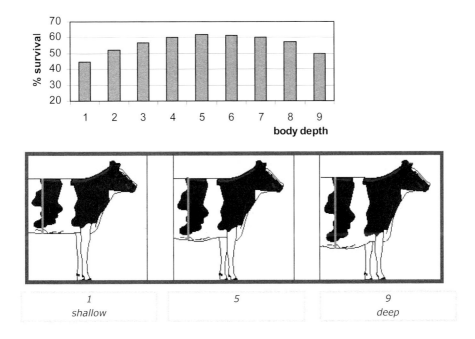

Figure 3. Relation Body Depth with survival percentage at third lactation (Dutch average 62%).

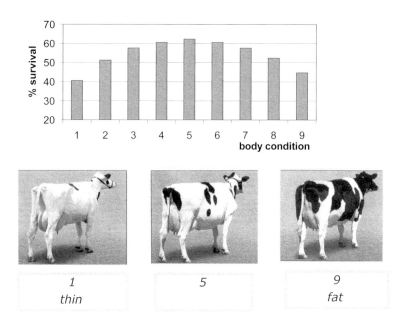

Figure 4. Relation Body Condition Score with survival percentage at third lactation (Dutch average 62%).

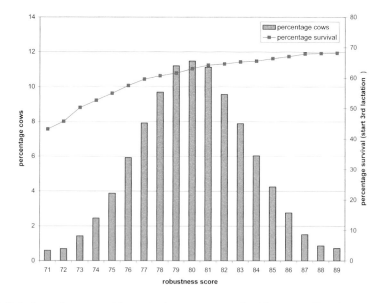

Figure 5. Relation robustness with survival percentage at third lactation (Dutch average 62%).

A balanced cow

The score for robustness is deduced from the linear traits and is therefore a so-called composite trait. In the Netherlands and Flanders, this way of determining scores for a type trait is completely new. The classifier reads the robustness score directly from his hand terminal after he has scored the

Breeding for robustness in cattle

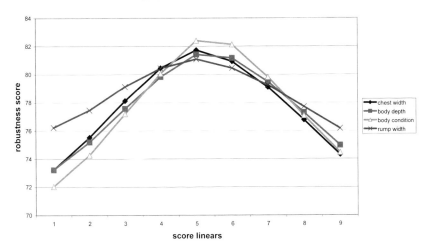

Figure 6. Effect of each linear trait on the robustness score.

linear traits. In the hand terminal, a module calculates the score. If a cow scores 5 points for each trait, it will have around 89 points for robustness and its chance of survival at the third lactation will be 67% (Figure 5). This is 5% higher than the national average. Two heifers, each with 80 points for robustness may look differently, because they both have been 'punished' or 'rewarded' on different traits. In fact, this also happens with the other three general characteristic traits. The udders of two cows, both with a score of 80 points, may also look differently because of differences in the suspensory ligament, front teat placement, etc.

The relation of the various traits with the robustness score is illustrated in Figure 6. The benefit of a composite trait is that a new trait is introduced faster, without any time-consuming harmonisation of classifiers. Changes can also be carried through quicker. On farms where body condition score has shortcomings, it is possible to carry through corrections, for example when resistance and/or health are strongly affecting robustness.

Using Stature as a parameter in the calculation has been a point of special interest. For a heifer of 1.40 meter, 3 points for chest width is less undesirable than for a heifer of 1.55 meter. This has been considered in the definitive version of the module that calculates robustness. The module considers correlations with other traits in the calculation. Besides stature, there is also angularity for example. Older cows are entered according to the same principle. The only difference is that classifiers have the possibility to round off robustness. The trait robustness is now part of the overall score for conformation. The overall score is determined by frame (20%), udder (35%), feet and legs (35%) and robustness (10%).

This paper describes how breeding for a more robust cow can be approached from a type classification point of view. So what can be done using just conformation traits of information to describe a robust cow (see Figure 7). But of course robustness and its relation with longevity also can be improved by selecting for other traits, for example traits which improve longevity or traits in the field of health and fertility. This last approach has resulted in using a selection index in the Netherlands and Flanders, NVI, which contains besides milk production traits, also longevity, somatic cell count, fertility and conformation scores of udder and feet & legs. All this efforts, using genetic selection, will result in cows that produce more milk but also are able to stay longer in the herd and have fewer problems.

Breeding for robustness in cattle 269

Figure 7. Robustness in practice. The cow on the right is desired: she is big enough and wide enough in the chest and rump, her body is deep enough and she has enough body condition. She scores for all traits around optimum. The cow on the left is too frail: she is small, narrow in the chest and rump, shallow in her body and she has almost no body condition. The cow on the right is the kind of cow farmers want, because of her good robustness she will be able to handle her production without problems and stay in the herd for a long time.

Breeding for robustness in cattle

Keyword index

Author index

About the editors

Marija Klopčič

Marija Klopčič works at the Department for Animal Science, Biotechnical Faculty, University of Ljubljana as a researcher and in transfer of knowledge to the field. She is also very much involved in different International projects. Recently she became Assistant Professor of Animal Science. For many years she has been professional secretary of the Slovenian Holstein Association, where she takes care of the breeding programme of the Holstein breed in Slovenia and of breeding work in the field. She is secretary of the Cattle Commission of the European Association for Animal Production (EAAP).

Reinhard Reents

Reinhard Reents is general manager of United Information Systems for Animal Production (Vit) in Germany. Vit is in charge of IT solutions - including breeding value estimation - for different species of farm animals. He has lengthy experience in genetic evaluation procedures such as the implementation of test day models in large-scale dairy cattle populations. On an international level he is a member of the Interbull Steering Committee (SC) and has been chairman of the Interbull SC since 2006. Beside this, he represents Interbull and the cattle breeding organisations of Austria, Germany and Switzerland on the Executive board of ICAR (International Committee of Animal Recording).

Jan Philipsson

Jan Philipsson is a professor in Animal Breeding at the Swedish University of Agricultural Sciences (SLU) in Uppsala, where he has in the past been head of the Department of Animal Breeding and Genetics. His research has to a large extent been devoted to functional traits, including calving, fertility, health and workability traits, in dairy cattle. He developed the Swedish Total Merit Index for dairy bulls back in 1975. He has been secretary of Interbull since its start in 1983. He also has broad experience of livestock breeding issues in developing countries.

Abele Kuipers

Abele Kuipers was director of the Research Institute for Cattle, Sheep and Horse Husbandry for 12 years in the Netherlands, including the experimental stations located around the country. He is currently director of the Expertise Centre for Farm Management and Knowledge Transfer, a foundation linked to Wageningen University and Research Centre. He is secretary of the Netherlands' Committee of the International Farm Management Association (IFMA-Holland), vice-president of the Central and Eastern European working group of the EAAP and president of the Cattle Commission of the EAAP.